Excellence in
Biology
NCEA Level 2

Martin Hanson

NELSON
CENGAGE Learning

Australia • Brazil • Japan • Korea • Mexico • Singapore • Spain • United Kingdom • United States

Excellence in
Biology

NCEA Level 2

Martin Hanson

NELSON
CENGAGE Learning

Australia • Brazil • Japan • Korea • Mexico • Singapore • Spain • United Kingdom • United States

Excellence in Biology Level Two
2nd Edition
Martin Hanson

Cover design: Book Design Ltd
Text design: Book Design Ltd
Illustrations: Martin Hanson
Proof reader: Sam Hill
Production controller: Siew Han Ong
Reprint: Jess Lovell

Any URLs contained in this publication were checked for currency during the production process. Note, however, that the publisher cannot vouch for the ongoing currency of URLs.

First published by New House in 2005 as Excellence in Biology: A textbook for Year 12 students.

© 2011 Cengage Learning Australia Pty Limited

For product information and technology assistance,
in Australia call **1300 790 853**;
in New Zealand call **0800 449 725**

For permission to use material from this text or product, please email
aust.permissions@cengage.com

National Library of New Zealand Cataloguing-in-Publication Data
National Library of New Zealand Cataloguing-in-Publication Data

Hanson, Martin.
Excellence in biology. NCEA level 2 / Martin Hanson. 2nd ed.
National Library of New Zealand Cataloguing-in-Publication Data

Hanson, Martin.
Excellence in biology. NCEA level 2 / Martin Hanson. 2nd ed.
Previous ed.: 2005.
Includes index.
Companion volume to: Level 2 biology workbook.
ISBN 978-0-17-021409-4
1. Biology. 2. Biology—Problems, exercises, etc.
I. Hanson, Martin. Level 2 biology workbook. II. Title.
570.76—dc 22

Cengage Learning Australia
Level 7, 80 Dorcas Street
South Melbourne, Victoria Australia 3205

Cengage Learning New Zealand
Unit 4B Rosedale Office Park
331 Rosedale Road, Albany, North Shore 0632, NZ

For learning solutions, visit **cengage.co.nz**

Printed in Australia by Ligare Pty Limited.
5 6 7 8 9 10 11 20 19 18 17 16

Contents

Part I: Ecology

Part II: Plant and Animal Adaptation

Part III: Genetics

Part IV: Biology of Cells

1 Who's Who in the Living World

Classifying living things

How many kinds of living things exist on Earth? A simple question, but there is no clear answer. Not long ago textbooks said the total was about one million, but now most agree on a number of over five million different species. Thousands more are being discovered every year, especially in rainforests and deep oceans.

Five million of anything — books, phone numbers, animals — needs to be classified in some way, or the list will become chaotic. Because there are many languages, alphabetic systems are not ideal. The preferred system for classifying animals and plants is based mainly on their physical features, not their size or what they eat.

The study of how organisms are classified is called **taxonomy**. Each group, or **taxon**, is divided into smaller groups. The main taxa, from smallest to biggest, are species, genus, family, order, class, phylum, kingdom. Animals that are only slightly similar are put in the same phylum; animals that are very similar (like dolphins) are put in the same family.

Using the above system, the classification of a tiger is:

Tiger

Kingdom:	Animalia
(sub)**phylum:**	Vertebrata (all animals with backbones)
Class:	Mammalia (all animals that produce milk)
Order:	Carnivora
Family:	Felidae (cat family)
Genus:	*Panthera* (also includes lions, leopards, puma)
Species:	*tigris*

ISBN: 9780170214094

It is not always necessary to write out all these taxa names. In most situations we use only the genus and species names, a custom known as the **binomial system**, invented more than 200 years ago by Carl Linneaus. Scientists always write the genus name with a capital letter, the species name with a small letter, and both words in italics. Examples: *Homo sapiens, Felis domesticus.*

The common names of species can cause confusion. In New Zealand a glow worm is the young stage of a small fly, but in Europe a glow worm is a beetle. The name 'snapper' is used for quite different kinds of fish in different parts of the world. Also, single species often go under different names, for example the same bird is known as kotare in Maori, tiotala in Samoan, and kingfisher in English.

What is a species? Why should we say that all dogs are one species, but sheep and goats are two different species? The short answer is that all dogs can interbreed, but sheep and goats cannot.

Kingfisher

A **species** is a group of organisms whose members are similar enough to mate with each other and to produce fertile offspring, but cannot or do not interbreed with other groups.

Horses and donkeys can interbreed to produce offspring, known as mules. Is this an exception? No. Mules are sterile and cannot themselves breed. So using the 'breeding test', we can say that donkeys and mules are separate species.

A number of species is put into a **genus** (plural, *genera*). Genera are put into **families** and families are grouped into **orders**. A number of related orders are placed in a class, and related classes are grouped into a **phylum** (plural, *phyla*). The largest group is the **kingdom**, and consists of a number of phyla.

The main groups of organisms

Scientists are not fully agreed as to how many kingdoms there should be. A common way is to divide organisms into five kingdoms.

Kingdom Prokarya ('prokaryotes')

The prokaryotes include the bacteria. Unlike all other organisms, they do not have a clearly defined nucleus (Fig. 1.1). This is because there is no nuclear envelope separating the chromosome from the rest of the cell.

All other living things do have a nucleus and are called **eukaryotes**. Some bacteria are *autotrophic* (make their own organic substances from carbon dioxide and water), while others are *heterotrophic* (use organic matter made by other living things). Of these, some are **saprobes**, feeding on dead matter and causing decay. Others are **parasites**, feeding on living organisms, and may cause disease.

ISBN: 9780170214094

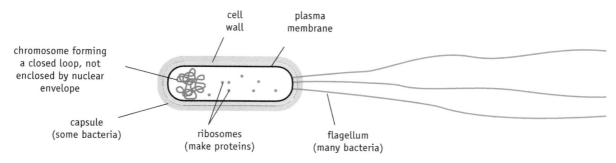

Fig. 1.1 Structure of a bacterial cell

Kingdom Fungi

Fig. 1.2 shows some of the 100 000 or so species of fungi. Fungi are fundamentally different from plants in that they lack chlorophyll and are thus *heterotrophic*, and their cell walls are made of chitin rather than cellulose. Many are saprobic, causing decay, and others are economically destructive parasites of plants. Many form mutualistic relationships with other organisms. Some are single cells (e.g. yeasts), but most have a body composed of thin threads called **hyphae**. The whole body of the fungus forms the **mycelium** (Fig. 1.3).

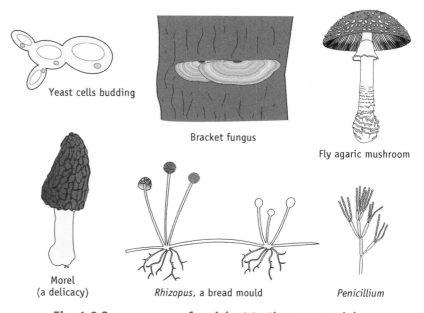

Fig. 1.2 Some common fungi (not to the same scale)

Kingdom Protoctista

This kingdom includes all those organisms formerly called 'algae' and 'protozoa' (Fig. 1.4). The algae were considered to be plants because they are photosynthetic, and protozoa were classified as animals because of their mode of nutrition.

The Protoctista are a ragbag group, some of which are actually more closely related to members of other kingdoms than they are to each other! For example the Chlorophyta ('green algae') are more closely related to the plant kingdom than they are to the 'brown algae' or to the 'red algae'.

Kingdom Plantae

All plants are multicellular and almost all are autotrophic. Their cells have cellulose walls and some have chloroplasts in which photosynthesis occurs. The life cycle is complicated, consisting of a diploid **sporophyte** (spore-producing) generation that alternates with a haploid **gametophyte** (gamete-producing) generation.

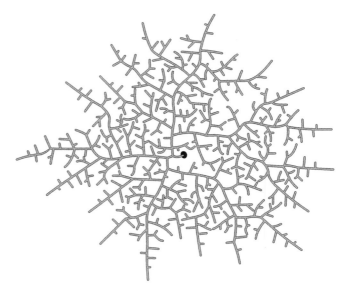

Fig. 1.3 Young mycelium of a mould showing hyphae growing and branching from germinated spore (after Kendricks)

ISBN: 9780170214094

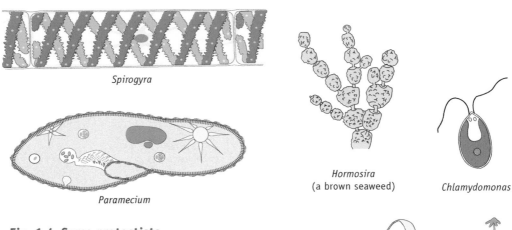

Spirogyra

Paramecium

Hormosira
(a brown seaweed)

Chlamydomonas

Fig. 1.4 Some protoctists

Phylum Bryophyta (Liverworts and mosses, Fig. 1.5).

These plants are poorly adapted for life on land for three reasons:

- Instead of roots they have microscopic thread-like *rhizoids* which function for anchorage rather than water absorption.

- They lack the highly specialised transport tissues of other plants.

- They require external water for fertilisation.

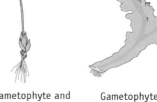

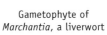

Gametophyte and sporophyte of *Funaria*, a moss

Gametophyte of *Marchantia*, a liverwort

Fig. 1.5 Bryophytes, the 'amphibians' of the plant kingdom

Because they cannot transport materials over long distances they are generally small. Most grow in damp, shady places. The gametophyte generation is a fully independent plant and lives from year to year, while the sporophyte obtains its energy from the gametophyte and dies after reproducing (see Chapter 13).

The plant phyla described below are collectively called *vascular plants*, so-called because they have conducting (vascular) tissues (xylem and phloem). The sporophyte is the dominant generation, the gametophyte being small and short-lived. In the seed-bearing plants it is vestigial and never lives an independent life.

Ferns and their relatives (Fig. 1.6)

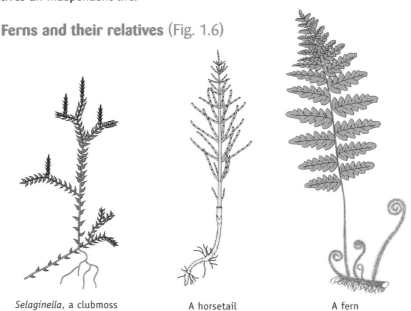

Selaginella, a clubmoss

A horsetail

A fern

Fig. 1.6 Vascular plants that do not produce seeds

ISBN: 9780170214094

Ferns are better adapted to life on land than bryophytes, for two reasons:

- They have true roots that can penetrate deep down into the soil to tap a more reliable water supply.

- They have specialised vascular tissues, so they can carry water rapidly up to the leaves to replace that which is lost.

Despite these adaptations, the life cycle still requires external water for fertilisation (see Chapter 13). Most grow less than a metre in height, but tree ferns reach several metres. In most ferns the spores are all the same and germinate to form a tiny, heart-shaped gametophyte. After fertilisation the zygote develops into a new sporophyte and the gametophyte dies.

Horsetails are all that remains of a major group of plants that flourished long before the time of the dinosaurs. Apart from one introduced species they are absent from New Zealand.

Clubmosses

Despite their name, these are not mosses. One group is particularly interesting because they produce two kinds of spore: tiny *microspores* that develop into male gametophytes, and large *megaspores* that develop into female gametophytes. In this respect they are a kind of link between the ferns and seed-bearing plants.

Phylum Coniferophyta (Conifers)

These plants include kauri and the podocarps such as totara, and also various introduced pines (Fig. 1.7). The gametophyte generation is microscopic and is entirely dependent on the parent sporophyte for energy (see Chapter 13). The seeds do not develop inside an ovary and are thus naked.

Fig. 1.7 A shoot of pine with female cone

Phylum Angiospermophyta (Flowering plants, Fig. 1.8)

In these plants the gametophyte generation is even more reduced. The seeds develop within the protection of an *ovary*, which grows into a *fruit*. There are two groups of flowering plants: monocotyledons and dicotyledons.

Dicotyledons These have the following features:

- the seed has two embryonic leaves (cotyledons);

- the veins in a young stem form a cylinder;

- flower parts are usually in fours or fives or multiples of four or five (e.g. kowhai has five petals and ten stamens);

- the leaf veins form a network;

- most undergo some growth in thickness (secondary thickening).

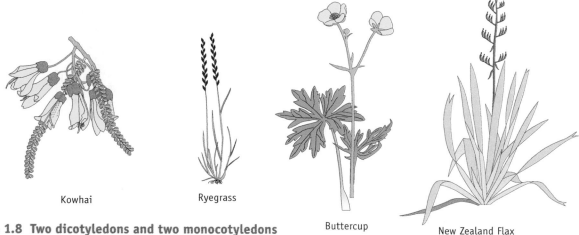

Kowhai

Ryegrass

Buttercup

New Zealand Flax

Fig. 1.8 Two dicotyledons and two monocotyledons

ISBN: 9780170214094

Monocotyledons These include grasses, lilies, orchids, palms, cabbage trees and supplejack. They have the following features:

- the seed has one cotyledon;
- the veins in the stem are scattered;
- the flower parts are in threes or multiples of three. For example a tulip has six stamens, in two layers (whorls) of three, and a three-part ovary;
- the leaves have parallel veins;
- secondary growth is rare.

Fig. 1.9 summarises these differences.

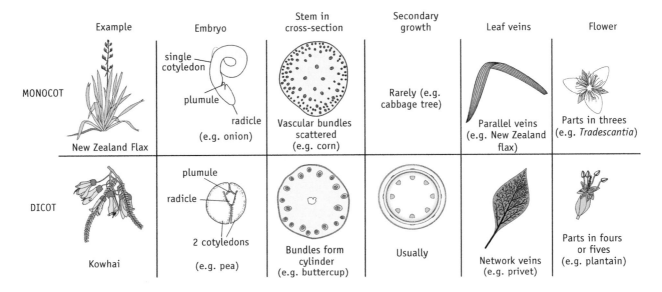

	Example	Embryo	Stem in cross-section	Secondary growth	Leaf veins	Flower
MONOCOT	New Zealand Flax	single cotyledon, plumule, radicle (e.g. onion)	Vascular bundles scattered (e.g. corn)	Rarely (e.g. cabbage tree)	Parallel veins (e.g. New Zealand flax)	Parts in threes (e.g. *Tradescantia*)
DICOT	Kowhai	plumule, radicle, 2 cotyledons (e.g. pea)	Bundles form cylinder (e.g. buttercup)	Usually	Network veins (e.g. privet)	Parts in fours or fives (e.g. plantain)

Fig. 1.9 Monocotyledons and dicotyledons compared

Kingdom Animalia

All members of the animal kingdom are multicellular (many-celled) and heterotrophic. Most move around in search of other organisms for food. Some are anchored (e.g. sea anemones), and many are parasites. The animal kingdom is divided into 33 phyla, but only a few are familiar. The most important are shown in Figs 1.10–1.30.

Phylum Porifera (Sponges, Fig. 1.10)

Sponges are about as unlike 'typical' animals as you can find. They live by filtering out small organisms from the surrounding water. Water is drawn into the body through the thousands of tiny pores from which the group takes its name. They have no nervous system, alimentary canal, blood system or excretory organs. Though they have different kinds of cells specialised for particular functions, these are not organised into tissues. They do, however, have a kind of skeleton, which in some species we use as a bath 'sponge'.

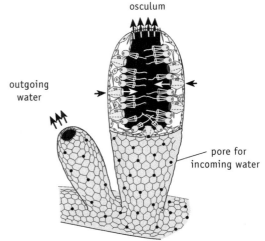

osculum

outgoing water

pore for incoming water

Fig. 1.10 A simple sponge

ISBN: 9780170214094

Phylum Cnidaria (formerly called Coelenterata)

These animals include the jellyfish, sea anemones, corals and freshwater *Hydra* (Fig. 1.11). They all have hollow, bag-like bodies. There is only one opening to the gut, so indigestible residue has to go out the same way the food entered. The mouth is surrounded by tentacles covered with stinging cells called **nematoblasts**. These are unique to the phylum and give it its name. The body is *radially symmetrical*, which means that it can be cut in many ways into two equal halves. Some are anchored (e.g. sea anemones and corals), and others swim weakly in the plankton (e.g. jellyfish).

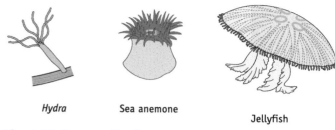

Hydra   Sea anemone

Jellyfish

Fig. 1.11 Some cnidarians

Phylum Platyhelminthes (Flatworms, Fig. 1.12)

These are so-called because of their flattened bodies and include many parasites such as flukes and tapeworms. Like cnidarians they have only one opening to the gut and lack a blood system. The phylum also includes non-parasitic animals called *planarians,* which live in streams and lakes.

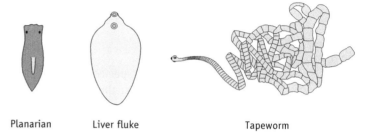

Planarian   Liver fluke   Tapeworm

Fig. 1.12 Some platyhelminthes

Phylum Nematoda (Roundworms, Fig. 1.13)

Though not familiar to most people, roundworms are an economically and medically important phylum. The body is round in cross-section and is not divided into segments. Most are less than 1 mm long, though some may be over a metre in length. Although only 80 000 species have been named, some nematologists estimate that there may be as many as a million species. Their numbers are vast. Over two million are present in each square metre of topsoil, and 90 000 were found in one rotten apple. Over 200 species were found in 7 cm³ of coastal mud. Some cause disease in humans (e.g. hookworms) or in crops (e.g. eelworms).

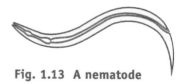

Fig. 1.13 A nematode

Phylum Annelida (Segmented worms, Fig. 1.14)

The annelids include the earthworms, leeches, and various marine worms. The body is divided into many repeating *segments*. They have two openings to the gut and a well-developed blood system. Between the gut and the body wall is a fluid-filled cavity, the **coelom**. This acts as a *hydrostatic skeleton*.

Phylum Mollusca (Fig. 1.15)

These include snails and slugs, mussels and oysters, octopus and squid. The body is unsegmented and typically there is a *shell* secreted by a fleshy *mantle*, and also a muscular *foot* by which the animal moves around. In all except bivalves there is a file-like tongue called a *radula*, used to rasp off food.

There are three major classes (and several smaller ones):

Class Gastropoda (snails and slugs). The shell is typically coiled, though slugs have lost it altogether in evolution. These include the only molluscs to colonise the land.

Class Bivalvia (mussels, oysters, pipi and scallops). These have two shells and most are filter-feeders.

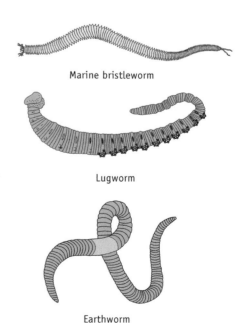

Marine bristleworm

Lugworm

Earthworm

Fig. 1.14 A selection of segmented worms

ISBN: 9780170214094

Class Cephalopoda (squids and octopus). The shell is greatly reduced or absent, and the mantle is modified for squirting out water in a kind of jet propulsion. The eyes and brain are highly developed.

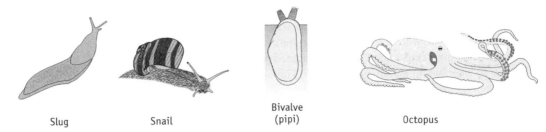

Slug   Snail   Bivalve (pipi)   Octopus

Fig. 1.15 A variety of molluscs

Phylum Arthropoda (Figs 1.16–1.19)

If the number of species is anything to go by, then arthropods are by far the most successful phylum, accounting for over three quarters of the animal kingdom.

- The skin secretes a **cuticle** consisting mainly of **chitin**. This is hardened in places, forming a jointed **exoskeleton** (external skeleton). Because the cuticle is non-living it cannot grow, so it has to be shed periodically as the animal grows.
- The body is segmented (though this is not always visible externally).
- Each segment typically bears a pair of jointed limbs (though some arthropods have lost limbs in evolution).
- There is an open blood system (see Chapter 7).

The most important classes are the following:

Class Insecta (Fig. 1.16) e.g. cicadas, weta, dragonflies, flies, beetles, ants, bees, wasps, butterflies and moths. Insects are by far the most successful arthropods, accounting for over three quarters of the phylum.

- The body consists of three parts: *head, thorax,* and *abdomen.*
- The head has a pair of *compound eyes* and a pair of *antennae.*
- The thorax has *three pairs of legs* and usually two pairs of wings.
- They breathe by means of long branching air-filled tubes called *tracheae.*

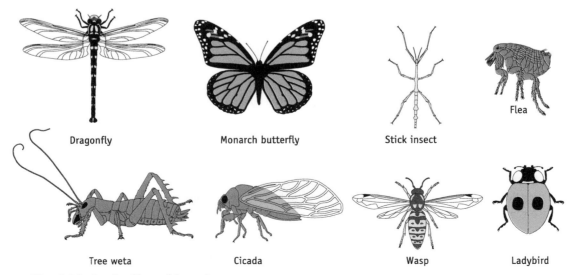

Dragonfly   Monarch butterfly   Stick insect   Flea

Tree weta   Cicada   Wasp   Ladybird

Fig. 1.16 A selection of insects

ISBN: 9780170214094

Class Myriapoda (Fig. 1.17) Centipedes and millipedes. These have many similar segments, each with a pair of legs. The head has a pair of antennae and they breathe by tracheae.

Centipedes (Chilopoda) are all carnivores and kill their prey by means of a pair of poison jaws. No centipede has a hundred legs as the name seems to suggest; most have fewer.

Millipedes (Diplopoda) are all herbivores. They do not have a thousand legs (though one species has 710!). Segments are fused in twos, so each apparent segment has two pairs of legs (hence the name Diplopoda).

Class Crustacea (Fig. 1.18) These include the barnacles, slaters, shrimps and crabs. Most are aquatic (live in water), and have two pairs of antennae.

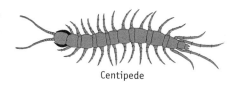
Centipede

Millipede

Fig. 1.17 Two myriapods

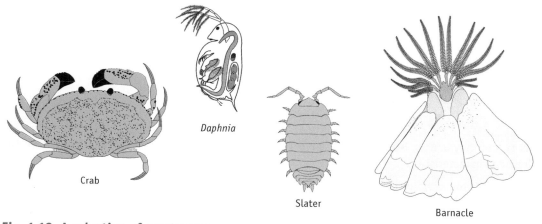

Crab

Daphnia

Slater

Barnacle

Fig. 1.18 A selection of crustaceans

Class Arachnida (Fig. 1.19) Spiders, ticks, mites and scorpions, with four pairs of legs and no antennae.

Tick

Spider

Scorpion

Fig. 1.19 A selection of arachnids

Phylum Echinodermata

These creatures all live in the sea and include starfish (Fig. 1.20), sea urchins, brittle stars and sea cucumbers (Fig. 1.21). The adults are radially symmetrical and have an internal skeleton of limy plates. These seem external but are actually covered by a thin layer of living tissue. They have no blood system but have a *water vascular system* filled with seawater. They move by means of hundreds of 'tube feet' that are outpushings of the water vascular system.

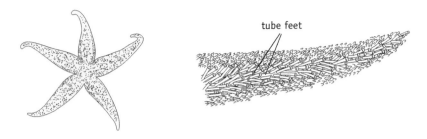

Fig. 1.20 A starfish, with underside of an arm showing tube feet

Sea urchin   Sea cucumber   Brittle star

Fig. 1.21 Echinoderm variety

Phylum Chordata (Fig. 1.22)

This is 'our' phylum. At least in the embryo (early stage of development), all chordates have the following characteristics:

- A stiff rod or *notochord* running down the dorsal (back) side. In nearly all chordates this is replaced early in embryonic development by a *backbone*. These chordates form the sub-phylum **Vertebrata**.

- A *hollow, dorsal* nerve cord. In the vertebrates this is protected by the backbone.

- The body is *segmented*. In amphibians, reptiles, birds and mammals this is only evident in the embryo, but in fishes the muscles are still segmentally arranged.

- The front part of the gut (pharynx) is also used in breathing. The pharynx connects with the outside by a number of *gill slits*. In fish these are used in gas exchange, but in other chordates they disappear during early development.

- A blood system in which the blood flows forward along the ventral (belly) side and backward dorsally.

- A tail extending behind the anus.

Only one group of creatures shows all these characteristics in the adult — small filter-feeding animals called lancelets. One species lives in shallow New Zealand waters.

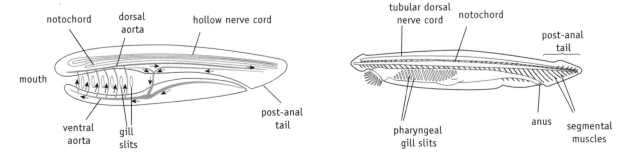

Fig. 1.22 Left: A generalised chordate. Right: Amphioxus, an animal that shows all chordate characters in the adult.

ISBN: 9780170214094

There are two sub-phyla; those without a backbone; and the **Vertebrata**, which have one. Besides the lancelet mentioned above, those without a backbone include the tunicates, or **Urochordata**. These bag-like creatures are filter-feeders and are often seen attached to rocks on the sea shore. They are also called sea squirts because if you poke them they squirt water out of the pharyngeal cavity. The adults have no notochord, which is only present in the tadpole-like larva (Fig. 1.23).

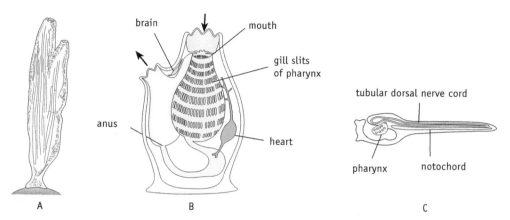

Fig. 1.23 A sea squirt (A) External features. (B) Diagrammatic view showing filter-feeding pharynx. (C) Free-swimming tadpole larva.

Sub-phylum Vertebrata

These animals include the familiar backboned animals as well as some weird, less well-known ones. There are two groups (super-classes); one without jaws (Agnatha); and the other with jaws (Gnathostomata). The jawless vertebrates include the lamprey (Fig. 1.24). It is actually a 'living fossil', a remnant of a group that flourished over 400 million years ago. The young live in New Zealand rivers and the adults live in the sea. Though loosely called a 'fish' because it has fins and breathes by gills, the lamprey has no jaws, paired fins, or scales. The adult feeds on fish by rasping flesh from them with a piston-like tongue covered with horny teeth.

Fig. 1.24 The lamprey, a primitive, jawless vertebrate

There are six classes of jawed vertebrates: cartilaginous fish, bony fish, amphibians, reptiles, birds and mammals.

Cartilaginous fish (Class Chondrichthyes, Fig. 1.25)
These are the sharks and rays.

- They have a skeleton of cartilage rather than bone.

- Their gill slits are visible externally.

- Their scales are like microscopic, backwardly-pointing teeth (with enamel, dentine and pulp cavity).

- The mouth is usually under the head.

- The blood passes only once through the heart in each circuit around the body (single circulation).

- Some lay large, yolky eggs protected by a horny shell. Others produce live young.

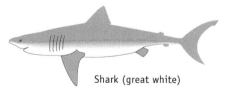

Shark (great white)

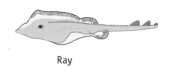

Ray

Fig. 1.25 Some cartilaginous fish

Bony fish (Class Osteichthyes, Fig. 1.26)

Though they are called 'fish', the bony fish are only very distantly related to cartilaginous fish.

- The gill slits are concealed beneath a gill cover.

- The scales are thin bony plates.

- They have a gas-filled *swim bladder*. This gives buoyancy, so the fish can 'hang' in the water without rising or sinking.

- The mouth is at the front of the head.

- There is a single circulation.

- Most lay large numbers of small eggs.

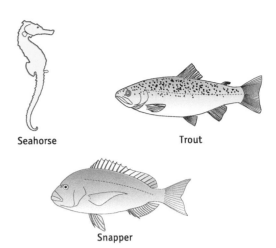

Fig 1.26 A selection of bony fish

Amphibians (Class Amphibia, Fig. 1.27)

Amphibians are so-called because, typically, the life cycle has both aquatic and terrestrial stages. There are two main groups – the salamanders, which have tails, and the frogs, which do not.

- Most have four limbs.

- The skin has no scales and is typically moist.

- The heart is partly divided, with two atria and one ventricle (see Chapter 7).

- Eggs are usually laid in water and fertilised outside the body.

- The eggs hatch into a fish-like tadpole that breathes by gills.

- The adults breathe by lungs and also through their skin.

Fig. 1.27 Some amphibians

The four species of New Zealand native frogs (*Leiopelma*) are primitive in that they have the remains of tail muscles. Also, in three of them the eggs are laid in moist places on land. The tadpole stage occurs entirely inside the egg membrane. Unlike other frogs the males cannot croak and neither sex has eardrums.

Reptiles (Class Reptilia, Fig. 1.28)

Reptiles are a far more diverse group than the amphibians. New Zealand reptiles include various lizards (geckos and skinks). The tuatara is not a lizard but is the only surviving member of a group that lived before the dinosaurs. Snakes are absent from New Zealand.

Reptiles resemble amphibians in having only one ventricle in the heart (though crocodiles have two). However, they are better adapted for life on land in a number of ways:

- The skin is dry (except in turtles) and scaly.

- The eggs are large and yolky and are protected by a leathery shell.

- Fertilisation occurs inside the body of the female.

- The eggs are laid on land and there is no tadpole stage. New Zealand lizards are unusual in that in all but one species the eggs develop inside the mother.

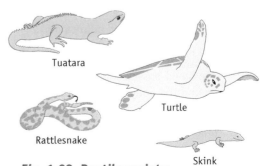

Fig. 1.28 Reptile variety

Birds (Class Aves, Fig. 1.29)

Birds are actually descendants of a group of dinosaurs.

Some of their characteristics are reptilian and others are adaptations for flight:

- They lay large yolky, shelled eggs (unlike reptiles the shell is made of lime).

- They have scaly legs and a horny beak, with no teeth.

- The skin has feathers and the forelimbs are modified as wings – though some are flightless, such as the kiwi and penguins.

- They maintain a high, relatively constant body temperature with a high metabolic rate.

- They have a *double circulation* with a *four-chambered heart*.

- With a high rate of activity, events move swiftly. This requires rapid information processing and hence a large brain. Some birds (e.g. parrots and ravens) are highly intelligent.

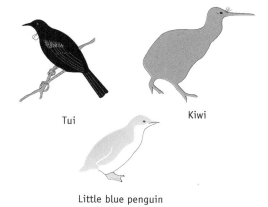

Tui

Kiwi

Little blue penguin

Fig. 1.29 A selection of birds

Mammals (Class Mammalia, Fig. 1.30)

Mammals are so-called because the females suckle their young on milk secreted by *mammary glands*. The young develop inside the mother (except two egg-laying species, the platypus and echidna). Most of their other characteristics are related to the fact that they maintain a high, relatively constant body temperature.

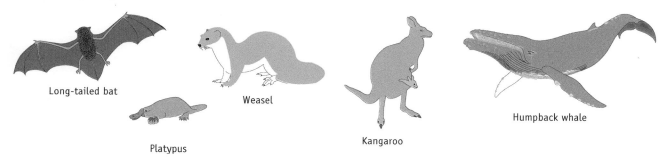

Long-tailed bat

Weasel

Platypus

Kangaroo

Humpback whale

Fig. 1.30 Mammal variety

- Heat is conserved by a coating of *hair* (except in whales in which hair has been lost).

- Excess heat is lost by evaporation of sweat, secreted by *sweat glands*.

- A high metabolic rate demands a high rate of delivery of food and oxygen to the cells. This is achieved by a number of adaptations:

  - There is a *double circulation* with a four-chambered heart (see Chapter 7).

  - A bony *palate* separates air and food passages, enabling food to be chewed without interrupting breathing, so it can be digested more rapidly. This is also helped by the specialisation of teeth into various types.

  - Gas exchange in the lungs is helped by the *diaphragm*, a muscular sheet between thorax and abdomen.

- As in birds, the brain is large.

The non-egg-laying mammals are divided into marsupial and placental mammals.

Marsupials. The young are born at an early stage of development. Each then moves to the pouch or *marsupium*, where it feeds from the nipple. This group includes all the native mammals of Australia, and the South American opossum.

Placental mammals. These are the more familiar mammals. The young become connected to the mother by a *placenta*, and are born at a much later stage of development than in marsupials.

Identifying living things – making and using keys

Because there are so many different kinds of organism, no one can ever recognise more than a tiny fraction of them. Biologists therefore have to specialise in particular groups. So, how do they identify unfamiliar animals and plants?

The answer is that they use a *key*, which enables organisms to be identified quickly. A key is really a way of dividing up a group of animals or plants according to easily visible features. For example, here is a list of eight common animals: gecko, brown rat, polynesian rat, wasp, spider, snail, earthworm, slug.

One way of sub-dividing them would be as in Fig. 1.31:

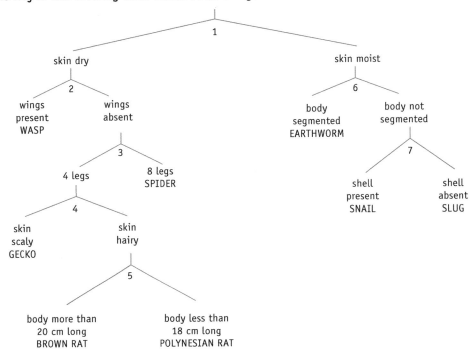

Fig. 1.31 A key written in the form of a classification

Notice that the classification branches like an upside-down tree. In each branch there are *two* choices, so it is called a *dichotomous key* (dichotomy means branching into two). Each choice is given a number.

It is more usual to re-write the classification in a different form, as shown below.

1. Skin dry ....go to 2
   Skin moist ....go to 6
2. Wings present .... WASP
   Wings absent ....go to 3
3. 4 legs ....go to 4
   8 legs ....SPIDER
4. Skin scaly ....GECKO
   Skin hairy..go to 5
5. Body more than 20 cm long............................... BROWN RAT
   Body less than 18 cm long .... POLYNESIAN RAT
6. Body segmented ....EARTHWORM
   Body not segmented ....go to 7
7. Body with shell ....SNAIL
   No shell..SLUG

ISBN: 9780170214094

Notice the following:

- This is only one of many ways in which this group of animals could be classified. The 'best' way is the one that gives a key that is most reliable and easy to use.

- Only clearly visible structural features are useful. For instance, 'animal active at night' is no use if you have a dead specimen!

- Size and shape are only useful if figures are given. For example, 'body more than 5 times as long as wide' is much more use than 'body long'.

- This key takes no account of young stages, which would make things more complicated in practice. Young wasps, for example, have no wings.

- A key will only work for the organisms for which it was designed.

Some practice

Use the key below to identify the order to which each of the sharks in Fig. 1.32 below belongs. Note that some sharks have a third eyelid that covers the eye and protects it when it is attacking prey and when it is near objects. This is shown in some of the illustrations.

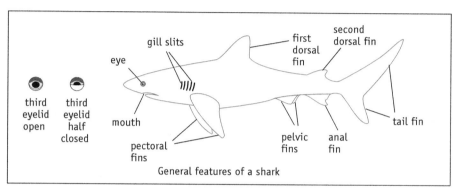

General features of a shark

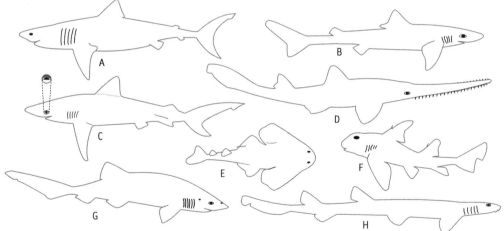

Fig. 1.32 Some of the orders of sharks

1. Anal fin present...2
   No anal fin...6
2. Five gill slits...3
   Six or seven gill slits .... HEXANCHIFORMES
3. Spine present at front of each dorsal fin.............. HETERODONTIFORMES
   No dorsal fin spines...4
4. Mouth well in front of eyes ................................ORECTOLOBIFORMES
   Mouth behind front of eyes 5
5. Third eyelid present .... CARCHARHINIFORMES
   No third eyelid .... LAMNIFORMES
6. Body flattened, ray-like .... SQUATINIFORMES
   Body not ray-like 7
7. Snout elongated and saw-like .... PRISTIOPHORIFORMES
   Snout not elongated, not saw-like .... SQUALIFORMES

Classification and evolution

If you think carefully about the classification of animals, you will realise that there is a pattern. For example, all animals that feed their young on milk also have a four-chambered heart, a diaphragm, and many other characteristics.

Why is this? Animals share features in common because they are *related*. If two or more animals are related, it means that they *share a common ancestor*. For example, in human families, brothers and sisters share the same parents, and cousins share two of their grandparents. If we could go back far enough in time, there would have been a human who was an ancestor of all humans alive today.

Going back about 60 million years, there would have lived an animal that was an ancestor of all primates. Biologists believe that about 130 million years ago there lived the ancestor of placental and marsupial mammals, and that about 400 million years ago lived the ancestor of all the vertebrates.

Fig. 1.33 shows two ways of representing the relationships between various groups within the mammalian Order Carnivora. The diagram on the left shows that the common ancestor lived about 55 million years ago, and that the common ancestor of seals and sea lions lived about 25 million years ago.

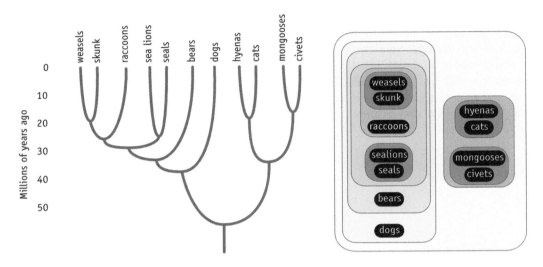

Fig. 1.33   Two ways of representing relationships in the Order Carnivora

Classification is based on the results of evolution. It is therefore not human-made, but existed long before humans were around, and is called the **natural classification**. Unlike the artificial classifications *invented* by humans to make keys, there is only one natural classification, which is being *discovered* by scientists.

Molecular taxonomy

The traditional classification of organisms into five kingdoms was based largely on anatomy. This works reasonably well when there is a lot of anatomical complexity, such as there is with plants and animals. But what about bacteria, which are structurally relatively simple? In recent decades the problem has been solved by comparing *molecules*, in particular DNA and RNA.

The new tree of life, based on these molecular studies, is shown in Fig. 1.34. It shows that there are actually three main branches, or *domains*: the Eukarya, the Archaea and the Eubacteria. The Archaea and Eubacteria are 'prokaryotes'.

The Archaea live in hostile environments such as hot springs, extreme pH and saltiness. These are the kinds of conditions that are believed to have existed early in the evolution of life.

The Archaea are actually more closely related to the eukaryotes than they are to the true bacteria. Thus an archaean living in a hot spring in Rotorua is more closely related to humans than it is to the bacteria living on the human skin! The term 'prokaryote' has therefore no real taxonomic significance. No doubt it will continue to be used, as will the old terms 'algae' and 'protozoa'.

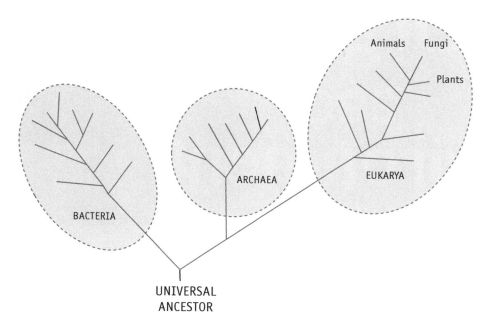

Fig. 1.34 The universal tree of life (most names of groups omitted)

Viruses

Viruses are so strange that many scientists consider them not to be living at all. For example they can be crystallised. They have one characteristic of life — they can reproduce — but they can only do so inside living cells. The reason is that they cannot carry out any life processes for themselves, but get their energy and raw materials from the host cell.

Viruses are not made of cells but are particles consisting of just two chemicals: a nucleic acid (either DNA or RNA) surrounded by a protein coat. The nucleic acid contains the genes ('instructions' for making new virus particles). The protein coat enables the virus to enter the host cell.

Summary of key facts and ideas in this chapter

○ Amid the immense variety of life there is a clear-cut pattern in which organisms fall naturally into groups, each of which can be divided into smaller groups, each of which in turn can be divided into smaller groups, and so on.

○ The smallest group is a *species*; its members are sufficiently similar to one another to interbreed and produce fertile offspring, but do not normally interbreed with members of other species.

○ Similar species can be grouped into a genus, genera can be grouped into families, families can be grouped into orders, which can be grouped into classes, which can be grouped into phyla, and phyla can be grouped into kingdoms.

○ This classification is the result of evolution, members of any given group having a common ancestor. Members of a species have a more recent common ancestor than members of a genus, which have a more recent common ancestor than members of a family, and so on.

○ Common names are unreliable, as one species can have different common names in different countries and cultures.

○ To get over this problem, scientists use *systematic* or scientific names.

○ Scientific names follow the *binomial* ('two-name') system, and are printed in italics.

○ The scientific name of an organism consists of two words; the first denotes the genus and is given a capital letter, and the second denotes the species and is given a small letter.

2 Organisms and Environment

Ecology is the study of organisms in relation to their surroundings, or *environment*. To begin with, some definitions.

Habitat

The *place* in which an organism lives is its habitat. The habitat of a trout, for example, is a lake or stream, and the habitat of the kowhai tree is in forest margins and the banks of streams and rivers. Many habitats are made up of patchworks of *microhabitats*, each with its characteristic inhabitants. A bark beetle and a leaf miner (an insect small enough to live and feed between the upper and lower epidermis of a leaf – Fig. 2.1) share the same woodland habitat, but they live in very different microhabitats.

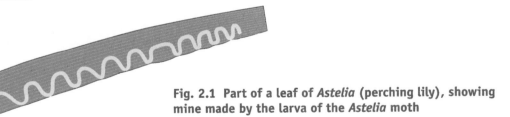

Fig. 2.1 Part of a leaf of *Astelia* (perching lily), showing mine made by the larva of the *Astelia* moth

Environment

This includes all the factors in the surroundings that affect an organism's chances of survival and reproduction. A trout may continue to live in the same habitat — a stream — but its environment changes with season and time of day.

Niche

Simply put, this is an organism's way of life, or role in the community. For example a centipede and a slater may share the same habitat, but they make their living in very different ways; a centipede is a carnivore and a slater is a herbivore (Fig. 2.2).

Fig. 2.2 A centipede (left) and a slater (right) may live under the same log, but have quite different ways of making their living

ISBN: 9780170214094

Ecology can be studied at a number of different levels:

- *Populations*. A population comprises all individuals of a given species in a particular habitat, such as the rainbow trout in a particular stream.

- *Communities*. A community comprises all the species in a given area, such as all the inhabitants of a pond. A community is thus a group of populations.

- *Ecosystems*. An ecosystem includes a community together with all the non-living components that interact with the community. A pond ecosystem thus comprises not only the community, but the water and its dissolved and suspended matter, and the mud at the bottom.

Adaptation

All organisms inherit characteristics that increase their chances of survival and reproduction. Such characteristics are called *adaptations*. Adaptations are commonly divided into three categories, which can be illustrated by the feeding of the female mosquito (Fig. 2.3):

- *Behavioural*. Female mosquitoes find their host by detecting carbon dioxide in the breath and also body heat.

- *Structural*. She has long piercing mouthparts, enabling her to pierce the skin and suck up blood.

- *Physiological*. Before sucking up the blood she injects an anticoagulant chemical that stops the blood clotting.

Only *inherited* features are adaptations. A suntan, for example, helps protect the skin, but is not inherited. The adaptation is not the suntan but *the ability to develop one*.

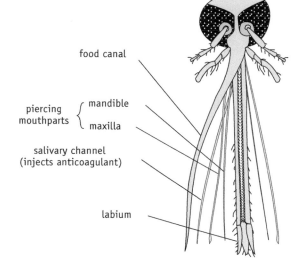

food canal

piercing mouthparts { mandible / maxilla

salivary channel (injects anticoagulant)

labium

Fig. 2.3 Mouthparts of a female mosquito

Kinds of environmental factors

These fall into two categories:

- *Biotic* factors include other organisms such as food, predators, parasites, and competitors.

- *Abiotic* factors are physical influences such as temperature and light intensity.

The distinction between biotic and abiotic factors is not quite so clear-cut as you might think. A kowhai tree, for example, is a biotic factor for insects feeding on it, but it is also a physical factor in the lives of organisms shaded by it.

Although it is easy to list abiotic factors separately, in many cases they interact with each other so that the effect of one depends on the level of another. For example a pond may freeze over, preventing herons from catching fish. Thus a physical factor (cold) may affect a biotic one (food availability).

Environmental factors can also be classified in a different way. *Resources* are factors that may be competed for, such as food, hiding places, mates, and territory in the case of animals, and light, water and mineral ions in plants. *Conditions* are factors that influence organisms without being used up in any way, such as temperature, soil pH, rate of water current flow, and so on.

These two classifications are independent of each other. Light, for example, is a resource for plants but a condition for animals. Some biotic factors, such as competitors, are neither resources nor conditions.

The physical environment

There are some places where physical conditions don't change, such as the ocean depths (Fig. 2.4). In Lake Taupo, the water temperature at a depth of 110 metres changes by no more than a fraction of a degree, at 11°C all the year round, and there is almost total darkness. Very deep underground caves are also physically monotonous.

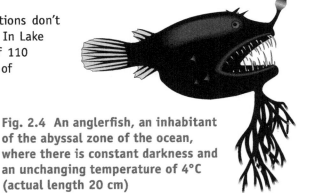

Fig. 2.4 An anglerfish, an inhabitant of the abyssal zone of the ocean, where there is constant darkness and an unchanging temperature of 4°C (actual length 20 cm)

In most places however, the physical environment is continuously changing, with changes of day and night and with the season.

Temperature

Temperature has an important effect on life because of its influence on the activity of enzymes. In homeothermic or 'warm-blooded' animals, enzymes function at near-optimum despite changes in environmental temperatures.

Though the effect of cold on enzymes is temporary, it may have a lethal effect on the whole organism. The Painted Lady butterfly, for example, migrates to New Zealand from Australia and breeds here every summer, but it cannot survive the cold of winter. The closely related Red Admiral on the other hand does survive the winter (Fig. 2.5).

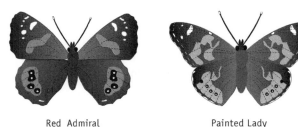

Red Admiral   Painted Lady

Fig. 2.5 The Red Admiral, a New Zealand butterfly that survives the winter cold, and the Painted Lady, an annual Australian immigrant, that cannot

Seasonal temperature changes influence almost every aspect of life. Most organisms, except birds and most mammals, become less active in winter. Because photosynthesis slows or even stops altogether, the supply of food for animals is greatly reduced. Birds and mammals are indirectly affected, since maintaining body temperature in winter requires increased food intake.

Organisms survive the winter in various ways. Some, such as tui and blackbirds, are able to find enough food to remain active throughout the year. Most, however, avoid the cold in ways briefly described below.

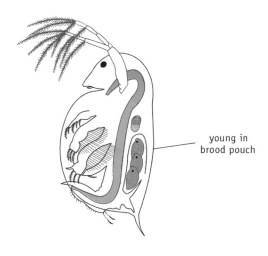

young in brood pouch

Dormancy

Many plants — and the eggs of many animals — survive adverse conditions in a state of greatly reduced activity called *dormancy*. This is not the direct result of low temperature on metabolism, because it requires some environmental factor

Fig. 2.6 *Daphnia* (a 'water flea') that does without sex in the summer, producing young without fertilisation

Excellence in Biology Level 2

ISBN: 9780170214094

besides favourable temperatures before activity can be resumed. A common stimulus is exposure to several weeks of winter cold.

Many animals pass the winter in the egg stage. In *Daphnia* ('water fleas'), reproduction in the summer is by *parthenogenesis*, in which all individuals are females and eggs develop without fertilisation. In response to shortening days of autumn, males are produced, and the resulting fertilised eggs are resistant to cold and other extreme conditions. The following spring the eggs hatch to give rise to another generation of parthenogenetic females (Fig. 2.6).

Hibernation and aestivation

Animals such as hedgehogs survive the winter by *hibernation*. This is a state in which metabolic rate and body temperature are very low. Before hibernating, the animal stores fat.

In aestivation the hazard is dryness rather than cold. Many land snails prepare for aestivation by secreting a thick layer of mucus over the entrance to the shell, which greatly reduces water loss (Fig. 2.7).

Fig. 2.7 A garden snail, an animal that survives hot, dry conditions by aestivating

Diapause

A special case of hibernation is *diapause*. This occurs in many arthropods and is a state in which development temporarily stops. It can occur in any stage of the life history. In the cricket it occurs in the egg stage. Only after a period of cold can the eggs continue development and hatch. This is an adaptive response that prevents the young emerging during unfavourable winter conditions (Fig. 2.8). In the white butterfly diapause occurs in the pupal stage, in response to shortening days during the last larval stage.

Fig. 2.8 The cricket, an insect that suspends development during the winter

Migration

Many animals make regular, usually seasonal, journeys. The shining cuckoo and many other birds migrate to warmer latitudes in the autumn (Fig. 2.9). This behaviour may not be just an escape from cold: by returning to cooler latitudes to breed in the spring, more daylight hours are available to feed the chicks. By the time they are nearly full size and are placing the greatest demands on their parents, days are longer than in the tropics. As with hibernation, seasonal migration occurs in response to cues such as day length.

Light

Fig. 2.9 The shining cuckoo, a migratory bird

Light is the energy that drives photosynthesis, so it is the ultimate source of energy for nearly all organisms. As a resource for plants it is competed for – plants producing the highest leaves get the most light.

Many animals use light as a source of *information*. Via their eyes, it can give information about food, predators and mates. In both animals and plants, day length or *photoperiod* can give information about time of year. This makes it possible to make adaptive changes in advance of unfavourable conditions.

Though visible light is not harmful to animals, it is usually associated with lower relative humidity. Animals that need damp conditions usually avoid light, which has the additional advantage of reducing the risk of being eaten.

Mineral ions

Mineral ions can influence organisms in two quite different ways. In low concentrations they are a *resource* for plants but in higher concentrations they influence the concentration of water, and thus the osmotic *conditions*.

As resources, certain minerals play a key role in the distribution of animals and plants. For example snails need calcium to make their limy shells, and are rare where the soil is low in calcium.

ISBN: 9780170214094

Salinity ('saltiness')

In the open sea the total salt concentration varies little, averaging 35 parts per thousand (35‰), compared with less than 1‰ in lakes. Most animals living in these conditions cannot withstand much change in the salt concentration of their surroundings. Freshwater animals have to get rid of water that enters by osmosis. Marine bony fish on the other hand, are constantly losing water by osmosis and are adapted to conserve it.

Estuaries and seashores are osmotically variable. In estuaries salinity varies with the state of the tide and with distance from the sea. Organisms living on bare rock on the seashore may alternately be exposed to rain and the drying effects of hot sunshine. Estuarine animals also have to cope with an environment that can vary between freshwater and seawater.

Some fish, such as salmon, migrate from saltwater to freshwater to breed, and make the return journey to the sea. Others, such as eels, migrate down rivers to the sea to breed (Fig. 2.10). In each case the fish have to make physiological adjustments.

Fig. 2.10 The salmon (left) migrates up rivers from the sea to breed; eels go in the opposite direction

Oxygen

Oxygen is a resource for animals, but on land above ground it is so concentrated (20.95%) that it is not competed for and is thus never a limiting factor.

The situation is quite different in water. Oxygen is only sparingly soluble in water, and even less so as temperatures rise. It also diffuses 10 000 times more slowly in solution than in air. In the cold, turbulent waters of shallow streams, the concentration of dissolved oxygen is high enough to support active fish such as trout. In ponds and slow-flowing rivers there is less oxygen and more sluggish fish such as carp and goldfish may be present.

In stagnant water, such as ponds and canals, oxygen concentrations may fluctuate considerably. Photosynthesis by pondweeds may enrich the water with oxygen during the day, but respiration reduces it at night. In the mud at the bottom of pond, conditions are usually permanently *anaerobic* (low in oxygen).

pH of soil and fresh water

Few plants can survive soil pH values below 3 or above 9. Extreme soil pH values can be hazardous to plants for two reasons:

- The direct effect of high or low H^+ concentration on metabolism itself.

- pH affects the solubility of certain soil ions.

Relative humidity

Fig. 2.11 An earthworm and a millipede - two animals that cannot survive for long in dry air

ISBN: 9780170214094

For terrestrial animals and plants, desiccation is always a potential hazard. How serious this is depends on three factors:

- The relative humidity of the air. At any given temperature, the lower the relative humidity, the more rapid is evaporation.

- The outer layer of the body. Insects and spiders have a waxy, waterproof outer layer in their cuticles and can survive for long periods in dry air. Many land invertebrates, mosses, and liverworts have relatively permeable body surfaces and cannot survive in dry air for long. Millipedes, centipedes, woodlice, earthworms and nematode worms control their internal water content by preferring dark places where relative humidity is usually high (Fig. 2.11).

- The surface/volume ratio of the body. Small organisms, or larger ones with thin or finely divided bodies, lose water more quickly.

Water flow

For an organism living in a stream, water flow has costs as well as benefits. It delivers resources such as oxygen and food, but it may also wash the organism away. In powerful currents, only organisms with low profiles such as some protoctists can hold on. The larvae of sandflies (*Austrosimulium*) live in streams and filter out suspended particles from the water. The abdomen has an adhesive tip that enables the animal to cling to the rock surface (Fig. 2.12).

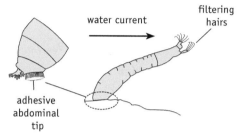

Fig. 2.12 Adaptation to water currents: sandfly larva holding on to rocky streambed

Tolerance

Every organism can only live within a certain range of physical conditions, such as temperature, and in the case of plants, soil water content and pH. The range of physical conditions in which an organism can survive is its range of *tolerance*.

An organism's tolerance in the laboratory is its *physiological* tolerance. In nature, it must not only withstand physical conditions — it must compete for resources, endure parasites, and in the case of animals, escape from predators. Not surprisingly, the *ecological* tolerances of most organisms are much narrower than their physiological tolerance.

The idea of a range of tolerance is sometimes expressed by a graph (Fig. 2.13), in which the vertical axis represents some measure of 'well being' or 'performance' such as rate of growth, number of offspring produced, and so forth.

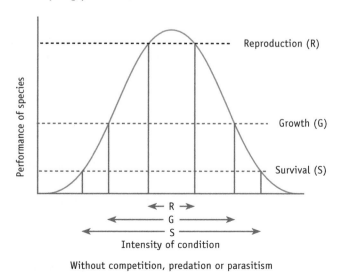

Without competition, predation or parasitism

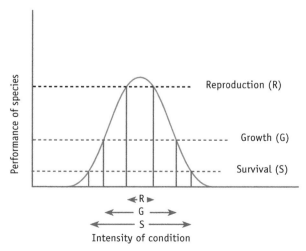

With competition, predation and parasitism

Fig. 2.13 Tolerance curve showing effect of a physical factor such as temperature on an organism in the absence (A) and presence (B) of adverse biotic factors factors

Factors interact with each other

Though environmental factors have been discussed separately, in reality they often interact with each other. They can do so in different ways:

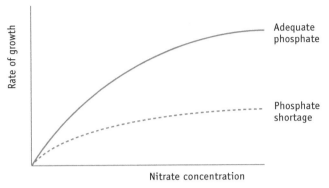

Fig. 2.14 Interaction between nitrate and phosphate supply on the growth of plants

- One factor directly influences another. For example, a rise in temperature of the air lowers its relative humidity. The solubility of oxygen in water decreases as it is warmed, and also with increasing salt concentration.

- One factor may influence the *effect* of another. For example, adding nitrate to a soil only increases plant growth if other minerals are present in sufficient amounts. Thus addition of nitrate has a much greater effect on plant growth if phosphate is adequate than if phosphate is deficient (Fig. 2.14). The mineral that is in shortest supply is said to be the one that *limits* growth. A similar kind of interaction occurs in photosynthesis, between light, temperature and CO_2.

Rhythms in the environment

In all habitats — except deep caves and ocean depths — physical conditions change rhythmically with day and night and with the seasons. On the seashore they also change with the tides.

Rhythmic changes can be predicted, and most organisms have an internal clock that enables them to change their behaviour before it gets dark or before the tide goes out. Nocturnal animals such as cockroaches become more active as evening approaches, even when kept in constant light, temperature and other conditions.

Most organisms use photoperiod (day length) to measure time of year; this is far more reliable than temperature and other seasonal cues.

The biotic environment

An organism's biotic environment is its relationships with other organisms, such as food supply, competitors, predators, parasites, mutualists and commensals.

Competition

A state of competition exists when the demand for a resource exceeds its supply. Thus plants may compete for light, water and minerals, and animals may compete for food, mates, or places to breed. *Intraspecific* competition occurs between members of the same species, and *interspecific* competition is between members of different species.

Competition has the following important features:

1. A resource can only be competed for if it is in limited supply relative to its requirements. Oxygen, for example, is never competed for by land animals since it is so abundant.

2. It operates in a d*ensity-dependent* manner (Chapter 3). Competition always results in a decrease in the reproductive rate of the population, though some individuals (the 'winners') may suffer very little. When competition is for mates or breeding sites, the link with reproduction is clear. If food is being competed for, then the effect on reproduction is indirect, the winners being more likely to become successful parents.

Intraspecific competition

Intraspecific competition is more intense than interspecific competition because individuals of the same species resemble each other more closely than members of different species. Resources that may be competed for by members of the same species include:

- Food, e.g. members of a pack of wolves after a kill.

- Mates, e.g. in red deer and seals, males compete for females.

- Breeding territories, e.g. blackbirds; or nest sites, e.g. gannets.

- Space, e.g. many sessile (attached) animals such as sponges, barnacles and mussels. In plants, competition for space as such is probably uncommon; most overcrowded plants are more likely to be competing for light above ground or for minerals or water below ground.

- Light, for example when plants grow close together.

- Minerals. This occurs in plants, particularly when other conditions such as light and temperature are favourable. In some situations, such as the upper layers of the sea and lakes in summer, mineral availability usually limits rate of growth of the phytoplankton.

Competition for mates

Competition by males for females takes two forms: *fighting* and *display*. In both cases there is strong selection pressure for competitive ability. This leads to *sexual dimorphism*, in which males differ markedly from females.

Competition for territory

Males do not always compete directly for females. In some species, males compete for possession of an area in which to mate. Such a *territory* is occupied and defended against others. In contrast, a *home range*, which is simply an area from which an animal does not usually stray, is not defended. Territorial behaviour is very widespread in the animal kingdom, especially amongst vertebrates. It is important to note that it is the area containing the resource that is defended, rather than the resource itself.

A territory may be:

- defended all the year round and represent a source of food, for example for the New Zealand robin.

- a site in which to build a nest and rear young, after which it is abandoned, as in the pukeko.

- very short-term, being simply a place for mating, as with red deer and seals.

Avoiding competition

Organisms may avoid competition in two different ways:

- By dispersal. In plants and fungi this is entirely passive, seeds and spores having little or no control over their destination. The larvae of many marine animals can actively move and often have some control over where they settle. Even so, the mortality of larval stages of marine animals is enormous.

- Different stages of the life cycle occupy different niches (Fig. 2.15). Leaf-eating caterpillars, for example, grow into nectar-feeding butterflies.

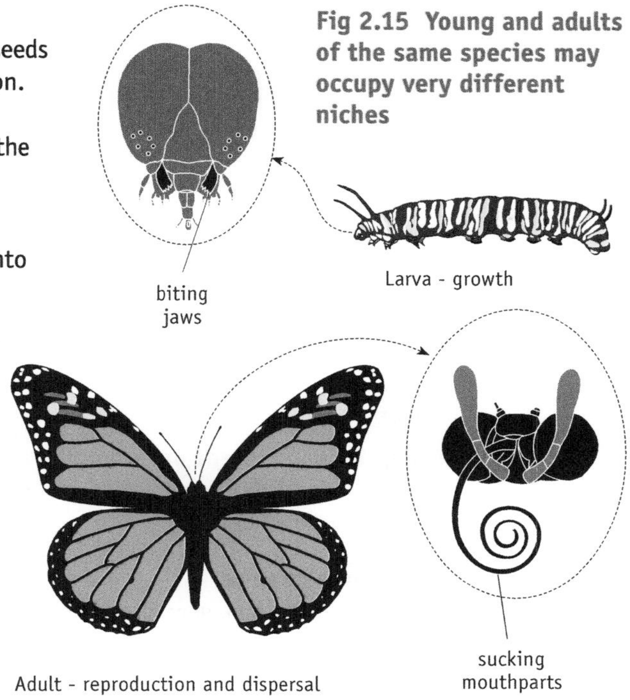

Interspecific competition – the Competitive Exclusion Principle

Since all members of the same population that are of the same age have the same requirements, they would be expected to compete strongly. It is less obvious that members of different species compete with each other. Just because caterpillars of the Red and Yellow Admiral butterflies can be seen on the same nettle plant, it does not follow that they are competing — there may be more than enough leaves to go round.

Some of the first experiments on competition were carried out by the Russian scientist G.F. Gause (pronounced 'Gauzer').

Fig 2.15 Young and adults of the same species may occupy very different niches

biting jaws

Larva - growth

Adult - reproduction and dispersal

sucking mouthparts

In one series of experiments he set up mixed and separate cultures of closely related species of unicellular protoctists, *Paramecium caudatum*, and a smaller species, *P. aurelia*. The paramecia fed on yeast that he regularly added to the cultures. The results are shown in Fig. 2.16.

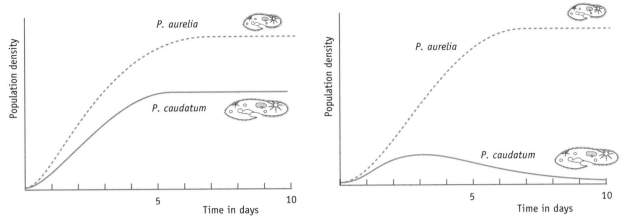

Fig. 2.16 Competition between *Paramecium* species. Left: single species culture. Right: mixed species culture.

Although both species flourished in isolation, in mixed culture *P. caudatum* was eliminated by *P. aurelia*. That competition was a two-way affair is shown by the slower rate of increase of *P. aurelia* in mixed culture.

Gause also found that by manipulating the conditions he could influence the outcome. For example when the water was regularly changed, *P. caudatum* won, suggesting that it was less able to tolerate wastes than *P. aurelia*.

Gause concluded from his results that two species cannot share the same habitat if they have similar requirements. This principle has become known as the *Competitive Exclusion Principle*.

The sum total of an organism's requirements constitutes its *niche*. This is often likened to its 'profession', in contrast to its habitat ('address'). The Competitive Exclusion Principle states that no two species can occupy the same niche indefinitely in the same habitat. A sea anemone and a mussel may live in the same rock pool but they clearly occupy different niches; the anemone feeds on small fish and crustaceans, and the mussel feeds on plankton.

There are many examples of animals which at first sight appear to live very similar lives but which are actually using different resources.

- Competition in various species of bird that feed on mud flats is avoided by differences in beak length (Fig. 2.17).

- Many flowers are so constructed that they can only be pollinated by certain kinds of insect. For example, flowers with nectar concealed at the end of a long tube are visited by butterflies, which have a very long, extensible proboscis and thus have an advantage over insects with short proboscides. The advantage to the plant is that it increases the chance that a visiting insect has previously visited the same species of flower.

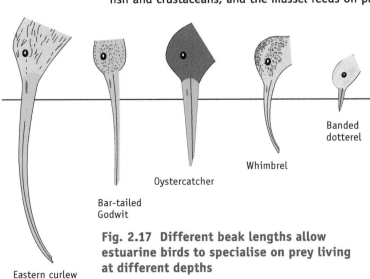

Banded dotterel

Whimbrel

Oystercatcher

Bar-tailed Godwit

Eastern curlew

Fig. 2.17 Different beak lengths allow estuarine birds to specialise on prey living at different depths

Niches change

An animal's diet may change considerably with its stage of development. This is especially true in species that undergo metamorphosis, as in amphibians and many insects. Even in the adult stage, there may be seasonal changes in diet.

Fundamental and realised niches

In captivity many animals can tolerate a much wider range of physical conditions and can exist on a wider diet than they can in nature. This is probably because, in the wild, animals can only survive on food that they can obtain in the face of competition and predation. A similar effect is seen in plants: most garden plants imported from overseas can only survive if protected from competition by 'weeds'; in their natural habitats in their countries of origin they are competitively superior to other species. In the wild, biotic factors force organisms to occupy narrower niches than they could under human protection.

The range of conditions in which an organism can survive under protected conditions constitute its *fundamental niche*. The narrower, more exacting conditions which it requires in nature constitute its *realised niche*. The difference between fundamental and realised niches is illustrated by the black rat. In Britain this species is more or less confined to cities, but after it was introduced into New Zealand it rapidly colonised most habitats, from forest to grassland. The house mouse has similarly spread throughout New Zealand, but in Britain it is far less numerous in woods and fields where it has to compete with field mice. Pohutukawa trees only grow naturally about as far south as Taranaki, but are grown in cultivation as far as Dunedin.

The difference between fundamental and realised niches is illustrated by Fig. 2.18, which shows the results of a famous study into competition in two species of barnacle, *Semibalanus balanoides*, and the smaller *Chthamalus stellatus*.

Barnacles have free-swimming larvae, which settle on the rocks and begin feeding and growing. Both species settle over a wide range, but reach adulthood over a narrower range. *Semibalanus* cannot survive desiccation and fails to survive on the upper shore. *Chthamalus* can survive desiccation but loses in competition. When *Semibalanus* was experimentally removed, *Chthamalus* became established, showing that it was competition rather than physical factors that was preventing it from living in the mid-tidal zone.

This study highlights an important ecological principle: the harsher the physical environment, the less intense are biotic factors such as competition. For example, mangroves can grow well in ordinary (low-salt) soils, provided that competitors are removed.

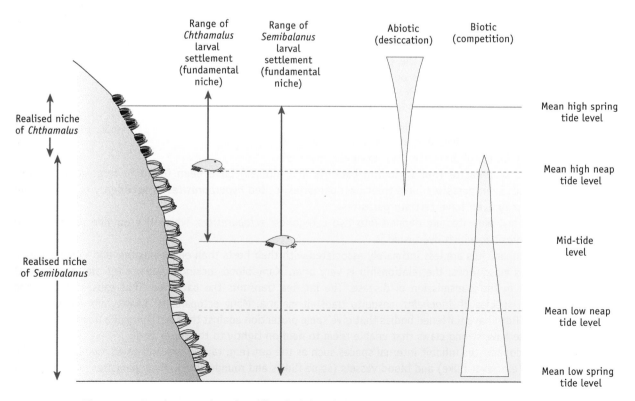

Fig. 2.18 Fundamental and realised niches in two species of barnacle on a Scottish shore

Exploitation

This is a relationship in which one species benefits at the expense of another. There are three different kinds:

- Predation. A predator feeds on another organism, the prey, killing it in the process. This definition therefore includes a squirrel eating acorns (killing the embryo plant in the process), a water flea feeding on phytoplankton, and a weasel feeding on a rabbit. The essential feature of a predator is that it exploits the prey as 'capital'.

- Grazing. One organism kills parts of many other organisms, leaving the rest to regenerate more tissue. A rabbit is a grazer; it feeds on 'interest' generated by a *population* of grass plants, which are not killed.

- Parasitism. A parasite feeds on another organism (the host). It feeds on the 'interest' generated by a *single* organism.

Predation

Predators and prey affect each other over two very different time scales:

- Over the short term they may influence each other's *numbers*, (see Chapter 3).

- Over the longer term they shape each other's *evolution*. Predators and prey are in a kind of evolutionary 'arms race'. Predators are under constant selective pressure to improve prey-catching ability. Consequently, prey are under selective pressure to improve escaping ability.

Parasitism

Whereas a predator feeds off 'capital' and must move on to find another prey, a parasite lives off 'interest'. It depends for its food on one organism, the host, which is not usually killed (although it may be weakened rendering it more likely to succumb to other hazards). In general, it is in the interests of the parasite to harm its source of food as little as possible.

Although parasitism is often considered to be a somewhat 'abnormal' way of life, there are actually more parasitic than free-living species because most free-living organisms are host to several kinds of parasite. For example, over 30

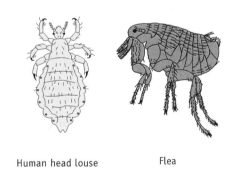

Human head louse   Flea

Fig. 2.19 Two ectoparasitic, wingless insects

species of parasite (not including bacteria and viruses) have been found in feral goats in New Zealand. Even parasites have their own parasites, called *hyperparasites* or secondary parasites, and these may even have tertiary parasites.

Animal parasites are divided into two categories: *ectoparasites*, which live outside the host, and *endoparasites*, which live inside it.

Ectoparasites are less intimately associated with their hosts than endoparasites, and in some cases, such as mosquitoes, the relationship is very brief. Many blood-sucking ectoparasites are important as *vectors* in the transmission of disease. The rat flea transmits the bacterium that causes plague, and certain species of *Anopheles* mosquito transmit malaria. Many ectoparasitic insects are wingless (Fig. 2.19). Both have flattened bodies that give some protection against the scratching of a mammalian host, and lice have strong claws that enable them to hold on tightly to the host's body.

Endoparasites inhabit internal spaces such as the gut (e.g. tapeworms and some roundworms), bile ducts (e.g. liver fluke) and blood vessels (some flukes and roundworms). Many parasites maintain their position in these places by holding on with hooks, suckers, or both.

Since they are protected and nourished by the host, an endoparasite does not need to spend as much energy on food-finding and escape. For this reason, *many endoparasites are structurally simpler* than their free-living relatives. This frees up resources for dealing with the biggest hazard for any endoparasite — *transmission*, or getting to another host. Adult parasites cannot travel from one host to another since they are not adapted to survive in the tough environment outside. Many blood parasites use a blood-sucking insect as a *vector*. Most other endoparasites leave the body as resistant eggs or specialised larval stages. The mortality in transmission is huge and is only offset by the equally high reproductive capacity of the adult.

Besides producing large numbers of offspring, many endoparasites have complex life cycles, often involving a succession of two or even three kinds of host, as in tapeworms.

An endoparasite that is potentially dangerous to humans is the hydatid tapeworm, *Echinococcus granulosus* (Fig. 2.20).

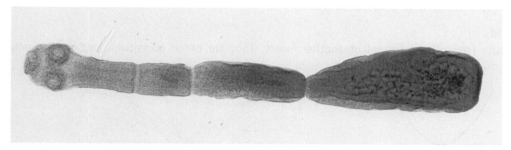

Fig. 2.20 The hydatid tapeworm (courtesy Dr Barry O'Brien, University of Waikato)

The adult worm lives in the small intestine of a dog and is only a few millimetres long. Its 'head' or *scolex* has hooks and suckers by which it anchors itself to the lining of the intestine. Whereas most tapeworms live solitary lives, up to 20 000 hydatid worms can be found in one dog. Tapeworms have no gut — they are surrounded by a soup of digested food, and protected from the host's enzymes by a thick cuticle.

Each 'segment' or *proglottid*, contains male and female organs and, like other tapeworms, the animal can fertilise itself. Once every few days a proglottid, containing an average of 500 eggs, is cast off and passed with the faeces (Fig. 2.21).

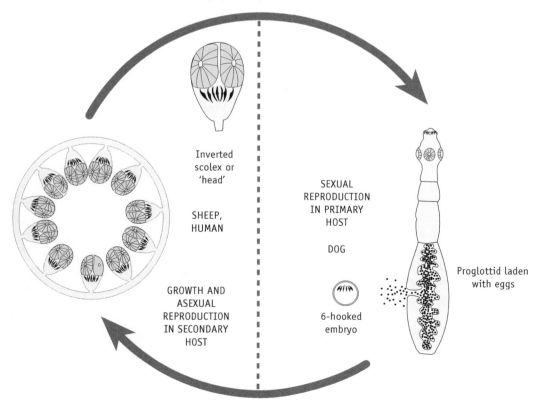

Fig. 2.21 The life cycle of the hydatid tapeworm

The proglottids rot away, leaving the 'eggs' (actually embryos) protected by a tough shell. The 'eggs' can survive for months, and if eaten by a sheep or other potential host, the shell is digested away. The embryo uses six hooks to bore its way through the gut lining. It enters the blood and is carried to the liver. Here most develop into *hydatid cysts*, though some are carried to the lungs or other organs. Each cyst buds off many tapeworm 'heads', each turned inside out. The cyst continues to grow, budding off 'daughter' cysts, which in turn produce further generations of cysts. Eventually the cyst contains hundreds of thousands of 'heads', each of which is capable of developing into a tapeworm.

To reach the next stage, the sheep must be eaten by a dog. A sheep infected with cysts in its liver, lungs or other organs is more vulnerable to predators. Thus, whereas it is in the interest of most parasites not to harm the host, the opposite is true of the hydatid worm. In the dog's intestine the inverted 'head' attaches itself and begins to grow into an adult worm.

Parasitoids

Some insects feed off the insides of another insect. They are called parasitoids, but are actually predators since they always kill the host.

Fig. 2.22 *Cotesia glomerata (=Apanteles)*, a tiny wasp (actual length 3 mm) that feeds inside the larvae of the cabbage white butterfly

Fig. 2.22 shows a tiny wasp that was introduced to control the cabbage white butterfly. The wasp lays its eggs inside the very young caterpillars. At first the wasp larvae feed off non-essential tissues and the caterpillar continues to grow normally. When the caterpillar has reached full size the wasp larvae start feeding off essential organs and then spin their cocoons and form pupae. The caterpillar is now dead. Another parasitoid wasp, *Pteromalus puparum*, attacks the pupal stage.

Mutualism

A surprising number of organisms enter into some form of partnership with another species in which both benefit. This is called mutualism. Lichens are examples, in which the hyphae of a fungus are interwoven with single-celled 'algae'. A few other examples are described below.

Root nodules

The roots of members of the legume family, such as gorse, lupin and clover, have swellings or nodules containing bacteria belonging to the genus *Rhizobium* (Fig. 2.23). The basis of the association is that each partner can do something the other cannot: the bacteria can fix nitrogen, and the plant can fix CO_2. The bacteria convert nitrogen into ammonia, which is used to make amino acids, the organic carbon part being supplied by the plant.

The bacteria live freely in the soil as saprobes, but when soil nitrogen is in short supply they penetrate the root hairs of the legume and stimulate development of a nodule. The legume is then able to compete very effectively with non-legumes.

Fig. 2.23 Nodules on root of clover

ISBN: 9780170214094

Mycorrhiza

One of the most widespread mutualistic relationships is mycorrhiza (meaning 'fungus-root'), a partnership between a fungus and the roots of a plant. The fungal hyphae form a meshwork in the cortex of the root and also extend for up to several centimetres into the soil. The hyphae are much better at absorbing minerals (especially phosphate) than the root hairs of the plant. This may be because they are finer than root hairs and extend further into the soil. So effective are the hyphae in absorbing minerals that root hairs may hardly develop, and in many trees the lateral roots are very short.

Experiments with radioactive tracers show that the fungus obtains sugar from the plant. The fact that mycorrhizal fungi do not live freely in the soil suggests that, so far as the fungus is concerned, the relationship is *obligate*, or essential for survival.

Mycorrhizae are more than just widespread — they are the rule rather than the exception. Some of the mushrooms commonly found beneath forest trees are actually the reproductive bodies of mycorrhizal fungi (Fig. 2.24).

Fig. 2.24 The fly agaric, a mycorrhizal fungus common under pines

Gut mutualists

Very few animals can produce a cellulase enzyme, and most herbivores rely on microorganisms in their guts to do the job for them. Many mammalian herbivores have a large part of the gut specialised for housing these microorganisms. Examples are the caecum of rabbits, rats and mice, and the rumen of sheep, cattle and deer (Fig. 2.25). In this protected environment the microorganisms obtain food, warmth and the anaerobic environment that they require. The herbivore derives a triple benefit:

- It obtains large amounts of carbohydrate from the digested cellulose.

- The microorganisms produce essential amino acids, which the animal uses when the microorganisms are eventually digested.

- They also produce vitamin B_{12}, which no animal can produce for itself, and also certain other B vitamins.

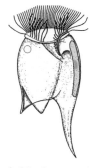

Fig. 2.25 A mutualistic proctoctist from the rumen of a cow

Fig. 2.26 A remora on a great white shark

Commensalism

This is a relationship between two species in which one benefits and the other neither benefits nor is it harmed. For example remoras are small fish that spend most of their time attached to sharks, anchored by a sucker that is actually a highly modified dorsal fin. They only detach themselves when the shark feeds, picking up small scraps of food (Fig. 2.26).

Humans have many commensals; many bacteria live harmlessly on our skin and in our colons, although some colon bacteria are potentially harmful. *Escherichia coli*, for instance, can cause infections of the urinary tract, and also food poisoning.

Antibiosis and allelopathy

This is a relationship in which one organism produces a chemical that inhibits the growth of potential competitors. Many fungi produce substances called *antibiotics* that inhibit the growth of bacteria. The first to be discovered was penicillin, produced by the mould *Penicillium* (Fig. 2.27). Some flowering plants also produce chemicals with similar effects on other plants. Though it is essentially similar, in plants the production of such chemicals is called **allelopathy**.

Fig. 2.27 Photograph of *Penicillium* on an orange

ISBN: 9780170214094

Summary of interspecific relationships

	Species A	Species B
Competition	–	–
Exploitation	+	–
Mutualism	+	+
Commensalism	+	0
Antibiosis	+	–

+ = benefits, - = harmed, 0 = no effect

Summary of key facts and ideas in this chapter

○ An organism's *habitat* is the place in which it lives.

○ An organism's *environment* is the sum total of all the factors in its surroundings that affect it. Unlike a habitat, environments are constantly changing.

○ An organism's *niche* is its way of life, and can be likened to its 'profession'.

○ Inherited features that increase an organism's chances of survival and reproduction are called *adaptations*.

○ Biotic environmental factors are the influences of other organisms, such as predators, competitors and parasites.

○ Abiotic factors include physical factors such as light intensity, temperature and humidity.

Test your basics

Copy and complete the following sentences. In some cases the first letter of a missing word is provided.

1. The place where an organism lives is its ___*___.

2. The ___*___ of an organism is all the factors in its surroundings that influence it.

3. An organism's way of life is its ___*___.

4. A ___*___ is all the members of a species in a particular area.

5. A ___*___ is all the populations in a particular area.

6. An ecosystem consists of a community plus all the ___*___ ___*___ components in a defined area.

7. Adaptations are i___*___ features of an organism that increase its chances of survival and ___*___. They may be s___*___, b___*___, or ___*___.

8. ___*___ factors in the environment are the effects of other organisms; physical factors such as light intensity are said to be ___*___.

9. Animals that can regulate their body temperature are said to be ___*___.

10. Many organisms survive adverse conditions by entering a state of ___*___, in which activity is greatly ___*___.

11. ___*___ in animals is a state of dormancy in which an animal survives a period of cold. When dormancy aids survival through a period of dry conditions, the animal is said to ___*___,

12. ___*___ is a special case of hibernation in which an arthropod survives harsh seasonal conditions by temporarily ceasing development.

13. Light is important to plants as a source of ___*___. In animals, it may be an important source of ___*___ about food, predators and other environmental factors. In both plants and animals, day-length or ___*___ may give information about the time of year.

14. Mineral ions are an important __*__ for plants; in aquatic animals the total concentration of ions ('salinity') determines the __*__ conditions.

15. In organisms that live above ground, oxygen is so abundant that it is never in short supply. This is not so in __*__ organisms since it is only sparingly __*__ in water.

16. Another important environmental factor in aquatic animals is its acidity/alkalinity, measured on the __*__ scale.

17. The rate at which a terrestrial organism loses water depends on the __*__ of the air, the __*__ of its body surface, and also the __*__ /__*__ ratio of the body. The last of these depends on the size and __*__ of the body.

18. Environmental factors may __*__ with each other, which means that the effect of one factor may depend on the level of another. For example, the oxygen supply in a stream-living animal may be affected by the __*__ of flow of water past the body.

19. Competition is a state of affairs in which the __*__ for a resource exceeds its __*__. __*__ competition is between members of the same species and is more severe than __*__ competition, which is between members of different species.

20. When male animals compete for females, it may be by physical combat or by __*__. In both cases the males differ markedly from females, a state called sexual __*__.

21. In some cases males compete for an area of habitat called a __*__, which is __*__ by the successful individual.

22. The __*__ __*__ Principle states that no two species can indefinitely occupy the same __*__ in the same __*__.

23. In __*__, one species benefits at the expense of another.

24. There are three kinds of __*__; in __*__ an organism feeds off another, killing it in the process. In __*__, one organism kills part of another, leaving the rest to grow again. In __*__, one organism, called a __*__, feeds off another, called the __*__, harming but not usually killing it.

25. __*__ is a relationship between two organisms of different species in which both benefit. An example is __*__, in which a __*__ lives in the roots of a plant. The __*__ obtains an increased supply of minerals, and the __*__ obtains organic matter from the plant.

26. __*__ is a relationship in which one species obtains benefit and the other neither benefits nor is harmed.

3 Populations

A population is all the organisms of a particular species living in a defined area; for example the lugworms in a particular stretch of shore, or the trout in Lake Taupo.

Population ecologists are concerned most of all with understanding how populations are *regulated*. Why, for example, are some species common and others rare? Why do some populations remain fairly stable over quite long periods of time, while others may fluctuate wildly?

The number of individuals per unit area is the *population density*. At any one time, population density is subject to two opposing influences: a tendency to *decrease* — due to **mortality** (deaths) and **emigration**, and a tendency to *increase* — due to **natality** (births) and **immigration** (natality includes hatching and germination).

Mortality actually refers to the death *rate*, and is expressed in terms of the number of deaths per thousand per year (or some other unit of time). Similarly, natality is expressed in terms of the number of 'births' per thousand per unit time.

In most populations the balance between natality and mortality is continually shifting, so that numbers are constantly changing. Fluctuations may be quite small, or they may be so large that only a logarithmic scale can accommodate them (Fig. 3.1).

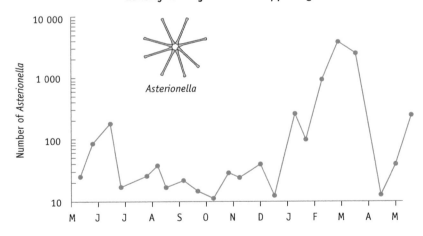

Fig. 3.1 Fluctuations in numbers of *Asterionella formosa*, a protist, in surface waters of Lake Rotorua

Estimating population densities

It is seldom possible to count the entire population — that is, to conduct a **census**. Most often, ecologists have to be content with estimates based on *samples*. There are various methods, the suitability of each depending on the species and situation.

Quadrats

With plants and sessile (attached) animals, ecologists use **quadrats** (*not* quadra**nt**s). A quadrat is a small part of the habitat whose dimensions are accurately known. The size of a quadrat is thus a

ISBN: 9780170214094

known proportion of the total habitat. If the number of individuals in each of a series of quadrats is accurately counted, the mean number per quadrat enables the population density to be estimated.

The size of the quadrat has to be adapted to the situation; when estimating the density of trees, it has to be much larger than when estimating the density of barnacles. When quadrats are being used to study the composition of a community, obviously it must be large enough to include most of the species present. To get an idea of a suitable quadrat size, it's best to do a preliminary survey using quadrats of different sizes, and record the number of species present in each case. The result for plant species in a lawn would be something like Fig. 3.2. Beyond about 1.5 m², increase in quadrat area brings little increase in the number of species, so 1.5 m² would be an optimum size.

Though quadrats are often square for convenience, they do not have to be any particular shape. In fact in some cases they are not even two-dimensional. For example, in estimating the density of cockles on a sandy shore, one would count the number of cockles in a certain area down to a certain depth i.e. in a certain *volume* of sand.

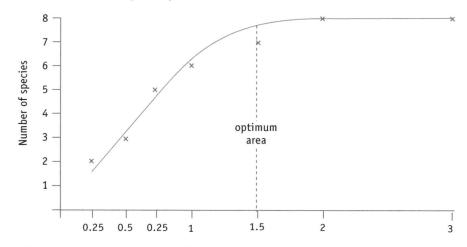

Fig. 3.2 Estimating the optimum size for a quadrat survey

Mark, release, and recapture

This method is most suitable for animals that are fairly mobile but can easily be caught, such as grasshoppers. A reasonably large number are caught, marked, and then released and allowed to mix with the rest of population. The following day another sample is caught. Some of these will be marked, and the proportion of these *recaptures* makes it possible to estimate the total number in the population (Fig. 3.3). The accuracy of this estimate depends on the extent to which certain assumptions, listed below, are valid.

MARKED SAMPLE

1. Catch and mark
a number of animals

POPULATION

2. Release into the population
and leave long enough to mix

SAMPLE

3. Catch another sample,
which should contain some
marked individuals

4. The proportion of marked animals in the second capture should be approximately equal to the proportion of released animals in the whole population.

Fig. 3.3 Mark, release and recapture method for estimating numbers of animals

ISBN: 9780170214094

Suppose that 100 individuals are marked and released. The larger the total population, the smaller the proportion of marked individuals will be. If, for instance, the total population is 1000 (a figure that we actually do not know), then about 10% of that population would be marked. The percentage of marked animals in the population can be estimated by taking a second sample a day or so later, a proportion of which would be marked. In the above imaginary case, about 10% would be marked.

It follows that:

$$\frac{\text{number of marked animals released (R)}}{\text{total number of animals in population (T)}} = \frac{\text{number of marked animals recaptured (M)}}{\text{total number of animals recaptured (N)}}$$

$$\text{Hence } T = R \times \frac{N}{M}$$

This estimate will only be reasonably accurate if the following conditions and assumptions hold:

- The number released must be large enough for a reasonable number to be recaptured. This number will vary with the species and the situation.

- The marked animals must mix completely with the unmarked ones. Some animals, such as slaters, live in local aggregations, and mixing is unlikely to be complete.

- Marked and unmarked individuals must be equally likely to be caught. If some are easier to catch than others, the marked ones are more likely to be recaptured. A higher proportion of marked individuals amongst the recaptures will lead to an underestimate of the total population, and vice versa.

- The mark must not reduce an animal's life expectancy in any way. It must be non-toxic and must not make it more likely to be seen by predators.

- The mark must persist at least until the day of recapture. In arthropods, which shed their cuticles, some marks will be lost in this way.

- There must be no immigration or emigration.

Population distribution – transects

Even in a fairly clear-cut area such as a lawn or pond, environments are rarely uniform, and as a result populations are not usually distributed evenly. There may be a clear gradient of physical conditions, such as on a sea shore. Distribution of plants or stationary animals (such as barnacles) along an environmental gradient is studied by recording the numbers along a line or **transect**. The line may be a length of string or rope, but it should be marked at suitable intervals, such as every 10 cm or every metre. At regular intervals the number of individuals in contact with the transect line is recorded. Samples can be taken in one of two ways (Fig. 3.4):

- Using a *line* transect. At regular, marked points along the transect string or rope, each species is recorded as being present or absent.

- Using a *belt* transect, in which counts are taken in a series of quadrats placed at regular intervals along the transect.

If the transect is up a rocky shore, it is the vertical height of each count that is important, rather than the horizontal distance along the transect. Vertical height is in relation to some fixed point such as the high tide mark.

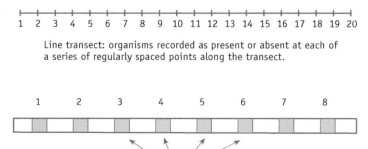

Line transect: organisms recorded as present or absent at each of a series of regularly spaced points along the transect.

Quadrat placings

Belt transect: quadrats placed at regular intervals along the transect. For each species the number of individuals is counted or percentage cover estimated.

Fig. 3.4 Line transect and belt transects

Population growth

Most populations fluctuate, but over a long period of time the rises and falls more or less cancel each other out. It follows that over a long period the tendency for the population to increase must be balanced by its tendency to decrease, in other words:

natality + immigration = mortality + emigration

In most populations the vast majority of individuals die without reproducing. Most will be weakened by parasites, eaten by predators or unable to obtain enough food. Many fail to reproduce because they cannot find a mating partner, and in plants, some ovules remain unfertilised. Because of this, death rates in natural populations are very high. To get an idea of how high mortality normally is, we only have to see what happens when a population grows without any of the usual checks.

Under favourable physical conditions and in the absence of predators, parasites or competition, a population grows **exponentially.** This means that numbers grow by the same *proportion* each unit time. Doubling every hour, tripling every ten years, or increasing by 1.5 times every century, would all be examples of exponential growth.

When numbers are plotted against time, the graph for an exponentially growing population gets steeper and steeper, and very soon numbers become too large to be plotted on an ordinary scale. The answer is to use a *logarithmic* scale, which allows very large as well as very small numbers to be plotted on the same graph.

Another advantage of using a log scale is that exponential growth is easy to recognise because it gives a straight line. In other words, the logarithm of the numbers of organisms is directly proportional to the time. This is why exponential growth is also called **logarithmic growth** (Fig. 3.5).

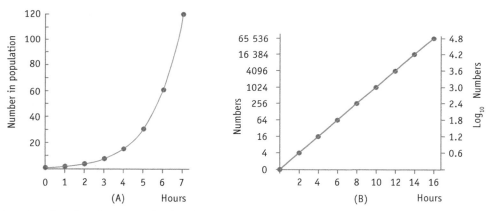

Fig. 3.5 Graphs showing growth of an imaginary population of bacteria, each bacterium dividing once every hour. (A) Plotted on a linear scale. (B) A logarithmic scale accommodates much higher populations. Note that both scales in (B) are logarithmic, but represented in different ways.

Though all populations have the capacity for exponential growth, the rate of increase varies greatly. One way of expressing the rate of increase is in terms of the **doubling time,** or time for the population to double. In organisms that reproduce by binary fission this is the same as the generation time, which for some bacteria under ideal conditions is less than 20 minutes. For fruit flies at 25° C it is just under two days, and for rabbits it is about three weeks.

Under ideal conditions the doubling time depends on three features of the organism's biology:

- The **generation time,** or the time it takes for an organism to become a parent. Generation times vary from less than 20 minutes in some bacteria to about 30 years in elephants.

- The average number of offspring produced by each parent in each reproductive event; this varies from one in apes and most monkeys to hundreds of millions in puffball fungi.

- The number of times an organism reproduces in its life. Some, such as eels, most insects, and annual plants, devote all their energy to one gigantic reproductive effort and then die. Others may reproduce more than once.

Of these factors, generation time is by far the most important. When a bacterium reproduces, it only produces two new individuals per generation, but in the time it takes for an oyster to produce millions of offspring a bacterium has had time to double its numbers thousands of times. A thousand doublings would give 2^{1000} offspring — far more than the estimated number of charged particles in the universe! It is no coincidence that most serious pests have short generation times.

Sigmoid (S-shaped) growth

Under natural conditions, logarithmic growth cannot continue for long; the larger the numbers, the greater the checks on further growth, for example:

- Increasing competition for food.

- Predators can catch more prey when the latter are more numerous.

- Parasites can spread more easily from one host to another when the hosts are closer together.

Collectively, these factors constitute the **environmental resistance**. When numbers have stabilised, the population has reached the habitat's **carrying capacity** and there is equilibrium between natality and mortality.

Suppose a million bacteria were to be introduced into a nutrient solution at a constant temperature of 20° C. If each bacterium divides every hour, the numbers would grow increasingly rapidly, but growth would eventually slow with the effects of overcrowding. Table 3.1 shows an imaginary set of figures to illustrate the important features of the growth of such a population, and Fig. 3.6 shows the figures plotted graphically.

The graph shows five distinct phases:

A The **lag phase**. If (as in this case) the new medium is chemically different from the previous one, there may be a period during which bacterial metabolism adapts to the new conditions, and the bacteria are not reproducing. (More complex organisms may have a lag phase for a different reason — when numbers are very low, it may be difficult for males and females to find each other.)

B **Exponential growth**. Each cell is dividing at a constant rate — every hour in this case, so the population doubles every hour.

C **Transition period**. Each cell takes longer to grow and divide due to the effects of crowding, but the graph is *still getting steeper*. This is because, although each individual bacterium is reproducing more slowly, this is more than offset by the increasing number of bacteria.

D **Slowing phase**. The gradient of the curve begins to decrease, because the increasing number of dividing cells is more than offset by their decreased rate of division.

E **Stationary phase**. Growth in numbers has ceased. Cells are either dead, or have formed resistant spores.

Growth period	Time (hours)	Numbers (millions)	Increase (millions)	% increase
	0	1	–	–
A	1	1	–	–
	2	1	–	–
	3	2	1	100
	4	4	2	100
	5	8	4	100
B	6	16	8	100
	7	32	16	100
	8	64	32	100
	9	128	64	100
C	10	232	104	81
	11	392	160	69
	12	496	104	26.5
	13	542	46	20
	14	564	22	4
D	15	582	18	3.2
	16	592	10	1.7
	17	598	6	1.0
	18	600	2	0.33
E	19	600	0	0
	20	600	0	0

Table 3.1 Growth of an imaginary population of bacteria

ISBN: 9780170214094

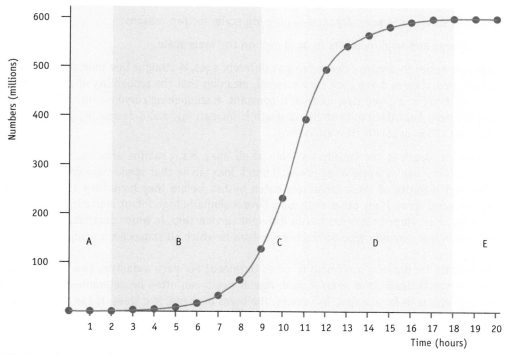

Fig. 3.6 Graph of the growth of an imaginary population of bacteria to show the five phases of growth

'J' curves

Not all populations grow in a sigmoid way. If a population has the capacity to grow faster than its food can regenerate, it may grow exponentially for a while before crashing, producing a 'J'-shaped 'boom and bust' curve (Fig. 3.7). This has happened on numerous occasions when herbivores such as deer have been introduced into a predator-free area such as an island. It almost certainly would have happened in New Zealand had deer numbers not been kept down by hunters.

Survivorship curves

In most species, death rates vary with age. This can be shown as a **survivorship curve**. It is a graph showing how the number of survivors of an original group born at approximately the same time changes with time. A group that started life at the same time is called a **cohort**.

Suppose that out of an original cohort of 10 000 individuals, half die every year. Then 5 000 will have survived after one year, 2500 after two years, 1250 after three years, 625 after four years, and so on (Fig. 3.8).

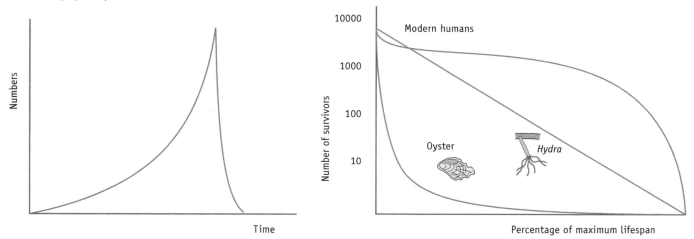

Fig. 3.7 A 'J' growth curve

Fig. 3.8 Three kinds of survivorship curve

The number of survivors is normally plotted on a log scale, for two reasons:

- It allows large and small numbers to be shown on the same scale.

- It makes it easier to compare death rates at different ages. A straight line indicates that a constant *proportion* is dying each time interval, meaning that the probability of any given individual dying in a given time interval is constant. A steepening curve means that life is becoming more hazardous (probability of death is increasing), and a decreasing gradient means that life is becoming less dangerous.

In most species, death is not equally probable at all ages. Many marine annelids, crustaceans and molluscs produce huge numbers of eggs which hatch into larvae that spend time drifting in the plankton. The vast majority of these larvae are eaten by fish before they have time to settle and develop into the adult form. Many other organisms have a similarly high infant mortality.

Humans living in developed countries have a different survivorship, in which death rates are quite low until old age. A less common type occurs in organisms in which all stages are equally vulnerable, as in *Hydra*.

How are the data for drawing survivorship curves obtained? For each organism, two things need to be known: its age at death, and when it died. Age at death can often be determined from clues such as growth rings, as in for example, fish scales, the horns of sheep, and trees. If the approximate date of death is also known, each dead individual can be assigned to its cohort.

Age group structure

Changes in mortality with age affect the age structure, or the relative proportions of each age group in the population. Fig. 3.9 shows the age structure of two human populations, Sweden and India in 1970.

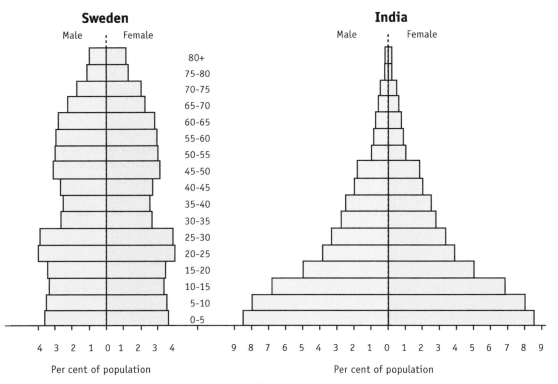

Fig. 3.9 Age structure for Sweden and India, in 1970

The age structure of a population results from the relationship between natality and mortality during previous years. Because these can vary independently, age structures need to be interpreted with caution. For example, a population with a high proportion of young individuals could result from two quite different causes:

Excellence in Biology Level 2

ISBN: 9780170214094

- High infant mortality, few reaching adulthood. If the population size is stable, mortality of each age group can be gauged by the decrease in the size of successive age groups up the pyramid.

- Even if infant mortality is low, a rapidly expanding population results in a high proportion of young individuals.

In some cases populations are divided into only three age groups: pre-reproductive, reproductive, and post-reproductive. In such cases one cannot draw any conclusions because the three phases may have very different durations. Cicada nymphs, for example, live for several years underground, but live for only one summer as adults. In cases like this there are bound to be many more pre-reproductive individuals than reproductives.

Population regulation

Every environment has a finite *carrying capacity*, so no population can grow unchecked for long. Sooner or later it either stabilises or falls drastically. Where physical conditions are only favourable for short periods, populations may undergo wild fluctuations, with a series of 'J'-shaped curves. In environments that are permanently favourable to life, most populations undergo only minor fluctuations because they are *regulated* by biological factors such as competition, predation and parasitism.

Checks on growth can act either by increasing mortality or decreasing natality — or more often, a combination of the two (Fig. 3.10). Since mortality is the number of deaths *per thousand* per unit time, an increase in mortality means an increase in the *proportion* of individuals dying per unit time. In other words, the average life expectancy of an individual must decrease. Similarly, a decrease in natality means that the average reproductive output *per individual* must decrease.

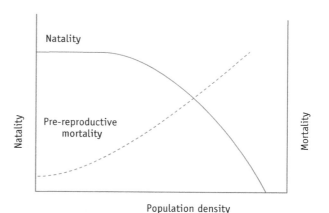

Fig. 3.10 Density-dependent effects on mortality and natality. As population density increases the probability that any given individual will die without reproducing increases, and the average number of offspring produced by those that reproduce, decreases.

It follows that population control must be **density-dependent**. It operates by a kind of **negative feedback** process, in which the greater the increase in numbers, the greater the checks on further increase. On the other hand, if population density falls below the carrying capacity, the intensity of density-dependent checks is relaxed, allowing population numbers to rise (Fig. 3.11).

Whereas density-dependent factors are usually biological, density-independent factors are generally abiotic. There is no special reason, for example, why a storm should kill a higher *proportion* of a dense population of birds than it would of a low population.

Control by predators

Every time a predator kills a prey, the number of prey is reduced. It might therefore seem obvious that predator populations control their prey populations. Although there is clear evidence for this in many cases, it is not always so. Some predators are simply acting as 'executioners' of animals that were doomed anyway. A gull chick that loses its parents will die even if adult gulls in the colony do not get it first. A wildebeest calf that loses its mother is doomed, whether or not it is killed by hyenas.

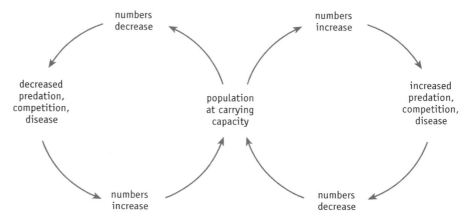

Fig. 3.11 Population regulation by negative feedback

There are cases where predators do control prey. An example is the small white butterfly (*Pieris rapae*) which was accidentally introduced into New Zealand about 1930. Within a few years it was devastating cabbage and other brassica crops. In an attempt to control the pest a small parasitoid wasp, *Pteromalus puparum*, was introduced. The wasp lays its eggs in the pupa of the butterfly, and the wasp larvae feed on it and eventually kill it. The result was a dramatic decrease in the numbers of the small white butterfly, which is now a nuisance rather than a major pest.

Predator-prey cycles

Some prey appear to be controlled mainly by their food supply, predators having only a minor effect. An example is the snowshoe hare and its chief predator, the lynx. In arctic Canada where these two animals live there are relatively few other species, so the biotic environment is simpler than in warmer latitudes. Both animals were trapped in large numbers and records of the Hudson's Bay Company give a picture of population fluctuations over more than a century (Fig. 3.12). Both populations undergo regular oscillations every nine to ten years. After a short delay, each hare peak is followed by a lynx peak.

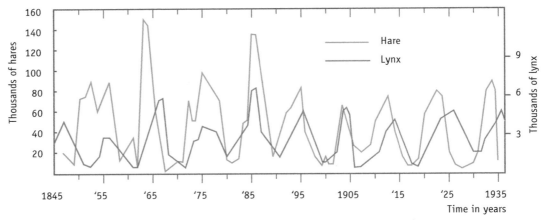

Fig. 3.12 Fluctuations in numbers of lynx and snowshoe hare in Canada

Lynx numbers were evidently fluctuating in response to changes in the supply of prey, but are the decreases in hare numbers caused by predation by the lynx? There are two reasons for believing that they are not the main factor controlling the hares:

- In parts of Canada where there are no lynx, hare numbers still undergo ten-year cycles.
- Hares have a higher reproductive potential than lynx, with an average of about ten young per year compared with two or three for the lynx. When food is adequate, the hares are always able to out-reproduce the lynx.

Research has suggested that the regular crashes in hare numbers is due to overgrazing of their food supply. Predation by lynxes, while the numbers of hares are low, delays the build up of hares. During this time the vegetation makes a full recovery, after which there is a rapid increase in hares.

Territory and population regulation

Predation and parasitism act by influencing mortality. In many territorial animals, competition for territory can act as a powerful population regulator by controlling natality. In many territorial birds, such as robins, only territory-holding males can attract a mate, non-territory-holders live as bachelors. The number of territories available therefore limits the size of the population. Since territory tends to prevent overcrowding, it results in the population size being held at a level that the food supply can support.

ISBN: 9780170214094

Summary of key facts and ideas in this chapter

- A population is all the organisms of a particular species living in a defined area.
- Populations have characteristics that individuals do not show, such as *natality* (birth rate) and *mortality* (death rate).
- The rate at which a population is growing or declining depends on the balance between natality + immigration and mortality + emigration.
- The *population density* is the number of individuals per unit area or unit volume of habitat.
- Population densities can be estimated by various methods, such as *quadrat* sampling and mark, release and recapture.
- Variation in population density in an area can be surveyed by taking samples along a line or *transect*.
- When a population grows without restriction, the increase in numbers is *exponential*, in which the population increases by the same percentage each unit of time.
- Invariably in nature, numbers become limited by factors such as shortage of food or increasing predation.
- Checks on numbers can act by decreasing natality or increasing mortality, or both.
- The life expectancy of an organism changes with age, and can be plotted graphically as a *survivorship curve*.
- Changes in mortality with age affect the *age structure* of the population.

Test your basics

Copy and complete the following sentences. In some cases the first letter of a missing word is provided.

1. In a population the 'birth' rate or ___*___ is the number of 'births' per ___*___ per ___*___.
2. The density of a population is expressed as the number of individuals per unit ___*___.
3. The density of a population of plants or of a population of attached animals is estimated by repeatedly counting the number of individuals in a portion of habitat of standard size, called a ___*___.
4. In the case of animals that move, population size can be estimated by c___*___, m___*___, and r___*___ individuals. The following day, another sample of the animals is captured, some of which are marked. The ___*___ of marked (recaptured) animals enables the total population to be estimated.
5. When there is a g___*___ of physical conditions (such as soil moisture or light intensity), the distribution of organisms can be mapped by making counts along a line or ___*___.
6. If numbers are increasing by the same percentage each year or other time unit, growth is said to be ___*___. If a population is increasing in this way and numbers are plotted using a ___*___ scale, the graph is a straight line.
7. When a population of microorganisms such as bacteria is grown in culture, the graph of numbers against time is typically 'S'-shaped or ___*___.
8. In most organisms, the probability of death changes with age. This can be shown graphically as a ___*___ curve.
9. If an environmental constraint on population growth acts by negative feedback, it is said to be ___*___-dependent.

4 Communities and Ecosystems

Communities

A community comprises all the organisms in a defined area, and is therefore a group of populations. A forest, a lake, and a coral reef all contain characteristic communities.

Members of a community are not living together simply because they are adapted to similar physical conditions; directly or indirectly, they depend on each other. Because of the network of interrelationships between them, a change in one part can affect many other parts. Like any other level of biological organisation, a community is thus more than the sum of its parts.

The most important interrelationships in a community are based on energy, nutrients, shelter, and reproduction:

- The trees and other plants of a forest provide food for a great many species of small leaf-eating, sap-sucking, wood-boring and root-eating animals. Long after a tree has died, its timber provides food for fungi and insects.

- The shade cast by the trees provides a cool environment for smaller plants such as ferns and mosses, and crevices in the bark provide shelter for a variety of small animals.

- Trees and shrubs provide safety for a variety of birds to build nests.

- In feeding on pollen, seeds and fruits, animals act as agents for pollination and seed dispersal.

- In bringing about decay, bacteria and fungi make inorganic materials (minerals and CO_2) available to plants.

Feeding relationships

The most important relationships in a community relate to nutrition, and are summarised in Fig. 4.1. Organisms can be divided into two groups: *autotrophs* and *heterotrophs*.

- **Autotrophs** convert carbon dioxide into organic compounds. Most are **photoautotrophs**, using light from the sun. A few are **chemoautotrophs**, for example nitrifying bacteria; they use chemical energy obtained by oxidising inorganic compounds in their environment.

- **Heterotrophs** obtain organic compounds from *other* organisms ('heterotrophic' means 'other feeders').

ISBN: 9780170214094

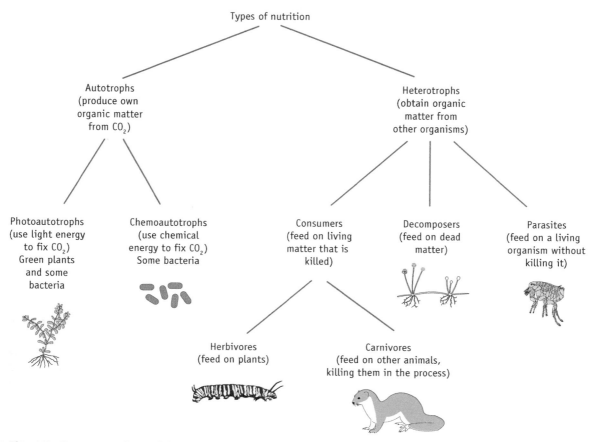

Fig. 4.1 Summary of nutritional types

Heterotrophs include all animals and fungi, most bacteria, and many protoctists. They are usually divided into three main categories:

- **Consumers** include most non-parasitic animals. They feed either on other organisms or on parts of them, killing living tissue in the process. A consumer thus directly reduces the ability of the food species to produce more food. Consumers that feed on plants are **herbivores** or **primary consumers**. Those feeding on herbivores are **carnivores** or **secondary consumers**. Animals feeding on both animal and plant food are **omnivores**.

- **Decomposers** feed on *dead* matter and include many fungi and bacteria. Unlike consumers, decomposers have no direct effect on the availability of their food. Instead, they depend on the rate at which some other process makes its food available, such as the rate at which leaves die. Decomposers that feed on detritus (fragments of dead plant and animal matter) are called **detritivores**. Whereas fungi and bacteria digest their food externally, detritivores digest their food inside a gut cavity. Decomposers are also known as saprobes or **saprotrophs**.

- **Parasites** feed on living matter without killing it, but harm it to some degree.

Fig. 4.2 shows some of the members of a pond community.

Although all communities contain heterotrophs, some lack autotrophs and depend on energy imported from elsewhere. Communities in deep underground streams, for example, depend on organisms and detritus washed down from sunlit areas above. Even in streams above ground, much of the animal life depends on detritus washed down from the surrounding land. A stream community is thus a *net importer* of energy.

ISBN: 9780170214094

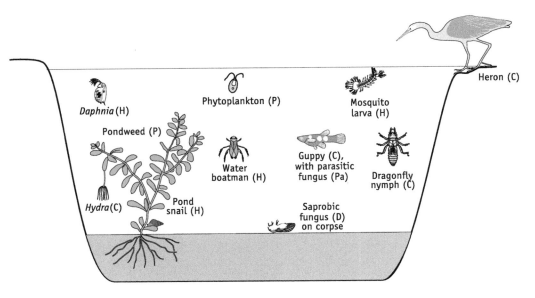

Fig. 4.2 A pond community (P = producer, H = herbivore, C = carnivore, Pa = parasite, D = decomposer)

What determines the membership of a community?

Ultimately, the composition of a community depends on physical conditions such as geology and climate, for it is these factors that determine the kinds of plant that can grow there. The plants in turn have a major influence on the animal life, because many animals depend on a particular kind of plant for their food, and many parasites are also specific to one kind of host.

Physical factors affect both animals and plants directly, but they also influence animals indirectly via their food. The distribution of many insects is determined by the soil requirements of their food plants. The rare helms butterfly, for example, is restricted to the forest in which its food plant (a sedge) grows.

Ecosystems

An ecosystem is a community plus all the non-living components of the habitat. Two processes are central to the functioning of any ecosystem: the flow of energy and the cycling of nutrients.

Fig. 4.3 shows a simple, artificial ecosystem. It consists of a sealed glass bottle containing some plants, together with soil and its bacteria, fungi, and small animals. Since the bottle is gas-tight the organisms inside are, so far as raw materials are concerned, independent of the world outside. No ecosystem, however, is independent of its surroundings for energy. This enters as light and, after

Fig. 4.3 A simple model ecosystem. In a bottle this size there would be insufficient plant growth to support any but very small animals.

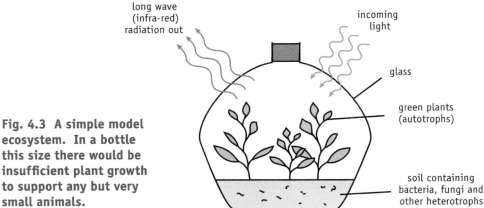

being used to drive living processes the energy is degraded to heat, which leaves the bottle as infra-red radiation.

Provided it receives enough light, a simple ecosystem like this can function indefinitely. The food supply of the consumers and decomposers is limited by the rate of growth of the plants. In adequate light this is limited by the rate at which the consumers and decomposers break down organic matter to the simple inorganic nutrients the plants need (Fig. 4.4).

In this simple system there is a balance between autotrophs and heterotrophs. The plants cannot grow faster than their supply of CO_2 and mineral ions allows, and the consumers and decomposers cannot outgrow their food supply. In this situation the total photosynthesis and respiration occurring over each 24-hour period is equal.

Of course, ecosystems are not sealed off from each other like the simple model described above. All ecosystems exchange both materials and energy with neighbouring ecosystems. Organic sediment is washed down from land to sea in rivers, and birds flying over sand dunes may deposit faeces. Many insects that live in fresh water as young stages fly from one pond to another as adults, and some animals migrate as adults over long distances. In all of these movements, matter and energy are transferred from one ecosystem to another.

In the biosphere as a whole, exchanges of energy and matter between ecosystems cancel each other out. The earth is simply a very much larger version of the sealed bottle in Fig. 4.3; energy entering as light and leaving as heat.

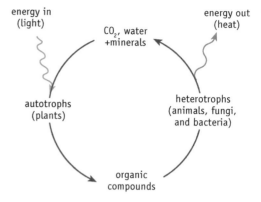

Fig. 4.4 Interdependence between autotrophs and heterotrophs

Food chains

The simplest way of expressing feeding relationships is by a **food chain**, in which transfer of energy from one organism to the next is represented by an arrow. Fig. 4.5 shows some examples of food chains.

Grazing and detritus food chains

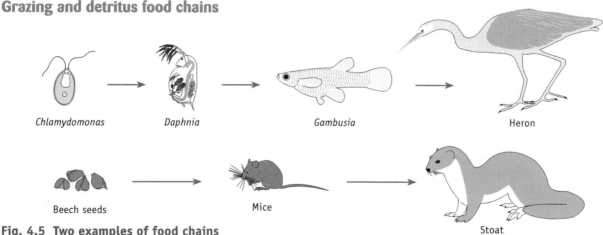

Chlamydomonas   *Daphnia*   *Gambusia*   Heron

Beech seeds   Mice   Stoat

Fig. 4.5 Two examples of food chains

There are two kinds of food chain:

- **Grazing food chains**, in which the primary consumers feed on *living* plant material.

- **Detritus food chains**, which begin with *dead* plant parts, which are food for decomposers such as fungi, and detritivores such as earthworms (Fig. 4.6). Many small arthropods play an indirect role by chewing up plant matter. As a result their faeces have a greatly increased surface area for attack by micro-organisms.

In terrestrial ecosystems most plant matter enters the detritus food chain as dead leaves, stems and roots; only a small proportion of live plant material enters the grazing food chain. In oceans the

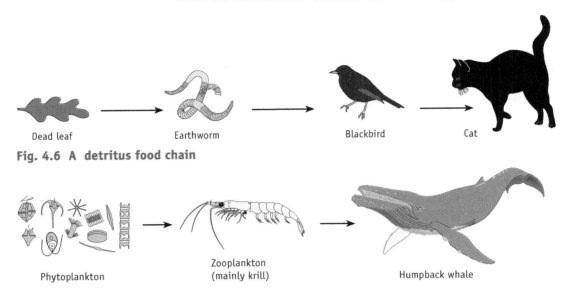

Dead leaf   Earthworm   Blackbird   Cat

Fig. 4.6 A detritus food chain

Phytoplankton   Zooplankton (mainly krill)   Humpback whale

Fig. 4.7 A marine food chain, in which detritus plays little part

reverse is true. The base of the ocean food chain consists entirely of microscopic floating algae, the **phytoplankton**. These are food for the tiny floating animals of the **zooplankton** (Fig. 4.7). Most of the phytoplankton is alive when eaten and thus enters the grazing food chain. A minority die without being eaten and begin to sink, entering the detritus food chain.

Food webs

A food chain obviously over-simplifies feeding relationships. Not many organisms feed on only one species and few, if any, are eaten by only one species. A spider eats many kinds of insect, and one species of tree may be food for hundreds of species of other organisms. In other words, food chains are branched and inter-connected to form **food webs** (Fig. 4.8).

Fig. 4.8 is highly simplified, in a number of ways:

- Many species are lumped together into single categories, e.g. more than one species of slater can often be found in a given habitat, and similarly for many other groups.

- Many animals change their diet as they get older; most caterpillars feed on leaves, but the adult butterflies and moths feed on nectar.

- Many animals change their diet with the seasons. For example, in spring and summer, possums may supplement their plant diet with birds' eggs.

- In ecosystems undergoing succession, the species composition slowly changes (see below).

Since every organism in a food web is, directly or indirectly, connected with every other, a change in one population may influence every other population in the community. In communities containing few species, a change in the numbers of one species is likely to have drastic effects on others. An example is the oscillations in the populations of the Canadian snowshoe hare and lynx (Chapter 3). In more complex communities there is much greater scope for animals to switch from one kind of food to another, so a disturbance in one population is more likely to be absorbed by the others. Probably for this reason, complex communities are generally more stable than simple ones.

A food web may contain thousands of species, and ecologists often simplify things by lumping together all producers, and similarly for all plant-eaters, all primary carnivores, secondary carnivores (animals that feed on primary carnivores), and so on. Each of these major feeding categories constitutes a **trophic level**. Thus all the producers together make up one trophic level, and all the herbivores form another, and so on.

As with many other biological terms, trophic levels are not mutually exclusive; omnivores belong to two or even more levels simultaneously, and the emphasis on membership of particular levels may change with the seasons. For instance tui feed as primary consumers on nectar and fruits, and as secondary consumers on insects, depending on the season.

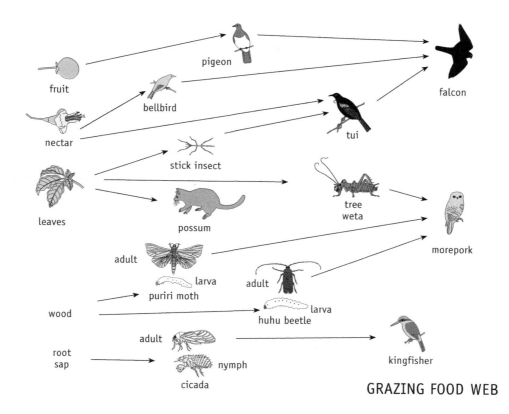

GRAZING FOOD WEB

A

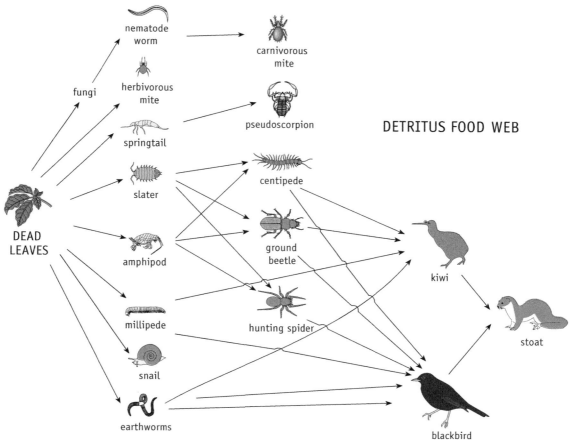

DETRITUS FOOD WEB

B

Fig. 4.8 Highly simplified food web based on a puriri tree. (A) Grazing food web. (B) Detritus food web.

How do we know what feeds on what?

Construction of a food web requires a detailed knowledge of the feeding habits of many animals, which may take considerable time and still leaves many details uncertain. There are several ways of studying feeding habits:

- Direct observation. Animals can sometimes be seen in the act of feeding, particularly in the case of specialist herbivores that spend all their time on their food plant. This is more difficult with carnivores, which feed intermittently.

- By keeping animals in captivity with a choice of food items. However, just because a predator eats a particular prey in captivity, this does not mean it does so in nature.

- A study of gut contents may reveal fragments of undigested food such as parts of exoskeleton and plant cell walls that in some cases can be identified.

- By immunological techniques. An animal such as a rabbit is injected with an extract of the suspected food, and allowed to develop antibodies against it. An extract of gut contents from the animal under study is then mixed with serum of the immunised rabbit. A precipitate indicates the presence of the same antigens in the food used to immunise the rabbit, confirming that the suspected material is part of the animal's diet.

- By treating plants with a radioactive tracer such as ^{32}P (as phosphate), and monitoring the radioactivity in members of the community at intervals afterwards. Radioactivity appears first in the phloem sap and in insects sucking the sap. Next to become radioactive are the youngest leaves, followed by leaf-eating animals, and then their predators. Members of the detritus food chain take longer to become labelled since leaves have to die before they become available as food. Fig. 4.9 shows the results of such a study.

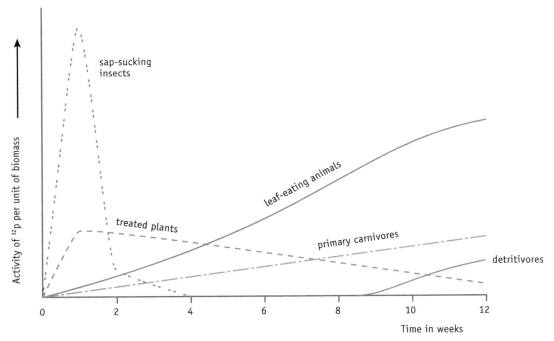

Fig. 4.9 Changes in radioactivity in members of a community after treatment of plants with radioactive phosphorus, ^{32}P

Organisms as energy converters

Few food chains have more than five links, and some have only three. This is because only part of the energy entering an organism is retained as chemical energy in new tissues; most is lost as heat in respiration. This is true of both animals and plants, but since their energy budgets are rather different we will deal with them separately.

Energy conversion in plants

Fig. 4.10 shows a typical energy budget of a crop plant under field conditions. The figures are only approximations, and the actual efficiency depends on light intensity, temperature, water supply, mineral supply, and other factors.

Fig. 4.10 may look complicated, but the important thing to remember is that only about 5% of the energy falling on the plant is converted into new plant tissue. This is in a crop plant, in which conditions for growth are artificially manipulated in the plant's favour. In many natural communities the efficiency is less than this.

Fig. 4.10 Energy flow through a crop plant

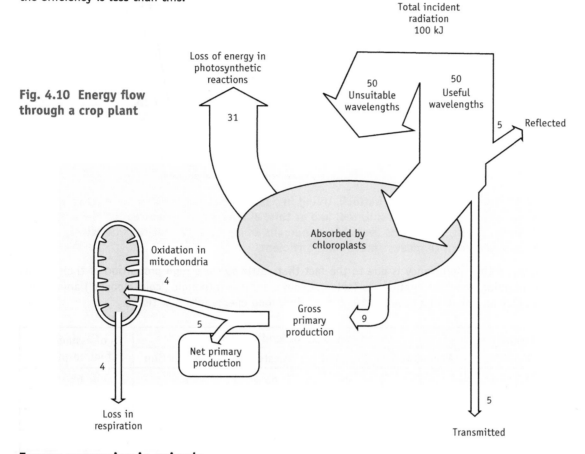

Energy conversion in animals

Fig. 4.11 shows the energy budget for a caterpillar. It shows that only about 23% of the energy taken in finishes up as new caterpillar flesh — and most ectothermic ('cold-blooded') animals convert about 10% of their food into new flesh. In endothermic ('warm-blooded') animals, the efficiency is closer to 2%.

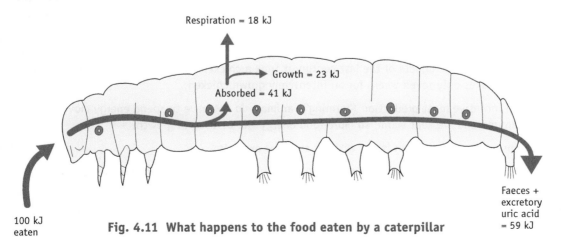

Fig. 4.11 What happens to the food eaten by a caterpillar

Fig. 4.12 shows the energy budget for a free-range bullock using traditional farming methods, and gives an idea of how inefficiently farm animals convert grass energy into meat energy. For every 100 kJ of energy entering the animal, about 4 kJ is potentially available to a predator, such as a human.

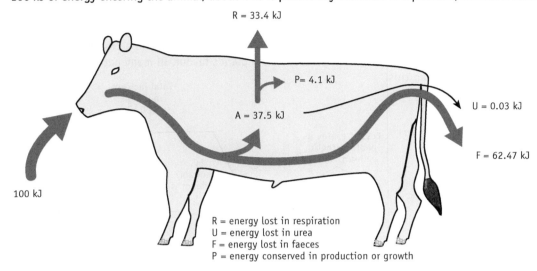

R = 33.4 kJ

P = 4.1 kJ

A = 37.5 kJ

U = 0.03 kJ

F = 62.47 kJ

100 kJ

R = energy lost in respiration
U = energy lost in urea
F = energy lost in faeces
P = energy conserved in production or growth

Fig. 4.12 Energy budget for a bullock, living in pasture. For every 100 kJ of energy eaten, little more than a third is assimilated, and of this, almost 90% is respired. Of the original food energy eaten, only 4% is available to humans as meat. Note that for intensively reared animals, conversion to meat is much more efficient.

Some of this inefficiency is due to the fact that cattle spend a high proportion of their energy in maintaining body temperature. As Table 4.1 shows, most ectothermic ('cold-blooded') animals are more efficient, converting between 10–25% of their food energy into new tissue.

Animal	Feeding category	Percentage of total food consumed				% of assimilated food respired
		Assimilated	Respired	Egested	Net production	
Grasshopper	herbivore	37	24	63	13	65
Caterpillar	herbivore	41	17.5	59	23.5	43
Spider	carnivore	92	57	8	27	62
Perch (a fish)	carnivore	83.5	61	16.5	22.5	73
Elephant	herbivore	33	32	66	1	97

Table 4.1 Energy budgets for five animals (modified from *Advanced Biology Learning Project*, Cambridge University Press)

The data shown in Fig. 4.12 are for a domesticated animal in the 1960s. In recent decades the efficiency of conversion of food into meat or milk has been enormously increased, for two reasons:

- As a result of selective breeding, growth rates have been greatly increased and a higher proportion of the growth is devoted to the production of meat rather than bone. For example whereas in the middle of the last century it took a chicken six months to reach full size, it now takes only seven weeks for an intensively grown chicken.

- In intensive meat production ('farming'), animals do not have to spend energy moving around and are kept warm, so more food energy can be used in growth.

Intensive meat production is not without cost, for example:

- Intensively farmed chickens and pigs are kept in conditions that many people consider inhumane.

- By selecting exclusively for growth rate, other characteristics, such as bone strength, may suffer, as a result of which the animals may suffer injury.

ISBN: 9780170214094

- The diet is often supplemented with antibiotics, which increase growth rate but may lead to antibiotic resistance to microorganisms, which may spread to bacteria that cause disease in humans. In some countries — for example the United States and the UK — growth is promoted by the addition of hormones to the animals' diet, and there is evidence that hormone residues may be getting into water supplies.

- For humans, efficiency of food production is becoming increasingly important. A hectare of land devoted to growing potatoes or wheat yields far more energy for humans than a hectare used for beef production, since the cattle waste about 96% of the energy in their food. From the energy point of view it therefore makes sense to use land for growing crops rather than for meat production.

There is another side to the energy question: we have to invest energy in order to obtain food energy. In traditional agricultural societies, the energy investment was in the form of muscle power. Since the Industrial Revolution the energy investment, in the form of farm machinery, fuel and fertiliser, has progressively increased, to the point where most food contains less energy than was used to produce it! Much of the energy used in intensive food production comes from an energy subsidy in the form of fossil fuels. Thus intensive agriculture may be efficient in terms of output per hectare, but in terms of output per unit of energy invested, it is highly *inefficient*.

Energy, however, is not the only consideration in food production. Not only is animal protein generally richer in essential amino acids than plant protein, but vitamin B_{12} is completely absent from plant food. Also, some herbivores can live on vegetation that humans cannot; by growing sheep on hills, humans can obtain food energy that would otherwise be unavailable.

Food pyramids

Because heterotrophs convert their food into flesh very inefficiently, only a small proportion of the energy entering each link is transferred to the next. By the time the energy reaches the top carnivores, a minute proportion of the solar energy trapped by plants remains; the rest has been lost as heat. This loss of energy along a food chain can be described by various forms of *food pyramid*.

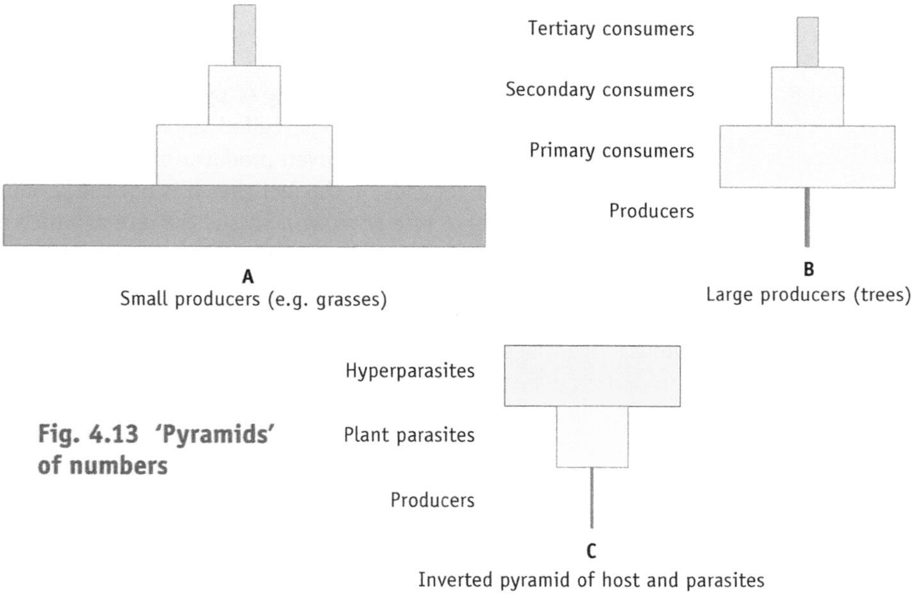

Pyramids of numbers

According to an old Chinese saying 'one hill cannot shelter two tigers'. The relative rarity of carnivores, especially large ones, is part of a

Fig. 4.13 'Pyramids' of numbers

A — Small producers (e.g. grasses)

B — Large producers (trees)

Tertiary consumers
Secondary consumers
Primary consumers
Producers

Hyperparasites
Plant parasites
Producers

C — Inverted pyramid of host and parasites

broader generalisation; in any ecosystem, herbivores are usually more common than primary carnivores, and primary carnivores are more common than secondary carnivores. If the total numbers of organisms at successive trophic levels in an ecosystem are represented by blocks of proportionate size, they often form a pyramid (Fig. 4.13). The lowest layer of the pyramid represents the total number of producers in the ecosystem, the next block represents the numbers of herbivores, and so on.

There are two reasons why the number of organisms decreases with successively higher trophic levels:

- Because of energy losses, each organism eats many times its body mass during its lifetime. Even if predators were the same size as their prey, each predator would eat many prey during its lifetime.

- More often than not predators are larger than their prey, increasing the difference between numbers of predators and prey.

The size disparity does not always work this way. Some animals are much smaller than their food organism. Thousands of insects, for example, can feed off a single tree, so that part of the pyramid may be *inverted*. A similar effect is created with parasites, which are always smaller (but often more numerous) than their hosts. Even amongst non-parasitic animals, some predators are smaller than their prey. A stoat, for example, is less than a quarter the mass of a rabbit.

Pyramids of biomass

The effect of body size is eliminated if total population *biomass* (in kg m^{-2}) is compared instead of numbers (Fig. 4.14). In a pyramid of biomass each block represents the **standing crop**, which is the total quantity of organic matter per m^2 present at one time. The result is usually a pyramidal shape, indicating that a large total mass of plant material supports a smaller total mass of herbivores, which supports a smaller mass of primary carnivores.

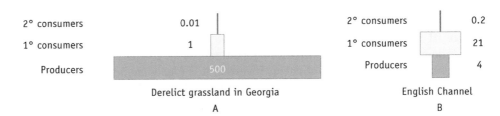

Fig. 4.14 Pyramids of biomass. (A) Derelict grassland in Georgia. (B) An anomalous pyramid, the English Channel in Spring.

Pyramids of productivity

Although biomass pyramids are usually of pyramidal shape, they do not have to be so, as Fig. 4.14B shows. Here the producers are single-celled 'algae', and the primary consumers are tiny filter-feeding crustaceans. In this example, a given producer biomass supports a greater biomass of herbivore.

The explanation depends on the fact that it is not the standing crop of the producers that matters, but its **productivity** or *rate of growth*. This is the rate at which new material becomes available to the primary consumers. Under optimal conditions members of the phytoplankton can divide once a day, so every day the producer biomass doubles — or it would do, if it were not being eaten by animals.

A more revealing kind of food pyramid is a **pyramid of productivity** or pyramid of **energy flow** (sometimes just called a pyramid of energy, Fig. 4.15). In this case the size of each block is proportional to the *growth rate* or productivity of the organisms in that trophic level. The units of productivity are kJ m^{-2} year^{-1}. A pyramid of productivity is *always* pyramidal, since some energy leaves each trophic level.

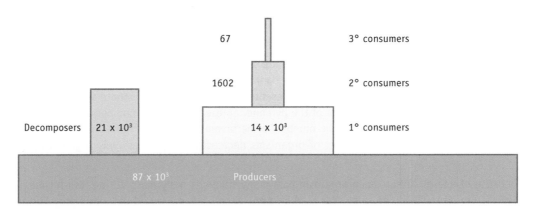

Fig. 4.15 Pyramid of productivity for a stream at Silver Springs, Florida. Note that the decomposers account for a considerable proportion of the energy flow through the ecosystem. Units are KJ m^{-2}yr^{-1}.

Excellence in Biology Level 2

ISBN: 9780170214094

Food chains and pesticide pollution

An important result of pyramids of productivity is that substances that are not excreted nor broken down in metabolism may become more concentrated along a food chain. In 1949, in an attempt to get rid of hordes of non-biting but troublesome midges, Clear Lake in California was sprayed with DDD, a close relative to DDT but less toxic to fish.

The initial results were spectacular, but repeated annual spraying became less and less effective. It was discontinued in 1954 when it was found that grebes (fish-eating birds) were dying in large numbers. Their fat was found to contain DDD at concentrations of 1600 parts per million (ppm), 32 000 times the original concentration in the lake water (Fig. 4.16). Since each animal eats many times its own body mass in its lifetime, the DDD became increasingly concentrated as it passes from one trophic level to the next. The grebes were at the end of the chain, so they suffered a much higher dose than the midges.

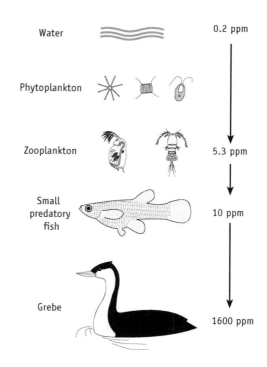

Fig. 4.16 How an insecticide can become concentrated along a food chain

Cycling of nutrients

Sooner or later all the energy that enters an ecosystem is radiated out into space as heat. Energy is therefore only used *once*. Matter, on the other hand, is recycled. Each element spends part of its time in complex organic molecules and part in simple, inorganic substances in the abiotic part of an ecosystem (Fig. 4.17). There are about 17 elements found in living matter, and each has its own cycle.

The details of each cycle vary with the element, but certain features are common to all. Nutrients are absorbed in the inorganic state by autotrophs, and passed on to the heterotrophs that feed on them. In a heterotroph, atoms either leave the body in inorganic waste, or they enter another organism in food. In every link in the food chain there are the same two alternative pathways, so sooner or later all elements are returned to the environment and become available for uptake by autotrophs.

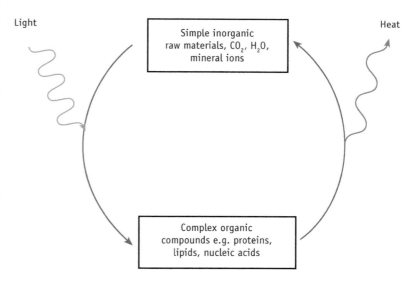

Fig. 4.17 A generalised nutrient cycle. The nutrient elements are recycled, but the solar energy that drives the cycle leaves as heat.

The carbon cycle

Carbon forms the 'skeleton' of all organic molecules. It enters green plants as CO_2 and is converted into carbohydrate in photosynthesis and, on a much smaller scale, by chemoautotrophic bacteria. Carbohydrates are then converted into all the other organic constituents of living matter, such as lipids, proteins and nucleic acids. Some of the organic matter made by plants is converted into CO_2 in respiration. As organic matter moves along the food chain it is converted back into CO_2 in respiration, which re-enters the atmosphere (Fig. 4.18).

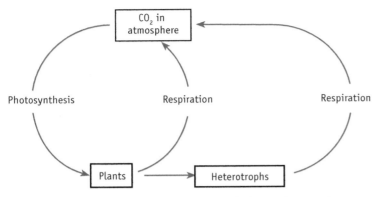

Fig. 4.18 Outline of the carbon cycle

Fig. 4.19 shows the carbon cycle in more detail. Each compartment represents a 'stop' on the cycle. In some compartments carbon is in the oxidised state, such as CO_2, carbonate (CO_3^{2-}), or hydrogen carbonate (HCO_3^-) ions. In organic compounds such as carbohydrates and proteins, carbon is in the reduced state (because at least some of the carbon atoms are directly bonded to hydrogen).

Until the Industrial Revolution, annual world photosynthesis more or less equalled respiration and combustion, so the amounts of carbon in each compartment changed little from year to year. In the summer, total photosynthesis slightly exceeds total respiration, so CO_2 is removed from the environment faster than it is returned, causing its level to fall slightly. During winter the situation is reversed.

Before the Industrial Revolution these changes more or less cancelled each other out over the year as a whole. More or less, but not exactly; for the following reason. In ecosystems where there is insufficient oxygen for decay, such as swamps, dead bodies of plants and animals accumulate and are slowly converted into coal, oil or natural gas. Although this represents less than 0.001% of the total annual photosynthetic product, when accumulated over millions of years this tiny annual deficit in the return of carbon to the atmosphere has resulted in the laying down of massive deposits of fossil fuel.

Since the Industrial Revolution the balance between the uptake and production of CO_2 has been disrupted on a much greater scale, for two reasons, both of which have contributed to a rise in the level of atmospheric CO_2:

- the burning of fossil fuels

- deforestation followed by decay and combustion of the timber.

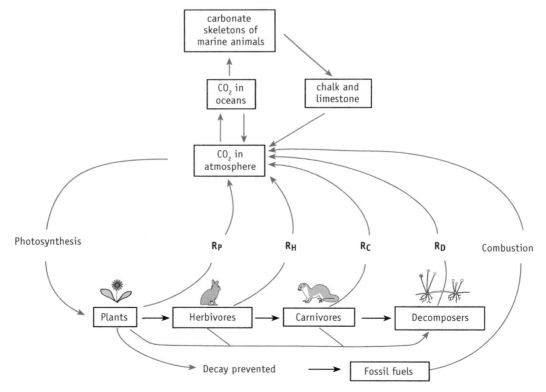

Fig. 4.19 The carbon cycle R_p, R_h, R_c and R_d = respiration by plants, herbivores, carnivores and decomposers

ISBN: 9780170214094

In the cycle we have so far been concerned with, carbon alternates between the oxidised, inorganic state and the reduced organic state. As Fig. 4.18 shows, there is another cycle, in which carbon remains oxidised throughout. In this cycle, CO_2 is used by planktonic protoctists to make their tiny skeletons of lime ($CaCO_3$). Over tens of millions of years, vast numbers of these skeletons are deposited on the ocean floor and eventually form *chalk*, or limestone. Over equally long periods this may undergo metamorphosis to become *marble*. If the ocean floor subsequently rises, chalk and limestone may be exposed to the weathering. Some dissolves in rainwater, freeing it to enter the other cycle via photosynthesis.

Of the two cycles, the purely 'inorganic' one turns far more slowly. This is partly because the amounts of carbon in the chalk and limestone compartment are so enormous, and partly because the rates at which carbon enters and leaves these compartments is relatively low. Because the carbon cycle involves geological as well as biological processes it is called a **biogeochemical cycle**.

The 'greenhouse' effect

Carbon dioxide is present in the atmosphere at a concentration of about 390 parts per million (in 2011). Besides being a raw material for photosynthesis, it also plays a vital role in acting as a kind of 'heat jacket' around the earth. Though CO_2 is transparent to visible light, it absorbs the longer wavelength infra-red rays. Sunlight absorbed by the earth's surface is re-emitted as infra-red radiation, which is trapped by atmospheric CO_2 instead of radiating out into space. The result is that the average temperature of the earth is 15° C instead of the −20° C it would be without CO_2 (Fig. 4.20).

Since the Industrial Revolution the burning of fossil fuels and the oxidation of forest timber (by burning or by decay) has led to a steady increase in atmospheric CO_2. As long ago as 1896 the Swedish chemist Svente Arrhenius predicted such a rise, and also that it would cause the mean temperature on the earth to increase. Only in recent decades has the possibility of global warming been taken seriously.

Evidence of a rise in atmospheric CO_2 comes from direct measurements made since 1957 at the top of Mauna Loa, a mountain in Hawaii (Fig. 4.21). At an altitude of over 4000 metres and in the mid-Pacific, the atmosphere there is relatively uncontaminated by proximity to human habitation. The graph shows not only a steady rise in CO_2 from 315 ppm in 1958, but a slight seasonal fluctuation as the balance between respiration and photosynthesis changes. Analysis of air trapped in ice thousands of years old in Antarctica has shown that before the Industrial Revolution the level of CO_2 was about 280 ppm.

The two essential ingredients in the situation are not in dispute: CO_2 traps outgoing heat radiation, and CO_2 levels are rising. What is uncertain is the extent to which climate will change as a result. Predictions are made more difficult because it appears that about half the net annual production of CO_2 is absorbed in the oceans and used in photosynthesis by phytoplankton. Some of the fixed carbon sinks to the ocean floor in the form of dead organisms, delaying or preventing its return to the atmosphere as CO_2.

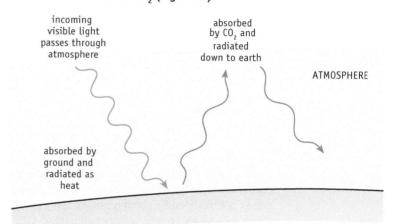

Fig. 4.20 How carbon dioxide keeps Earth warm

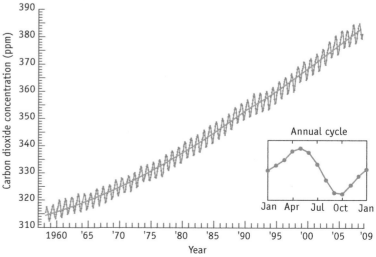

Fig. 4.21 Changes in CO_2 concentration measured at Mauna Loa, Hawaii

The magnitude of this effect is itself likely to be influenced by warming of the oceans. Since the solubility of CO_2 decreases with rise in temperature, the more the seas warm, the less CO_2 they can absorb.

Other feedback effects complicate the picture still further. For example, a rise in temperature will increase evaporation of water from the oceans. The resulting increase in cloud cover would raise the reflectance of solar energy by the earth. This would be a *negative* feedback effect and would tend to stabilise world temperature.

On the other hand there may be *positive* feedback effects. For example, a melting of the permanently frozen ground (permafrost) in North America and Asia, where there are enormous quantities of dead organic matter, would bring about the release of very large amounts of CO_2, causing further warming and production of more CO_2.

Despite these and other uncertainties, the overwhelming majority of climate scientists agree that if present trends continue, mean annual temperatures are likely to rise by about 2–4 °C in the next century. To put this seemingly small rise into perspective, mean temperatures during the most intense ice age were about 5° C lower than they are today.

The situation is further complicated by the fact that certain other gases, such as water vapour, methane, and nitrous oxide are also 'greenhouse' gases. Methane is produced naturally by anaerobic microorganisms in swamps, and also by mutualistic microorganisms in the guts of many herbivorous animals. Increases in the numbers of cattle are having a significant effect on world methane production. Nitrous oxide is produced in denitrification, which has increased with the large-scale use of nitrate fertilisers.

The possible effects of a rise in world temperatures are several. Some of the most obvious include the following:

- A rise in sea level due to the thermal expansion of the oceans and the melting of terrestrial ice caps (melting of *floating* ice has no effect on sea level).

- A change in the wind patterns and distribution of rainfall, with more extremes of weather. In particular, the frequency of tropical cyclones, which develop over warm ocean, is likely to increase.

- An increase in the rate of photosynthesis, both as a direct result of increased CO_2 and also as a result of warmer conditions.

If photosynthesis increases, could the increased rate of removal of CO_2 counteract the effect of fossil fuel burning and deforestation? No, unless extra plant material is allowed to remain as biomass, and is not oxidised by combustion or respiration. Any increased photosynthesis is likely to be in crops, which will be eaten and rapidly oxidised to CO_2.

At various times in the geological past, atmospheric CO_2 concentrations and global temperatures have been higher than they are now, and species have adapted to them. Is it not likely that similar adaptation will occur in response to human-induced changes that are occurring now? Studies of preserved pollen have shown that many times in the past, plants have shifted their distribution in response to changes in climate. However, while species with long-range dispersal mechanisms and short life histories probably would be able to adapt, many trees (and the animals that depend on them), would not.

The nitrogen cycle

Nitrogen is a constituent of amino acids and nucleotides, and therefore of proteins and nucleic acids. In all these compounds, nitrogen is bonded to hydrogen and is therefore in the reduced state.

The nitrogen cycle differs in an important way from the carbon cycle, in that several stages can only be carried out by bacteria. Like the carbon cycle, it actually contains two interlocking loops, one of which 'turns' much more rapidly than the other (Fig. 4.22). The rapid cycle involves organisms from all five kingdoms, whilst the slow one involves bacteria only.

Like the carbon cycle, the nitrogen cycle involves oxidation and reduction, with associated energy changes. The main cycle can be divided into three phases. These occur respectively in plants, in consumers and decomposers, and in chemosynthetic bacteria. Interlocked with this cycle are two other natural processes, *denitrification* and *nitrogen fixation*. Also affecting the nitrogen cycle are agricultural harvesting and addition of fertilisers, and lightning.

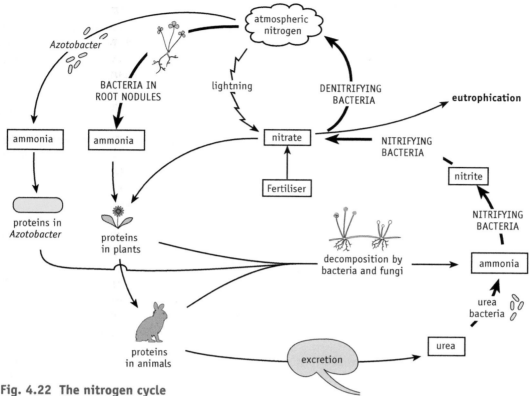

Fig. 4.22 The nitrogen cycle

How plants use nitrogen

Although molecular nitrogen (N_2) accounts for about 80% of the atmosphere, it is extremely unreactive and cannot be used by plants. Instead, plants absorb their nitrogen in the form of nitrate ions (NO_3^-), which they then reduce to ammonia:

$$NO_3^- \rightarrow NH_3$$

This is an energetically 'uphill' process, using energy from photosynthesis. Ammonia is then used to make nitrogenous organic compounds such as amino acids and proteins:

$$NH_3 + \text{organic acids} \rightarrow \text{amino acids} \rightarrow \text{proteins}$$

Ammonia formation by heterotrophs

When herbivores feed on plants they digest the plant proteins, converting them to amino acids. Some of the amino acids are used to make herbivore protein, the rest are *deaminated* to release ammonia:

$$\text{amino acids} \rightarrow \text{organic acids} + NH_3$$

A similar process occurs in every heterotroph, with the result that all the organic nitrogen is sooner or later converted to ammonia.

In many animals the ammonia produced in deamination is converted to a less toxic substance such as *urea* or *uric acid*. The end result is the same, however, since these excretory products are converted into ammonia by bacteria.

Nitrification

The ammonia produced by deamination of amino acids is oxidised by bacteria to nitrates. This process is called **nitrification** and can only be carried out by bacteria. It is essentially the reverse of the reduction of nitrate to ammonia by plants. It occurs in two steps, carried out by different kinds of **nitrifying bacteria**. *Nitrosomonas* oxidises ammonium ions to nitrite, and *Nitrobacter* oxidises nitrite to nitrate. Both reactions release energy that is used to convert CO_2 to carbohydrate. These bacteria are thus *chemoautotrophs* — instead of using light to make carbohydrate from CO_2, they use chemical energy.

ISBN: 9780170214094

Denitrification

Under anaerobic conditions, bacteria such as *Pseudomonas denitrificans* reduce nitrate to nitrogen gas or to nitrous oxide, N_2O. This is called **denitrification** and obviously tends to lower soil fertility.

Denitrifying bacteria are *facultatively anaerobic*. This means that they use oxygen if it is available, but if the soil becomes depleted of oxygen they switch to anaerobic respiration (using nitrate instead of oxygen).

Anaerobic conditions can occur after flooding. In well-drained soils denitrification is seldom an agricultural problem.

One of the environmental problems of denitrification is that nitrous oxide is contributing to the destruction of ozone. Though present in minute traces in the upper atmosphere, it absorbs the most dangerous wavelengths of ultraviolet radiation. On the other hand, swampy land provides habitats for specialised and diverse communities of ecological interest.

Nitrogen fixation

Denitrification would eventually lead to serious depletion of nitrate were it not for the opposing process of **nitrogen fixation**, in which nitrogen gas is converted to combined nitrogen. To combine nitrogen with other atoms the very strong triple bond in the nitrogen molecule must first be broken. The large amount of energy needed can come from two sources: heat generated in combustion or lightning, and the metabolism of living organisms.

Non-biological fixation occurs when nitrogen and oxygen are heated to high temperatures; for example in car engines and in lightning flashes they form oxides of nitrogen, which react with water to form nitrite and nitrate.

Of the total natural annual nitrogen fixation, about 90% occurs by biological mechanisms. The only organisms that can fix nitrogen are prokaryotes. Some nitrogen-fixing bacteria live freely in the soil, for example *Azotobacter*. They get their energy for fixation from organic matter in the soil.

Economically, the most important nitrogen fixers are mutualistic bacteria. By far the most important are members of the genus *Rhizobium*, which live with members of the legume family, such as clover, peas and beans. A number of non-legumes, such as tutu (a native shrub), also have nitrogen-fixing nodules.

The human impact

In most natural ecosystems nitrogen is recycled with little loss; but in agriculture, harvesting removes large quantities of nitrogen as protein. To some extent this may be replaced by growing clover and other legumes, but considerable quantities of inorganic fertiliser in the form of ammonium nitrate also need to be added.

The disadvantage of inorganic fertiliser is that it leads to a short-term rise in soil nitrate. The surplus may be *leached*, or washed out into streams and lakes. The effect is that lakes become **eutrophic**, or enriched with mineral nutrients. As a result plant growth is stimulated and when the plants die off in the autumn their decay may cause deoxygenation.

Interactions between the cycles

All nutrient cycles are interlocked because many organic molecules contain four or more elements. Nucleotides contain nitrogen and phosphorus as well as carbon, and proteins contain sulfur also. A change in the rate of cycling of one nutrient must therefore be accompanied by a change in the others.

Nutrient cycling in lakes

Although the fundamental principles of nutrient cycling are similar in aquatic and terrestrial environments, there are some important differences. During late spring and summer, heat from the sun warms the upper layers of the water. Since warm water is less dense, it does not mix with the colder water below (Fig. 4.23). The boundary between the upper, warm water and the cold water below is called the **thermocline**.

At the same time as the water has been warming up, the light intensity has been increasing and there is a population explosion of phytoplankton (Fig. 4.24). Because the deeper, nutrient-rich water can no longer mix with the warmer water above, the upper layers soon become depleted of minerals

ISBN: 9780170214094

by the phytoplankton. The phytoplankton population crashes, as does the zooplankton that depends on it.

In autumn the upper layers of the lake cool somewhat. As a result there is some mixing with nutrient-rich water below and there may be a second, smaller surge of phytoplankton. By wintertime, reduced light greatly slows the growth of phytoplankton. At the same time there is now mixing of water, and the mineral content is restored.

Similar layering of the water occurs in the sea. In temperate waters a thermocline occurs only in summer, but in the tropics it is a permanent feature. This is why tropical seas are generally poor in nutrients and are much less productive than polar seas are in summer.

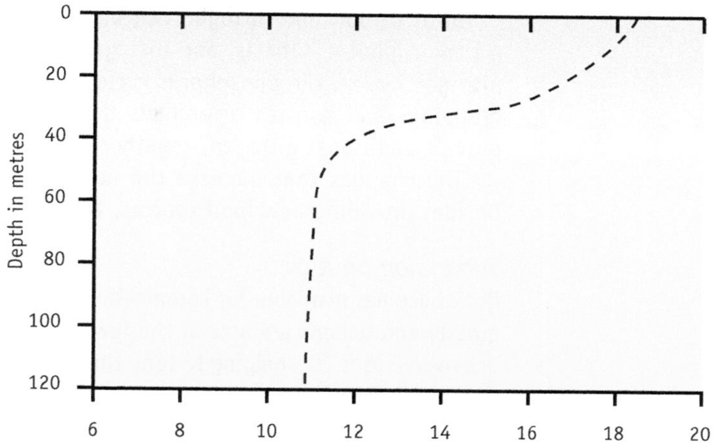

Fig. 4.23 Thermal stratification in Lake Taupo in Summer

Nutrient cycling in New Zealand forest

As in most terrestrial ecosystems the return of nutrients to the soil is mainly via litter that falls onto the forest floor. Unlike deciduous forests of Europe and North America, leaves in New Zealand forests fall all the year round, with a slight surge in March. Beetles, cockroaches, slaters, spiders, weta and a host of other creatures depend on this harvest that falls from the trees.

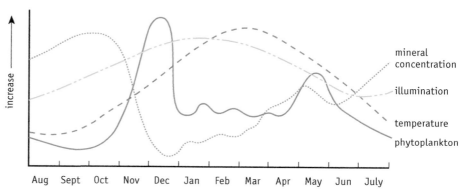

Fig. 4.24 Seasonal changes in biotic and abiotic factors in a lake

Most of this detritus consists of leaves, which are much richer in protein (and hence nitrogen) than wood. In one study in the Orongorongo forest near Wellington, an average of 143 tonnes of litter lay on each hectare of ground, with about 7 tonnes being added annually. The leaves decompose much more rapidly than the twigs and branches. Thus a carbon atom may spend centuries as part of a cellulose molecule in a rimu tree trunk, but only a few years as part of a cellulose molecule in a leaf.

Succession

Succession is a gradual, progressive and predictable change in the composition of a community over a period of time. There are two kinds: primary and secondary.

Primary succession

Primary succession occurs when conditions at the beginning are unfavourable for plant growth, but soil is gradually made fertile by plants. To begin with, physical conditions are harsh and only a few **pioneer species** can survive. As roots of pioneer plants die they leave humus. This slowly raises the water-holding capacity of the ground and thus its fertility.

Pioneer species are adapted to a tough physical environment but are weak competitors. As other plants begin to get established they compete with the pioneers, eventually displacing them. As the process continues, stronger competitors displace less competitive species until there is no further change. When the final or **climax** community has developed, any further change is very slow and is due to climatic change. The kind of climax vegetation depends on the climate and soil conditions. In most parts of New Zealand it is evergreen forest.

During primary succession there is an increase in species diversity, both of plants and the animals they support. There is also an increase in biomass. This means that nutrients must be imported from outside the ecosystem.

Both carbon and nitrogen can be imported into an ecosystem from the air. Carbon is fixed by plants in photosynthesis, and nitrogen is fixed by root nodules of legumes. Unlike the carbon and nitrogen cycles, the phosphorus cycle has no gas phase. Phosphorus is imported in the form of bird droppings and corpses of animals that have come in from outside before they died. Faeces also provide additional nitrogen, together with other minerals.

The changes that increase the number of plant species also result in greater animal diversity. Besides providing new food sources, the plants also provide shelter and nesting places for birds.

Succession on rock

Rock becomes available for colonisation after landslips and volcanic eruptions. Bare rock dries out very quickly and lichens are among the few producers that can survive there. The fungal component of the lichens releases CO_2, helping to form tiny cracks in the rock surface. Some mosses can withstand complete drying for quite long periods, becoming active during brief rainy spells. Later, a few small flowering plants may get established in cracks, such as bird's foot trefoil, an introduced legume. Its ability to fix nitrogen enables it to flourish in areas poor in nitrogen.

As a result of the growth of the pioneer plants, enough humus develops to allow slightly larger plants such as gorse (another legume) and manuka to establish. These out-compete the pioneers and eventually replace them. As the process continues, each new species adds to the dead matter and increases the water-retaining capacity. Eventually, trees become established.

Succession in sand dunes

Sand dunes are parts of the coast where the land is advancing as a result of the activity of plants. Unlike succession on rock, the changes in time in a sand dune system are reflected by changes in space — the youngest dunes being nearest the sea (Fig. 4.25). So in walking inland from the high tide mark, we encounter progressively later stages in the succession.

Sand dunes are a tough environment characterised by high winds, frequent very dry conditions, salt spray, and heat in summer. The chief pioneer plants are the introduced marram grass (*Ammophila arenaria*, the native sand sedge pingao (*Desmoschoenus spiralis*), and in Northern and Central New Zealand, the native sand dune grass *Spinifex hirsutus*. A few other plants are able to tolerate these conditions, such as bird's foot trefoil. As a result of the activity of the pioneer plants, the sand gradually becomes stabilised. Some humus begins to form, increasing the water-holding capacity of the soil. The resulting increase in fertility allows other species to become established. Further inland, pohuehue (*Muehlenbeckia complexa*) may form dense tangles over wide areas. Tree lupin (*Lupinus arboreus*), an introduced legume, becomes very common and its nitrogen-fixing ability enables it to raise soil nitrogen content. In northern parts, kikuyu grass may become dominant. Other plants that

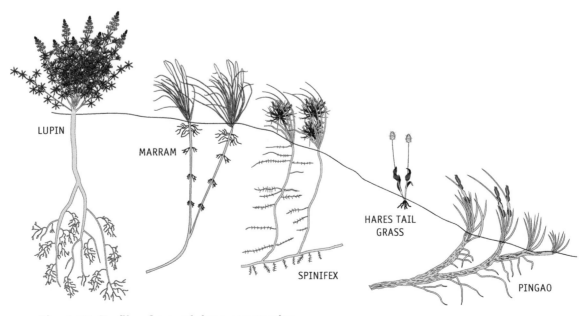

LUPIN

MARRAM

SPINIFEX

HARES TAIL GRASS

PINGAO

Fig. 4.25 Profile of a sand dune community

ISBN: 9780170214094

begin to appear are shrubs such as cottonwood (*Cassinia retorta*), and trees such as karo (*Pittosporum crassifolium*) and on northern coasts, pohutukawa (*Metrosideros excelsa*).

Not only do the plants provide food for animals, but they also provide shelter, and the increase in diversity of plants is accompanied by increasing number of animal species. For example, the caterpillars of the native copper butterfly feed on *Muehlenbeckia*.

Secondary succession

A quite different kind of succession occurs after existing vegetation is removed, for example by fire. The early stages of this **secondary succession** can be seen every time a garden is left unweeded. The soil is fully fertile, and so it is not long before plants become established. The first to appear are either those that can produce large numbers of wind-dispersed seeds or spores, or those which were already present in the soil as seeds, tubers or other dormant stages. Since the soil is fertile from the start, secondary succession is much faster than primary succession.

Forest stratification

The vegetation in a forest occurs in layers or *strata*. Each stratum consists of trees of characteristic species, though the details vary with the latitude (Fig. 4.26).

- The tallest trees are called **emergents**, because they protrude from the continuous canopy layer just below. These are the forest giants such as totara, rimu, matai, miro, and (in swampy areas) kahikatea and (in the North) kauri.

- The **canopy layer** consists mainly of angiosperm species such as beech, hinau, kamahi (except in the far North), pukatea (especially in wet soils), tawa, pigeonwood, nikau. In the South Island beech dominates. Included in the canopy layer are various species of climber or liane, such as lawyer, passion vine and *Clematis*.

- Below the canopy layer are various species of tree fern.

- Next is the **shrub layer** consisting of hangehange, fivefinger, various species of *Coprosma*, and others.

- The lowest layer of vegetation is the **ground layer**. This consists of small ferns, mosses and seedlings of tall trees.

Not belonging to any specific layer are the various species of **epiphyte**. These germinate on another plant, often high up, thus saving the energy needed by other plants to gain height. Though they are anchored to another plant by their roots, they do not obtain any nutrients from them. They are, in effect, commensals.

1. Emergents: e.g. rimu, totara, kahikatea, rewarewa
2. Canopy: e.g. kamahi, puriri, taraire
3. Sub-canopy: e.g. mahoe, mapou, lemonwood
4. Shrub layer: e.g. rangiora, kawakawa
5. Ground layer: mosses, ferns

Fig. 4.26 Stratification in New Zealand forest

ISBN: 9780170214094

In moving down from the canopy, there is a gradient of light intensity and quality, temperature, wind speed, and humidity. At any given level, plants are adapted to those particular physical conditions. Canopy trees are adapted to high light intensities and can withstand higher rates of transpiration. The ferns and mosses of the ground layer can grow in dimmer light. Tree seedlings also must be able to survive in these conditions, although they grow very slowly until a light gap is created by a large tree falling over.

Zonation on rocky shores

More than any other habitat, a rocky shore shows the effects of physical factors. Within a very short distance there is a change from organisms adapted to living in the sea to those adapted to life on land. Except for the inhabitants of rockpools, all shore organisms spend periods of exposure to the air (emersion), alternating with periods of submergence.

Emersion carries several hazards:

- Desiccation (drying).

- Rain can cause osmotic uptake of water.

- Temperatures fluctuate far more in the air than in water.

- Even in rock pools — especially small ones — the environment changes drastically when the tide retreats. On hot days the temperature may rise by several degrees, raising the metabolic rates of the inhabitants. The oxygen supply, on the other hand, is reduced since an isolated pool is cut off from fresh supplies of water.

Most organisms cease activity when the tide is out: seaweeds cease photosynthesis, and most animals retreat into shelter and stop feeding. On a rocky shore there is the additional hazard of wave action, and all the inhabitants must be able withstand being washed away.

The rhythmic changes in physical conditions are complicated by the fact that the tidal cycle is just over 12 hours, so the time of low tide is a little later each day. Dangers of desiccation are obviously much more severe when low tide comes at mid-day than when it occurs at dawn or dusk.

Superimposed on the daily cycle are two other rhythms. Once a fortnight, at around the time of New Moon and Full Moon, the tides come up higher and retreat lower. These *spring tides* alternate with the *neap tides*, when the tidal range is least, and which occur during the 'half moon' phase of the lunar cycle.

On a much longer time scale is a twice-yearly rhythm. At around 21 March and 21 September, the spring tides are even greater than usual. These are called the *equinoctial* tides because they occur at the equinoxes, when day and night are of equal length.

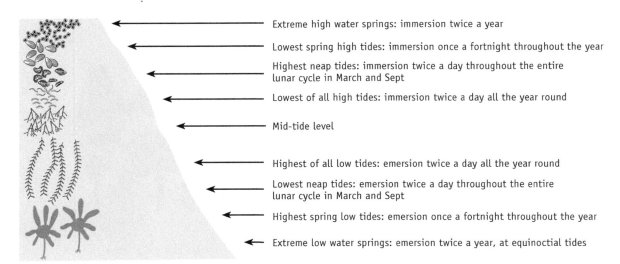

Fig. 4.27 Ecologically important zones on a rocky shore

ISBN: 9780170214094

These twice-daily, twice-monthly and twice-yearly cycles have great significance for the inhabitants of the shore (Fig. 4.27). At the extreme low water level, organisms are uncovered only about two days a year, at the equinoxes. A little higher up is the mean low water of spring tides where, throughout the year, organisms are exposed about once a fortnight. Higher still is a level reached by the highest low tides of the year, below which organisms are exposed part of every day, even during neap tides.

Above mid-tide is the lowest high tide level of the year, where organisms are submerged at least once a day throughout the year. Higher up is a zone that is submerged about once a fortnight all year round, and at the very top of the shore organisms are submerged only at the equinoctial tides, twice a year. This zone is not clearly defined because it varies greatly with exposure, merging with the splash zone.

Though the gradients of emersion and submersion are continuous, the distribution of some of the shore organisms may show quite sharp boundaries. Experiments have indicated that competition (Chapter 2) is a probable factor. For example, when brown seaweeds are artificially removed, the red seaweeds that normally are confined to the lower shore grow in abundance. This suggests that competition intensifies the effects of small changes in physical conditions.

Zonation with altitude

As one walks up a mountain, such as Mt Taranaki, it gets progressively colder, wetter and windier. Paralleling these changes in physical conditions, the vegetation changes too. At the boundary of Mt Taranaki National Park the altitude is about 400 m and the forest is typical podocarp. Rimu, rata and kamahi dominate, and the tallest trees are over 20 m high. With increasing altitude the height of the canopy decreases and by 1300 m the trees have given way to shrubs (Fig. 4.28). Higher still there is tussock and herbs such as alpine daisies and mosses.

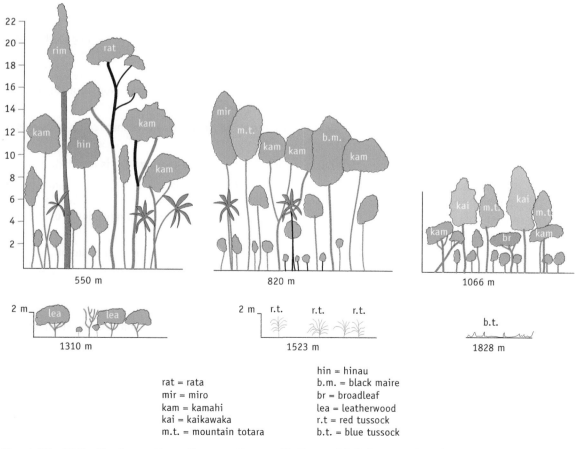

rat = rata	hin = hinau
mir = miro	b.m. = black maire
kam = kamahi	br = broadleaf
kai = kaikawaka	lea = leatherwood
m.t. = mountain totara	r.t = red tussock
	b.t. = blue tussock

Fig. 4.28 Altitudinal zonation of vegetation on Mt Taranaki (After Clarkson)

Summary of key facts and ideas in this chapter

○ A community is all the organisms living in a defined area.

○ *Autotrophs* or producers make organic compounds from inorganic raw materials.

○ *Heterotrophs* obtain organic materials from other organisms.

○ *Consumers* feed on other living organisms or on parts of them, killing living tissues in the process.

○ *Herbivores* are consumers that feed on plants, and carnivores are consumers that feed on animals.

○ *Omnivores* feed on both animal and plant material.

○ *Decomposers* feed on dead matter, and include many bacteria, fungi, and some animals. Whereas bacteria and fungi digest their food externally, animal decomposers digest it in a gut cavity and are called *detritivores*.

○ *Parasites* feed on living matter without killing it, though harming it to some extent.

○ An *ecosystem* is a community plus all the non-living components of the habitat.

○ In an ecosystem, autotrophs and heterotrophs are interdependent. The autotrophs depend on the heterotrophs for inorganic raw materials, and the heterotrophs depend on autotrophs for energy and organic raw materials.

○ Organic compounds pass from organism to organism along a *food chain*. The first member of a food chain is an autotroph or dead autotrophic material.

○ *Grazing food chains* begin with living plant material; *detritus food chains* begin with dead plant material.

○ Food chains are interconnected to form *food webs*.

○ Energy flow through a food web can be simplified by grouping all plants together as a trophic level, and similarly for herbivores, primary carnivores (carnivores that feed on herbivores), secondary carnivores, and so on.

○ Of the energy entering an animal, only a small proportion is retained in new tissue in growth; most is lost in respiration, and in the organic matter in faeces and urine.

○ Of every kJ of energy in the producer trophic level, only a fraction enters the herbivore level, and so on. When the total biomass of each trophic level are represented as blocks of proportionate size, a pyramid of biomass is usually the result (for an exception, see text).

○ Pesticides that are not excreted or broken down in metabolism become concentrated to dangerous levels along a food chain.

○ Whereas energy is lost along a food chain because it cannot be re-used, matter is recycled.

○ The carbon cycle involves organic compounds, and is described in detail in the text. Human activities have severely altered the carbon cycle. One of the most obvious effects has been a pronounced increase in CO_2 concentration in the atmosphere. The overwhelming majority of climate scientists think that this is warming the planet.

○ The nitrogen cycle can also be disrupted by human activity and can lead to local pollution.

○ When bare ground is colonised by plants, a series of changes in species composition occurs over time, called *succession*. These are of two kinds, primary and secondary, and occur over very different time scales and involve different processes.

○ In a forest, plant foliage occupies different levels. This layering of vegetation is called *stratification*.

○ On a rocky shore, there is a gradient of physical conditions extending from low tide to high tide. The change in species composition that occurs along this gradient is called *zonation*.

○ Zonation also occurs in the vegetation with increase in altitude up a mountain.

Test your basics

Copy and complete the following sentences. In some cases the first letter of a missing word is provided.

1. Organisms that make their own organic materials from carbon dioxide and water are ___*___.

2. Organisms that cannot make their own organic compounds from carbon dioxide and water are ___*___.

3. A ___*___ feeds on living material, killing it in the process. A ___*___ ___*___ feeds on plants, and a ___*___ ___*___ feeds on herbivores. ___*___ feed on living matter without killing it, and ___*___ feed on dead matter.

4. In an ecosystem, ___*___ is recycled, but ___*___ can only be used once.

5. In a ___*___ food chain, the primary consumers feed on living plant matter; in a ___*___ food chain the primary consumers feed on dead plant matter.

6. In a marine food chain the producers are single-celled photosynthetic organisms, the ___*___. These are eaten by tiny animals, the ___*___.

7. Food chains are normally interconnected to form food ___*___s.

8. In a given area, all the producers collectively form the lowest ___*___ level of a food ___*___,

9. The ultimate source of energy in nearly all ecosystems is ___*___ from the ___*___. Eventually, all energy leaves ecosystems as ___*___.

10. Most ___*___ic ('cold blooded') animals convert about _*_% of the energy in their food into flesh. Most is lost as heat in m___*___; the rest is lost as ___*___ energy in f___*___ and ___*___.

11. In a given ecosystem the various trophic levels can be represented by 'boxes'. If the area of each box is proportional to the total mass of organisms in that particular trophic level, the boxes collectively nearly always form a ___*___ shape.

12. In recent times the concentration of atmospheric CO_2 has risen rapidly as a result of the oxidation of timber following ___*___, and the burning of ___*___ ___*___. The vast majority of climate scientists believe that this is causing the earth's ___*___ to rise.

13. Nitrogen is a key constituent of ___*___ and ___*___ acids such as DNA.

14. Plants absorb most of their nitrogen in the form of ___*___ ions. These are converted to ___*___, which is then combined with carbon compounds to form ___*___ acids and then ___*___.

15. When organisms die, their proteins are digested by heterotrophs to form ___*___ ___*___. Most of these are then ___*___ed to release ammonia. In the soil, ammonia is oxidised by ___*___ bacteria, first to ___*___ and then to ___*___.

16. Some bacteria are able to convert nitrogen gas into ammonia, which is then combined with carbon compounds to form ___*___ ___*___ and then ___*___. Most of these ___*___ ___*___bacteria live in the roots of plants of the ___*___ family.

17. In anaerobic conditions, ___*___ bacteria use ___*___ in ___*___ respiration to oxidise organic matter.

18. In p___*___ succession the first plants to get established are the ___*___ species. These are tolerant of harsh ___*___ conditions but are weak ___*___. Later, as soil conditions become more ___*___, other species replace the earlier ones. Eventually, after there is no further species change, a ___*___ community develops.

19. In s___*___ succession, the soil conditions are favourable to begin with, such as after a forest fire.

20. The layering of vegetation in a forest is called ___*___.

21. The change in the species composition of a community along an environmental gradient is called ___*___.

5 Adaptation in Animals

All living organisms are equipped or *adapted* in various ways to make a living. To do this they must obtain energy and raw materials, avoid being eaten, and finally, reproduce.

Any inherited feature that increases an organism's chances of survival and reproduction is called an **adaptation**. The light skeletons of birds and streamlined shapes of fishes are obvious examples.

So well adapted are organisms to their ways of life that it seemed they must have been the product of deliberate design by an intelligent Creator. Then in 1859 Charles Darwin published *The Origin of Species*. It was a bombshell, because it challenged the orthodox view. Darwin maintained that organisms have become adapted to their environment by a very gradual process, which later became known as **evolution**.

Darwin was not the first to suggest that organisms are the result of gradual change rather than Divine Creation. The reason why his book made such an impact was that he presented a massive body of evidence, and suggested a plausible mechanism. He called the mechanism **natural selection**. It depended on a number of simple observations (O) and deductions (D):

- Organisms produce far more offspring than are needed to replace the parents. (O)

- Over long periods, the numbers of any given species remain approximately stable. (O)

- Therefore, over long periods, birth rates and death rates must be equal. (D)

- Therefore the majority of offspring do not survive to reproduce. (D)

- Organisms *vary*; no two members of a species are exactly the same. (O)

- Some individuals must therefore be more likely to survive than others, so there must a process of *selection*. (D)

- Variation is, at least to some extent, *inherited*; successful individuals will therefore hand on their advantageous characteristics to their offspring. (D)

The strength of Darwin's proposal lay in the fact that selection had been proved to work with domesticated organisms, the difference being that selection was by humans rather than by nature.

A case study of adaptation: The mosquito

Mosquitoes are important insects because some species spread diseases. There are 12 species of mosquito native to New Zealand, and four introduced species (Fig. 5.1). The details of the life

Excellence in Biology Level 2

ISBN: 9780170214094

cycle vary slightly with the species; the description below applies mainly to *Culex pervigilans,* a common New Zealand species.

Like other insects, mosquitoes belong to the phylum Arthropoda, which also include crustaceans (e.g. crayfish), arachnids (e.g. spiders), centipedes and millipedes. Their most defining character is a **cuticle** of *chitin*, hardened in places to form a jointed **exoskeleton** (external skeleton).

Like all organisms, mosquitoes are adapted in far more ways than can be described in a few paragraphs; the following is a selection.

Feeding

Adult mosquitoes are fluid feeders. Both sexes feed on nectar (a sugary solution secreted by flowers), but females must also take a blood meal to gain the protein necessary for egg production. They are attracted to the warmth and CO_2 produced by the host. The mouthparts of the female (Fig. 5.2) are highly adapted for piercing and sucking, without attracting the attention of the host animal, which might swat them with its tail (or human hand).

The piercing mouthparts (*stylets*) are like two pairs of microscopic scalpels: the **mandibles** and the saw-toothed

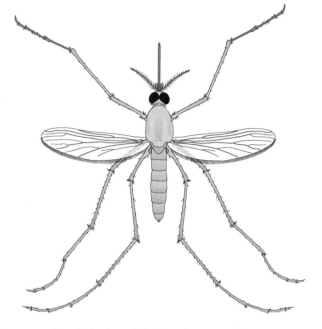

Fig. 5.1 An adult female mosquito

maxillae. These are so fine that in most cases the host does not feel the puncture. Before blood is sucked up through the food channel, saliva is injected into the wound. It contains an anticoagulant protein that stops the blood clotting, but also causes itching. The injection of saliva is also the route by which some parasites (such as the malaria parasite), enter the host.

The stylets are supported by the **labium** or 'lower lip'. This is too wide to enter the wound and bends in the middle as the stylets enter the skin. Blood is sucked into the mid-gut by a pump in the pharynx (front part of the gut). A female can take more than her body mass of blood, and to accommodate this huge meal her abdomen has to be very distensible. Receptors in the abdomen cause the pharyngeal pump to stop when she is fully distended.

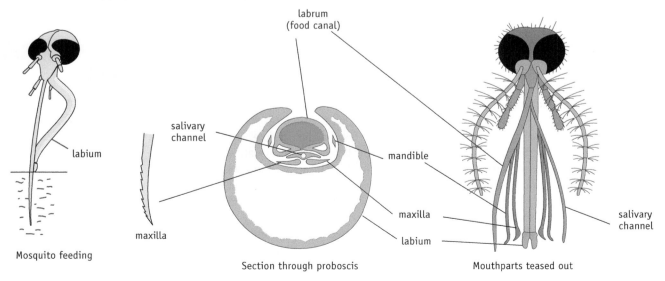

labrum
(food canal)

labium

salivary
channel

maxilla

mandible

maxilla

labium

salivary
channel

Mosquito feeding

Section through proboscis

Mouthparts teased out

Fig. 5.2 Mouthparts of female mosquito

The ability of the abdomen to swell so much enables the mosquito take fewer, larger, meals. This is of survival value because with each feed the insect risks being swatted.

The main organic foods for the mosquito are proteins (haemoglobin and plasma proteins) and lipids (in the plasma membranes of the blood cells). These require protease and lipase enzymes to be digested.

Summary of key nutritional adaptations:

- Ability to detect host from a distance.
- Mouthparts that can pierce flesh without the host feeling it.
- Anticoagulant in saliva.
- Highly distensible mid-gut and pharyngeal pump.
- Production of protease and lipase enzymes.

Gas exchange

In insects the blood or *haemolymph* is colourless and plays no part in oxygen or CO_2 transport. Instead these gases are transported by a system of branching, air-filled tubes called **tracheae**. These open at the sides of the body by holes called **spiracles**. The finest 'twigs' of the tracheal system are microscopic **tracheoles**. These reach to within a fraction of a millimeter of every cell in the body. A more detailed description of the insect tracheal system is given in Chapter 8.

Osmoregulation

As a small, terrestrial (land-living) animal, a mosquito must avoid desiccation (drying). Water loss is reduced in several ways:

- The outer layer of the cuticle is *waxy*, making it waterproof.

- Because the tracheae are *intuckings* into the body, the air next to the cells lining them is very humid, so water is lost more slowly. The rate of water loss is reduced still further by the ability to partially close the spiracles when the animal is less active. The rate of ventilation of the tracheae is regulated by the concentration of CO_2 in the tissues. During flight, the increased rate of CO_2 production causes the spiracles to open, but when respiration is slower, the spiracles close, reducing water loss.

- Very little water is lost in nitrogenous excretion. The excretory organs are the *Malpighian tubules*, which open into the hindgut. Like other terrestrial (land-living) insects, adult mosquitoes excrete **uric acid**. This is non-toxic, almost insoluble, and is passed out with faeces. Since it is excreted as a crystalline solid, very little water is lost.

Life cycle

A male mosquito detects the female by the high-pitched hum of the female's wings. The 'ears' of the male are their extremely bushy antennae (Fig. 5.3).

Mating takes place in mid-air. The male introduces sperm into the genital opening of the female, but before the fertilised eggs can develop, she must take a blood meal. A few of days after a blood meal she lays her eggs in groups ('rafts') on the surface of small freshwater pools rich in decaying organic (Fig. 5.4). Suitable pools are recognised by their smell.

The egg rafts float on the surface because all except the lower surface of each egg is covered with wax, which is not wettable. After a couple of days the young emerge from the eggs. Each looks completely different from the adult and is called a **larva** (Fig. 5.5).

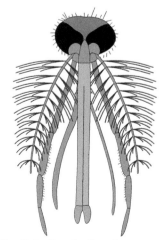

Fig. 5.3 Head of male mosquito showing bushy antennae

Excellence in Biology Level 2

ISBN: 9780170214094

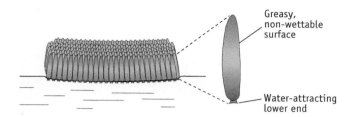

Fig. 5.4 Egg raft of a *Culex* mosquito.

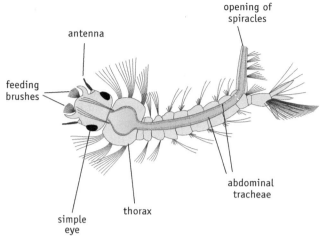

Fig. 5.5 A mosquito larva, side view

The larvae live in freshwater pools and feed on bacteria which they filter out of the water using a pair of *feeding brushes* (Fig. 5.6).

Water rich in bacteria is usually poor in oxygen, but the larvae are unaffected because they breathe at the surface. The two long tracheae open by spiracles at the tip of the abdomen. These, together with some of the smaller tracheae, can easily be seen in the living animal when viewed under the microscope. The spiracles are protected by flaps that are greasy on the inside but wettable on the outer surface (Fig. 5.7). While breathing they hang from the surface. The flaps are held open by the surface tension of the water. When the insect dives, the flaps close, sealing the spiracles. If oil or detergent is added the surface tension is destroyed and the flaps cannot open, so the insects drown.

Fig. 5.6 Head of a mosquito larva, showing feeding brushes

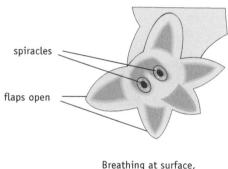

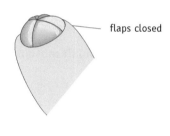

Fig. 5.7 Mosquito larva, showing spiracles open at surface, and closed when animal is submerged

Larvae spend most of their time feeding while hanging from the surface, but in response to vibration or a shadow falling over the water, they dart down to the bottom by rapid wriggling movements of the abdomen.

As time goes on, the cuticle can stretch to some extent, but cannot grow. As in all insects, a new cuticle is secreted by the epidermis (skin) under the existing one. The old cuticle splits and the 'new' larva emerges, rapidly swelling as a result of absorption of water. This shedding of the cuticle during growth is called **ecdysis**. There are four ecdyses, and each larval stage is called an **instar**.

The role of the larva is simple: to feed, grow, and accumulate sufficient energy and materials to develop into the adult. The adult, however, is utterly different with regard to movement, feeding, gas exchange and sense organs, to mention but a few.

In the last ecdysis, the last larval instar gives rise to a **pupa**, which resembles neither larva nor adult (Fig. 5.8). The pupa breathes at the surface through two 'breathing trumpets' in a similar way to the larva. The pupa does not feed, but when danger threatens it can swim rapidly by movements of its abdomen.

Inside the pupa, many of the larval organs are broken down, and adult organs grow from tiny buds of cells. The energy and raw materials for this were laid down and stored in the larva.

After three or four days (depending on temperature) the adult emerges from the pupa while it is hanging from the surface (Fig. 5.9). The adult stands on the pupal case as blood is pumped into the veins of the wings, causing them to expand. The adult cuticle hardens, and after an hour or so the mosquito flies away.

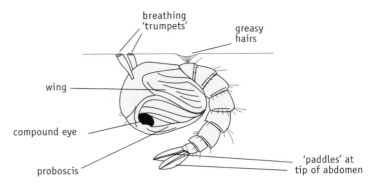

Fig. 5.8 Pupa of mosquito

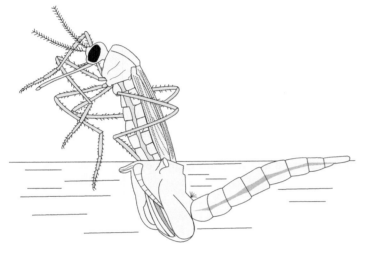

Fig. 5.9 Adult mosquito emerging from the pupal case

Types of adaptation

Adaptations are often described under four headings, all of which can be illustrated by the mosquito.

- Structural adaptations. The extremely thin, sharp mandibles and maxillae of the female enable her to penetrate the skin with minimum irritation to the host – thus minimising the risk of being swatted. This potential hazard is further reduced by her distensible abdomen that enables her to take larger, and therefore fewer, meals.

- Physiological adaptations. Thermoreceptors that detect heat and chemoreceptors that detect CO_2 enable the female mosquito to locate the host.

- Behavioural adaptations. Mosquito larvae respond to any vibration or shadow of a potential predator by wriggling down to the relative safety of the bottom of the pool.

- Biochemical adaptations. The anticoagulant protein in the saliva enables the blood to continue flowing, and the protease and lipase enzymes she secretes in her gut enable the blood to be digested.

These distinctions are somewhat artificial. Take, for example, the escape reaction of a mosquito larva as it dives down to the bottom of the pool. Though we call this a behavioural adaptation, it involves anatomical structures, physiological processes, and biochemical changes.

Structures called receptors detect the stimulus of a sudden change in light intensity. This starts a chain of physiological processes such as electrical signals (impulses) travelling along structures called nerve fibres. Nerve impulses involve biochemical changes in protein molecules in the plasma membranes of the nerve fibres. This is followed by contraction of structures called muscle fibres, in which certain protein molecules slide past each other, causing the muscle fibres to shorten.

Adaptations work together

In the above treatment of the mosquito, various life processes have been treated separately. But of course, animals are not collections of separate adaptations; all working parts and processes evolve together. For example, the ability to fly could not have evolved without an efficient gas exchange system to deliver oxygen rapidly to the muscle cells. Also, the environment of a flying animal can change quickly. Insects therefore need sophisticated sense organs such as compound eyes, and a brain to process the information.

Acclimatisation vs adaptation

The word 'adaptation' can be a noun or a verb. As a verb, the *process* of adaptation is an evolutionary one and thus involves genetic change. What this means is that *individual* organisms do not adapt; populations do, over many generations. Adaptive change takes place *between* generations rather than within an individual organism's lifespan.

This does not mean that individuals are totally at the mercy of environmental change. An organism may adjust to new circumstances, for example:

- Fair-skinned people can develop a suntan in response to increased UV radiation.

- Skin on the hands and soles of the feet get thicker in response to wear.

- Heart and skeletal muscles become stronger in response to increasing demands.

- Bones becomes stronger in response to increasing stresses placed upon them.

- Over a period of several weeks, a person can adjust to high altitude.

- During periods of prolonged cold, a person's metabolic rate (and therefore heat production) rises.

Changes like this are called **acclimatisation** and are not inherited — children of weightlifters do not inherit powerful muscles!

Does this mean that such changes are of no long-term significance? Not necessarily, because the *ability* to acclimatise *is* inherited, and the speed and extent to which individual organisms can acclimatise varies. For example:

- People with similar skin colour may inherit the ability to tan to different degrees even if they are exposed to the same solar radiation.

- Different people may have different (inherited) tendencies to store fat even if other environmental factors are the same.

- People of similar age may inherit different abilities to respond to exercise.

Acclimatisation of individuals can therefore give time for more permanent, genetic adaptation to evolve.

Extension: Comparing animals from different groups

Adaptation can also be studied by comparing the same process in a range of different animals, as described in Chapters 6–8. One can compare feeding and digestion in closely related meat-eating and plant eating animals. For example sheep and dogs are both mammals, so they have fundamentally similar anatomies. Their basic mammalian anatomy has been modified in evolution adapting them to feed in very different ways.

When studying adaptation to a particular way of life in animals that are *not* closely related it is important to remember that many of the differences one observes have nothing to do with differences in the present day environment. Rather, they are the result of evolutionary changes that took place hundreds of millions of years ago.

For example, a weta and a human are both adapted to exchanging oxygen and CO_2 in air rather than water, but they do so in completely different ways. This is because when the remote ancestors of weta and humans were becoming adapted to life on land, their anatomies were already completely different. Natural selection works by making piecemeal modifications to genes controlling *existing* structures — it can't 'think ahead' to future needs. The insect tracheal system evolved by the development of deep intuckings of the exoskeleton, whereas the vertebrate lung evolved as an intucking of the front part of the gut, with totally different results.

ISBN: 9780170214094

Summary of key facts and ideas in this chapter

○ The word 'adaptation' can be a noun or a verb. As a noun, an adaptation is an *inherited* feature that increases the probability of an organism's survival and reproduction. As a verb it means the *process* by which, over successive generations, such inherited features develop in the course of evolution.

○ The mosquito is used to illustrate the idea that adaptive features evolve *together* as an integrated ('working together') whole. For example, the evolution of a jointed, waxy cuticle makes possible the evolution of:
  - rapid movement by flight,
  - the ability to pierce skin with sharp mouthparts,
  - the ability to greatly reduce water loss.

○ The evolution of such a highly energetic movement as flight could not have occurred without an efficient method of delivering oxygen to the cells; the tracheal system. Flight also demands a relatively (for such a small animal) sophisticated nervous system to process rapidly-changing sensory information.

○ The tracheal and excretory systems, with the waxy cuticle, all act together to reduce water loss in the adult.

○ Adaptation should not be confused with *acclimatisation*, which is a change in the body of an individual organism that increases its ability to survive new environmental conditions (for examples see text). Whereas adaptation involves genetic change, acclimatisation involves purely phenotypic change and is not inherited.

Test your basics

Copy and complete the following sentences. In some cases the first letter of a missing word is provided.

1. Adaptations are ___*___ features that increase the chances of an organism ___*___ and ___*___.

2. Adaptations are the result of the process of ___*___, in which the best- ___*___ ('fittest') individuals tend to have more surviving offspring than the less well-___*___ individuals.

3. Individual organisms do not become better-adapted during their lifetimes; p___*___s do, over many g___*___s.

4. Adaptation is quite different from ___*___, in which individual organisms develop characteristics that increase their chances of survival. For example after repeated vigorous exercise, the heart responds by becoming ___*___.

6 Nutrition in Animals

Unlike plants, animals are **heterotrophic**, meaning that they obtain their energy and raw materials from other organisms. In most animals the food consists largely of starch, proteins, fats and water, with small quantities of vitamins and minerals.

Before the food can be used by the cells it has to be broken down into simpler substances such as glucose and amino acids. This process is called **digestion** and occurs in a cavity called the **gut**, or **alimentary canal**. Digestion is brought about by enzymes secreted into the gut. In most animals this is a tube with two openings, the *mouth* and *anus*.

Digestion is one of five stages of food processing. These are:

- **Ingestion**, or the taking in of food at the mouth. This often involves physical breakdown of the large food lumps into smaller bits, by **chewing**.

- **Digestion**, or the chemical breaking down of the complex foods into simpler, soluble substances.

- **Absorption**, or the passing of the simple products of digestion through the walls of the gut. Until now, the food has not strictly been inside the body. This is because the cavity of the gut is an extension of the outside world.

- **Egestion**, or the passing out of the gut of the indigestible remains as *faeces*. Egestion is quite different from *excretion*, since faeces are not produced in metabolism.

- The final stage is **assimilation**, or the absorption of the digested foods by the cells.

Though these processes are common to most animals, the details of food capture, ingestion and digestion vary considerably. **Herbivores** feed on plants and **carnivores** feed on other animals. Animals that feed on both animals and plants are called **omnivores**.

Hydra

Hydra is a small freshwater animal found attached to weed in ponds and lakes (Fig. 6.1). It is a member of the phylum Cnidaria. Other members of the phylum include sea anemones, corals and jellyfish.

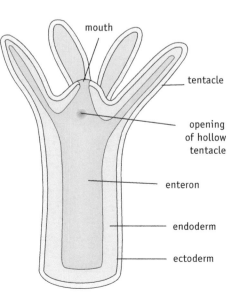

Fig. 6.1 *Hydra* **cut longitudinally**

ISBN: 9780170214094

These creatures have a number of distinctive features:

- The body wall consists of only two layers of cells, an outer **ectoderm** and an inner **endoderm**, with a jelly-like **mesogloea** between. In *Hydra* the endoderm is green because of the tiny photosynthetic organisms called *zoochlorellae* living in a mutualistic relationship inside the endoderm cells.

- The body is like a hollow bag, with only one entrance (the older name for the phylum, Coelenterata, means 'hollow inside').

- The mouth is surrounded by tentacles.

- The ectoderm has thousands of **nematoblasts** or **cnidoblasts**, from which the phylum takes its name. These are particularly abundant on the tentacles.

- They are *radially symmetrical*, meaning that the body can be cut into equal halves in several ways.

Hydra is *sessile*, meaning that it is anchored to one spot, so it waits for food to come to it. It is carnivorous, feeding mainly on tiny animals such as *Daphnia* (Fig. 2.6). The mouth is surrounded by hollow tentacles, each covered with nematoblasts. Each nematoblast contains a **nematocyst**, a poison sac leading to a hollow thread that can be shot out (Fig. 6.2).

When the prey blunders into a tentacle, thousands of nematocysts are discharged into it, injecting paralysing poison. The tentacles slowly move the prey into the mouth and into the digestive cavity or **enteron**. This is like a bag, with only *one* opening, so indigestible food is egested the way it came in. Gland cells in the endoderm secrete enzymes and slowly digest the food, reducing it to a soupy liquid. (Fig. 6.3). Some small food particles are taken into endoderm cells by *phagocytosis* and enclosed in food vacuoles where they are digested. Flagella on some of the cells help to bring semi-digested particles into contact with endoderm cells. Digestion is thus partly *extracellular* (outside cells), within the enteron, and partly *intracellular* (within cells). The prey's indigestible remains are then egested.

Having a gut with a single opening has two important consequences:

- Each prey item has to be fully processed before any more can be eaten.

- Since the indigestible remains are egested the same way they came in, food traffic is *two way*. It is therefore not possible for different parts of the gut to be specialised for different stages of food processing. In animals with tube-like guts (two openings), the gut is divided into different regions specialised for different stages of digestion and absorption.

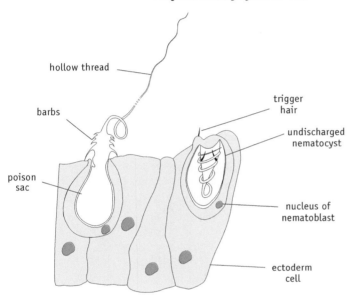

Fig. 6.2 A nematocyst, shown discharged (left) and undischarged (right)

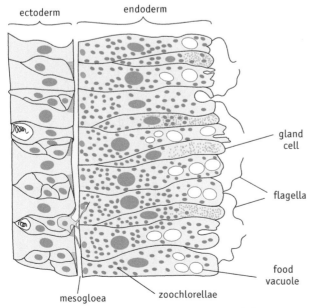

Fig. 6.3 *Hydra*: longitudinal section through body wall

Earthworm

Earthworms are herbivorous, feeding on dead leaves which they drag down into their burrows at night. They also ingest quantities of soil as they burrow through it, digesting the humus (semi-decomposed organic matter) it contains.

ISBN: 9780170214094

Unlike *Hydra*, earthworms have two openings to the gut, allowing food to travel in one direction as it is processed. This has two advantages:

- Feeding can be more or less continuous, enabling processing of large quantities of low energy food.

- Different parts of the gut can be specialised for different stages in food processing, such as mechanical break-up, digestion, and absorption.

The gut or alimentary canal is divided into a number of regions, each with a particular function (Fig. 6.4).

Food is sucked into the mouth by the **pharynx**, which has large muscles attached to its outer wall. It is then propelled by muscular waves in the **oesophagus** to the **crop**, where it is temporarily stored. It then passes to the **gizzard**, whose muscular walls grind it up. It then passes into the long **intestine**, in which digestive enzymes are secreted. Starch is converted into *glucose*, proteins into *amino acids*, and fats into *fatty acids* and *glycerol*. The surface area of the intestinal lining is increased by a longitudinal ridge called the **typhlosole** (Fig. 6.5). Unlike *Hydra*, all the food is digested extracellularly (outside the cells), and this is true of all other animals in this chapter.

The digested food is absorbed into the blood by *active transport*. This is an energy-requiring process by which substances are pumped across cell membranes (see Ch. 21).

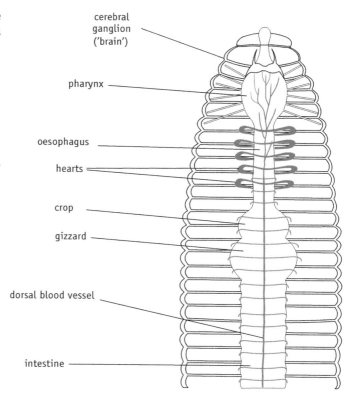

Fig. 6.4 Earthworm dissected to show the anterior part of gut from the dorsal side

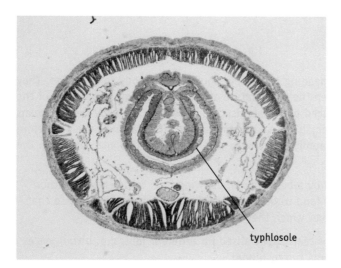

Fig. 6.5 Transverse section of an earthworm in region of the intestine

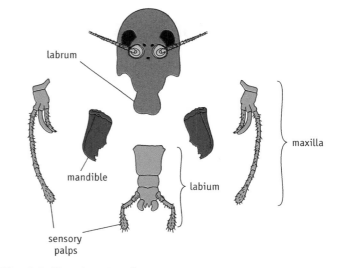

Fig. 6.6 Mouthparts of a weta

Weta

The common tree weta is a member of the class Insecta in the phylum Arthropoda. As in other arthropods the epidermis ('skin') secretes a **cuticle** consisting mainly of **chitin** and protein. Except at the joints the cuticle is hardened and acts as an **exoskeleton** (external skeleton). Hard mouthparts have enabled arthropods to exploit a wide range of foods requiring biting, piercing and/or sucking. Weta are herbivores, feeding at night on leaves of various kinds of tree. They also eat some wood as they enlarge the holes in the branches in which they spend the daytime.

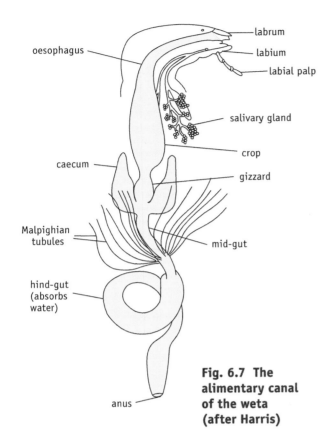

labrum

labium

labial palp

oesophagus

salivary gland

crop

caecum

gizzard

Malpighian tubules

mid-gut

hind-gut (absorbs water)

Fig. 6.7 The alimentary canal of the weta (after Harris)

anus

As in earthworms and most other animals, arthropods have a tube gut, with two openings. Associated with this, different regions of the gut are specialised for different functions.

Unlike earthworms, weta have jaws which break up food before it enters the mouth. This increases its surface area, making it easier for enzymes to digest. Behind the upper lip or **labrum**, are three pairs of mouthparts (Fig. 6.6). Food is first bitten off using the large **mandibles**. The **maxillae** (singular, *maxilla*) cut it into smaller pieces. Each maxilla has a sensory **palp** that tastes the food. The last pair of mouthparts are joined together to form the **labium** which acts as a lower lip. It helps to manipulate the food before swallowing.

A pair of **salivary glands** secrete saliva onto the food (Fig. 6.7). Though the composition of weta saliva has not been studied, in most leaf-eating insects the saliva contains a starch-digesting **amylase**. The food then passes down the **oesophagus** into the **crop**. Here it is temporarily stored before being passed into the **gizzard** where it is ground into smaller pieces. It then passes into the mid-gut where it receives more digestive enzymes secreted by the **caeca** (singular, *caecum*). These enzymes complete the digestion, yielding amino acids, fatty acids and glycerol. These are absorbed by the mid-gut.

Leaves contain considerable amounts of water and this, together with the water in the digestive juices, is absorbed by the hind-gut. As a result the faeces are dry, and water is conserved.

Mammals

Like birds, mammals are **homeothermic**, meaning that they regulate their body temperature. They are also **endothermic**. This means that when the temperature of the body is higher than that of the environment, the 'extra' body heat is produced by metabolism — i.e. *within* the body ('endo' means 'inside'). As a consequence, endothermic animals have a high metabolic rate. Mammals must therefore take in, digest, and metabolise food very rapidly. To achieve this, they have two adaptations in the beginning of the alimentary canal:

- Food and air passages are separated by a bony *palate*, allowing food to be chewed without interrupting breathing. This enables food to be finely broken up before swallowing, so that enzymes have a large surface area over which they can act on the food.

- They typically have different kinds of teeth, specialised for different jobs, e.g. biting pieces off food, followed by cutting, grinding, or crushing.

The raised body temperature also enables the digestive enzymes to work rapidly. Though not usually considered as part of nutrition, mammals also have an extremely efficient *double circulation* that enables food and oxygen to be transported rapidly to the cells.

Mammals show considerable variety in their dietary habits. **Carnivores** feed on other animals. **Herbivores** feed on plants, while **omnivores** have a mixed diet.

These distinctions are not absolute. For example, possums eat eggs as well as leaves, and dogs eat some vegetable material.

Human (omnivore)

The human alimentary canal is shown in Fig. 6.8.**Teeth**

ISBN: 9780170214094

Before digestive enzymes can act on the food they have to mix with it. Solid foods are eaten in large lumps, which must first be broken up to increase the area of contact with the enzymes. This is the function of the teeth (Fig. 6.9). Humans use tools to break up the food before it enters the body, and the teeth and jaws are correspondingly small.

Humans and other mammals have different kinds of teeth, adapted for special tasks. From front to back these are:

- **Incisors**, with chisel-like edges for cutting off pieces of food.

- **Canines**. They are only slightly pointed in humans and have a similar function to incisors.

- **Premolars** have one root and two cusps (projections on the surface of the tooth). They are used for crushing food into smaller pieces.

- **Molars** differ from premolars in that they are *first* teeth — they do not replace any earlier 'milk' or *deciduous* teeth. They have several roots and four or five cusps. They are larger than the premolars and, like them, crush the food.

Teeth are anchored in sockets in the jawbone by

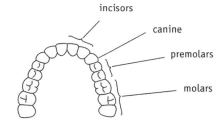

Fig. 6.9 Teeth of the human upper jaw

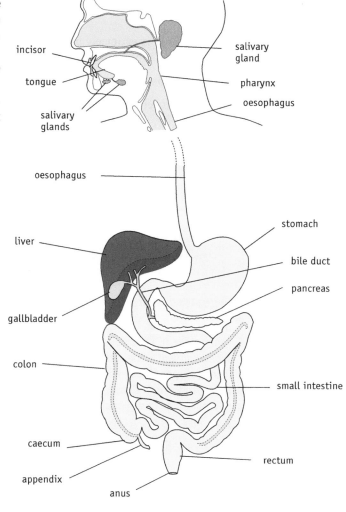

Fig. 6.8 The human alimentary canal

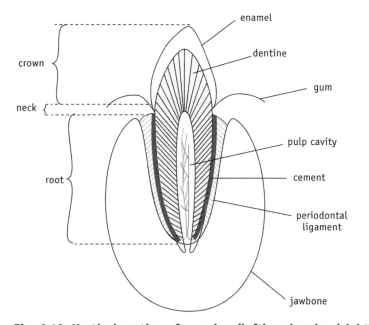

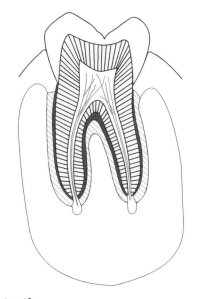

Fig. 6.10 Vertical section of a canine (left) and molar (right) tooth

ISBN: 9780170214094

tough collagen fibres of the **periodontal ligament**. Each tooth consists of a number of parts (Fig. 6.10):

- **Enamel** is extremely hard and covers the crown, forming a hard, biting surface. It consists almost entirely of calcium phosphate.
- **Dentine** forms the bulk of the tooth and is not as hard as enamel.
- **Pulp cavity** contains blood vessels and nerve endings.
- **Cement** covers the root of the tooth and resembles bone.**Chewing**

Aided by saliva secreted by **salivary glands**, chewing breaks up the food into a soggy lump or *bolus*. At the same time the enzyme *salivary amylase* converts starch into the complex sugar, *maltose*.

Swallowing

Swallowing is the muscular propulsion of food from mouth to stomach. First, the tongue pushes the bolus to the **pharynx** where air and food passages merge. The food is prevented from entering the trachea (windpipe) by the flap-like **epiglottis**. The bolus is squeezed down the oesophagus by waves of muscular contraction called **peristalsis** (Fig. 6.11).

Digestion in the stomach

Gastric juice is secreted onto the food, and at the same time it is mixed by the action of muscle in the stomach wall. The enzyme **pepsin** converts proteins into **polypeptides**, and as a result the food is slowly converted into liquid **chyme**. Gastric juice contains hydrochloric acid, which not only activates pepsin but also kills most bacteria. The exit to the stomach is guarded by a ring-like **sphincter** muscle, which controls the rate at which food leaves the stomach.

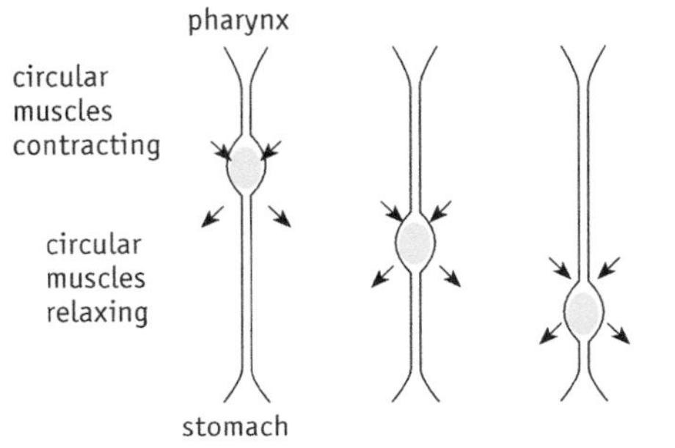

Fig. 6.11 How food passes down the oesophagus by peristalsis

Digestion in the small intestine

As the food is liquefied it is slowly released into the **duodenum**, where it receives two digestive juices:

- **Bile** is a yellow-green, alkaline liquid. It is secreted by the liver and temporarily stored in the **gallbladder**. Though it contains no enzymes it is essential for the digestion of fat. It contains *bile salts* which break up large fat globules into an *emulsion* of millions of tiny droplets (Fig. 6.12). This greatly increases its surface area, speeding up the action of the fat-digesting enzyme **lipase**.

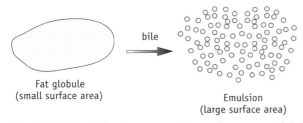

Fat globule
(small surface area)

bile

Emulsion
(large surface area)

Fig. 6.12 How bile increases the surface area of fat

- **Pancreatic juice**, secreted by the pancreas, contains enzymes that digest fat, proteins and starch.

Both these juices contain *sodium hydrogen carbonate*, which makes them alkaline. The digestion of proteins and carbohydrates is completed by other enzymes in the small intestine. The result is a mixture of simple sugars, amino acids, fatty acids and glycerol (Fig. 6.13). While the food is in the intestine it is mixed and propelled by muscular action.

Absorption of the digested food

Glucose and amino acids are absorbed into the blood by **active transport**, which is one reason why the gut uses more energy after a meal. The energy for active transport is in the form of **ATP**. This is produced in the abundant mitochondria in the cells lining the small intestine.

The surface area of the intestinal lining is greatly increased by many finger-like **villi**, each about 1 mm long. The area is further increased by millions of **microvilli** on the individual cells lining the

ISBN: 9780170214094

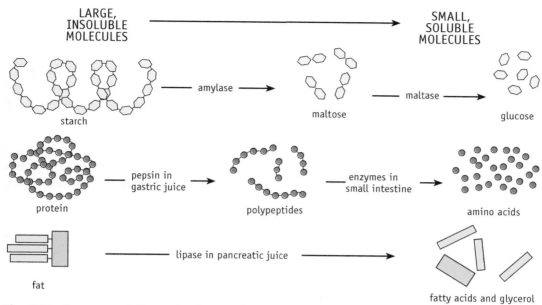

Fig. 6.13 Summary of the main changes in digestion

intestine (Fig. 6.14). Each villus contains a network of **capillaries**. These eventually join to form the **hepatic portal vein**, which takes blood to the liver.

Fatty acids and glycerol cross the lining of the intestine and recombine to form tiny fat droplets. These enter the tiny lymph capillary in each villus as an emulsion. The lymph capillaries in the villi

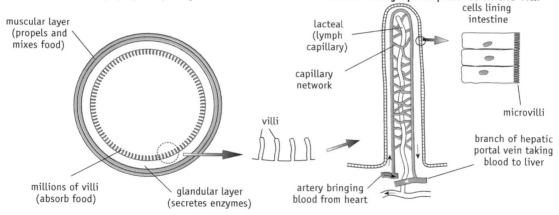

Fig. 6.14 How the surface area of the gut lining is increased

are called **lacteals** because after a meal they appear white due to the milky emulsion of fat.

The absorption of water

The digestive juices contain large amounts of water, which is too valuable to waste. About 95% is reabsorbed in the small intestine and most of the rest is reabsorbed in the colon.

The large intestine

By the time the undigested food has reached the large intestine, all that remains is cellulose and some water. The bacteria in the colon convert the material into **faeces**, which are temporarily stored in the rectum before being **egested** via the anus.

The cat

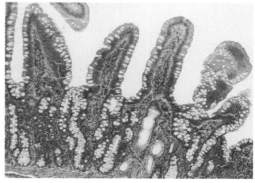

Villi in duodenum

ISBN: 9780170214094

Cats are adapted for catching and digesting animal food. Meat is rich in energy but carnivores have to work hard to get it. Most members of the cat family stalk their prey and then capture it in a short burst of intense activity. The final leap on to the prey depends on knowing how far away it is. This *stereoscopic vision* is made possible by forward-facing eyes, so the fields of view overlap. By comparing the image seen by the two eyes, the brain can judge the distance to the prey (Fig. 6.15).

Having made contact with the prey, the next job is to kill it. The pointed canines concentrate great force on a small area and their length enables them to penetrate vital organs. The incisors are used to scrape flesh away from the bone. The last upper premolar and first lower molar work like scissor blades to cut meat, and are called **carnassials** (Fig. 6.16). The jaw joint is hinge-like, allowing only up-and-down movement. This also makes the joint strong enough to resist dislocation when the prey struggles.

The main biting muscle is the **temporal**

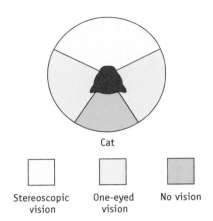

Cat

Stereoscopic vision   One-eyed vision   No vision

Fig. 6.15 Forward-facing eyes gives a cat stereoscopic vision

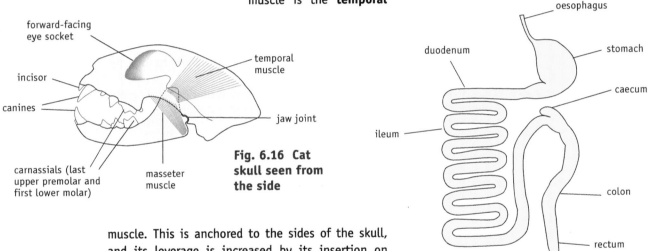

forward-facing eye socket

incisor

canines

carnassials (last upper premolar and first lower molar)

masseter muscle

temporal muscle

jaw joint

Fig. 6.16 Cat skull seen from the side

oesophagus

duodenum

stomach

caecum

ileum

colon

rectum

Fig. 6.17 Alimentary canal of the cat

muscle. This is anchored to the sides of the skull, and its leverage is increased by its insertion on an upward projection of the jaw. The other biting muscle is the smaller **masseter** muscle.

The rest of the gut (Fig. 6.17) is adapted to processing foods rich in protein and fat, the only carbohydrate being glycogen stored in the liver and muscle of the prey. Protein-rich food is relatively quickly digested and the gut is relatively short. The digestive enzymes are similar to those of a human except that the pancreatic juice contains relatively little glycogen-digesting amylase.

The small amount of carbohydrate in the diet means that glucose (needed by the brain) must be obtained from protein by deamination of amino acids in the liver.

Sheep

Sheep are *herbivores*, feeding on grass, clover and other pasture plants. Leaves have three drawbacks as food:

- Very few animals can produce a *cellulase* enzyme that digests the cellulose cell walls.

- Leaf tissue is relatively low in nutrients and energy, so it has to be eaten in large amounts. Most herbivores spend a high proportion of their 'working day' feeding. At the same time, they have to maintain vigilance against predators. Early warning against approach by a predator is made possible by their wide field of view (Fig. 6.18).

- Leaves are less rich in essential amino acids than most animal food, and do not contain all the necessary vitamins (for example, vitamin B_{12} is completely absent from plants).

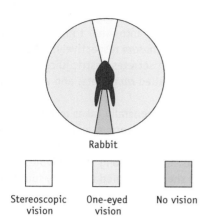

Rabbit

Stereoscopic vision | One-eyed vision | No vision

Fig. 6.18 Eyes on each side of the head gives a herbivore a wide field of view

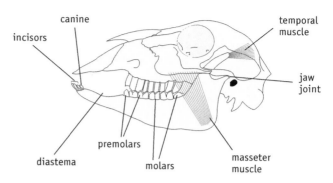

Fig. 6.19 Sheep's skull

Most herbivores have overcome both these disadvantages by having huge numbers of mutualistic microorganisms living in their guts, as explained below.

In sheep there are no incisors or canines in the upper jaw (Fig. 6.19). In their place is a horny pad against which the large chisel-shaped lower incisors and canines bite when the sheep is cropping grass — rather like a kitchen knife cutting on a chopping board. Behind the canines is a long gap in the dentition called the **diastema**. This provides space for the tongue to keep freshly cropped grass away from grass that is being chewed.

The premolars and molars are large and ridged. When the teeth first emerge from the gum in a young lamb the crowns are completely covered in cement, but this wears down as the animal chews. Enamel is harder than dentine, so it wears more slowly, forming ridges that run predominantly in the front to rear direction. These act like a file, cutting open the cellulose walls of plant cells. The lower jaw moves in a side-to-side fashion, at right angles to the ridges on the premolars and molars. This side-to-side jaw action is possible because of the flat jaw joint surface on the skull.

Unlike most other plants, grass has another disadvantage as food: the cell walls are strengthened by *silica*, a mineral that wears teeth rapidly. Herbivorous mammals have evolved a counter-measure: their teeth grow continuously. This is made possible by the rich blood supply to the pulp cavity, which has a wide opening (Fig. 6.20).

The jaw muscles are the same as in the cat, but their relative sizes are reversed, the masseter being much larger.

Sheep, like goats, deer and cattle are called **ruminants** because the oesophagus leads into a huge chamber called the **rumen** (Fig. 6.21).

The rumen contains vast numbers of cellulose-digesting microorganisms (Fig. 6.22). They produce large quantities of methane that has to be regularly belched. They also make a number of vitamins and essential amino acids that are not present in the food. The food then passes to the next chamber where it is worked into small lumps called 'cuds'.

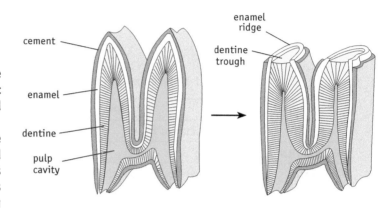

Fig. 6.20 A sheep's molar before and after wear

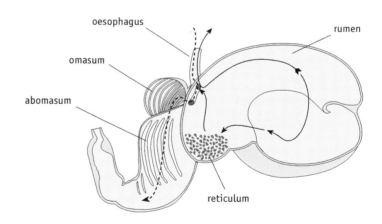

Fig. 6.21 The stomach compartments of the sheep

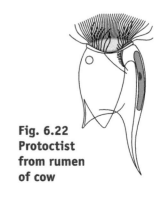

Fig. 6.22 Protoctist from rumen of cow

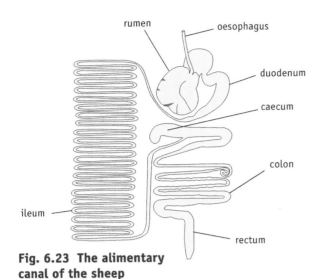

Fig. 6.23 The alimentary canal of the sheep

These are regurgitated and chewed again before being swallowed, but this time the food passes into the third and fourth compartments (*omasum* and *abomasum* respectively). The last of these is the true stomach, which secretes gastric juice.

Animals with a rumen are called *ruminants*, and are *fore-gut fermenters*.

Feeding and food processing generally takes much longer in herbivorous mammals than in carnivorous mammals. A sheep may graze for eight hours a day and ruminate (chew the cud) for a similar period. The gut is relatively long in herbivores — 27 times the body length in a sheep compared with only four times in a cat (Fig. 6.23).

Summary of differences between sheep and cat nutrition

	Sheep	Cat
Eyes	On sides of head	Forward-facing
Feeding time	Long	Short
Jaw joint	Allows horizontal (sideways) movement	Hinge-like, allowing up-down movement only
Incisors	Upper incisors absent	Present in both jaws
Canines	Absent in upper jaw	Long and pointed
Diastema	Present, allowing manipulation of grass	Absent
Tooth growth	Continuous throughout life	No growth after eruption
Premolars and molars	Ridged for grinding	Last upper premolar and first lower molar specialised for cutting (carnassial teeth)
Gut length	Long	Short
Time for food to pass through gut	Long	Short

Extension: The rabbit

Fig. 6.24 shows an incisor tooth, skull and alimentary canal of a rabbit.

Like sheep, rabbits are herbivores, with teeth that grow continuously from open roots. The enamel on the incisors is confined to the anterior (front) surface, and since this wears away more slowly than the dentine behind, the incisors maintain a sharp chisel shape. The premolars and molars have side-to-side ridges, enabling the food to be ground up. The ridges are at right angles to the front-rear movement of the lower jaw in chewing. This contrasts with ruminants, in which the ridges extend in front-rear direction and the lower jaw moves from side to side. In both types the lower jaw moves at right angles to the direction of the ridges on the teeth.

As in sheep and cattle, cellulose is digested by anaerobic prokaryotes and protoctists, but in rabbits these live in a greatly enlarged caecum. The microorganisms also synthesise essential amino acids and certain vitamins. Because rabbits digest their cellulose in the lower part of the gut they are called *hind-gut fermenters*, in contrast to ruminants, which are fore-gut fermenters.

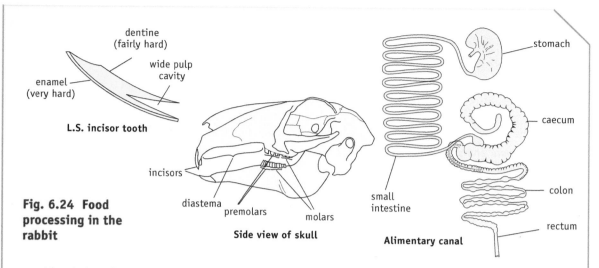

Fig. 6.24 Food processing in the rabbit

Labels on diagram:
dentine (fairly hard)
wide pulp cavity
enamel (very hard)
L.S. incisor tooth
incisors
diastema
premolars
molars
Side view of skull
stomach
caecum
colon
rectum
small intestine
Alimentary canal

After being chewed, grass passes straight through the stomach and small intestine to the caecum without any secretion of gastric or pancreatic juices or bile. In the caecum the grass is partly digested, the cellulose being converted to glucose, some of which is used by the microorganisms for growth. The food is now 'downstream' of the small intestine, where absorption of digested food occurs. To reach the small intestine, the food has to go around again. The 'first time around' faecal pellets are produced at night, and are eaten. In its second time around, the food is subjected to the full action of the rabbit's digestive juices. This time, the food bypasses the caecum and the resulting faeces are not eaten.

Summary of key facts and ideas in this chapter

○ Animals obtain their raw materials and energy in the form of other organisms or parts of other organisms.

○ Much of an animal's food consists of complex organic substances that must be broken down into simpler substances before it can be absorbed.

○ This process is called *digestion* and is accomplished by *enzymes*.

○ In most animals digestion occurs in a cavity called the *alimentary canal* or *gut*.

○ Undigested food is got rid as faeces. This process of *egestion* is quite different from excretion.

○ Animals such as sea anemones and the freshwater *Hydra* have sac-like guts, with a single opening. As a result the mouth also serves as an anus, so there is two-way movement of food.

○ In most other animals there are two openings to the gut, so food moves in one direction. This allows different parts of the gut to be specialised for different stages of food treatment.

○ In arthropods (such as insects) and vertebrates, a jointed skeleton makes it possible to exploit a wide range of foods using hard mouthparts that can bite, pierce, chew, or suck food.

○ As *homeothermic* animals, mammals are adapted to achieve a high food intake by their ability to chew food and breathe at the same time — unlike other vertebrates.

○ Mammals may be specialised for meat-eating (*carnivory*), plant-eating (*herbivory*) or a mixture (*omnivory*). Adaptations for different diets are evident in the teeth and jaws, and in the length and complexity of the alimentary canal. Herbivores generally have longer alimentary canals than carnivores.

○ Very few animals produce a *cellulase* enzyme, and many herbivorous mammals have parts of the gut specially adapted to contain cellulose-digesting microorganisms.

○ In fore-gut fermenters, cellulose is digested in enlargements of the oesophagus. In hind-gut fermenters, cellulose is digested in the large intestine.

Test your basics

Copy and complete the following sentences. In some cases the first or last letters of a missing word are provided.

1. Animals are ___*___ic, meaning that they cannot make organic compounds from CO_2 and water. Their food consists of complex organic molecules such as p___*___, ___*___ and c___*___. Before they can be ___*___ into the body they have to be broken down into ___*___er substances by catalysts called ___*___. This process is called ___*___ and occurs in an intucking into the body called the ___*___ canal or gut.

2. In *Hydra* and sea anemones, the gut consists of a simple sac with a s___*___ opening. Food and undigested residue therefore enter and leave by the same route.

3. In an earthworm there are ___*___ openings to the gut; food enters via the mouth and ___*___ leave by the ___*___. The ___*___-way movement of food through the gut enables different regions to be ___*___ for performing different stages of food processing.

4. In insects such as a weta the ___*___ is hardened in parts to form an ___*___. This has enabled the evolution of biting jaws that can process hard food, breaking it up into smaller pieces. This increases the ___*___ ___*___ for the action of ___*___.

5. In most mammals there are two sets of teeth; the 'milk' or ___*___ teeth, and the adult set they replace. Mammals have different kinds of teeth specialised for different functions. ___*___ at the front cut pieces off the food. Behind these are the ___*___ which, in ___*___ (meat-eaters) are usually long, pointed and used for ___*___. Behind these are the ___*___. The rear-most teeth are the ___*___ which, unlike all the others, do not replace 'milk' teeth.

6. The part of a tooth that contacts the food is the ___*___. Under this is a thick layer of ___*___, surrounding a ___*___ cavity containing ___*___ vessels and sensory ___*___ endings. The tooth is anchored in the socket by a tough layer of ___*___ fibres, the ___*___ ligament.

7. As the food is chewed into smaller pieces, ___*___ is secreted on to it. In humans this contains the enzyme ___*___, which converts ___*___ into ___*___.

8. In the first stage of swallowing, a lump or ___*___ of food is pushed into the ___*___ by the ___*___. The food is then squeezed along the ___*___ by muscular waves called ___*___, until it arrives in the ___*___.

9. Glands in the stomach wall secrete ___*___ juice. This contains the enzyme ___*___ which breaks down ___*___ into ___*___. It is only active in acid conditions; these are provided by ___*___ in the ___*___ juice. Mixed by peristaltic contractions of the stomach wall, the food is slowly converted into a creamy liquid called ___*___. The exit of the stomach is guarded by a ___*___ muscle, which slowly relaxes enough to allow the ___*___ to slowly enter the first part of the ___*___ intestine, the ___*___.

10. Two digestive juices enter the ___*___. ___*___ is secreted by the ___*___ and stored in the gall bladder, and ___*___ juice is secreted by the banana-shaped ___*___. Both juices contain s___*___ h___*___ c___*___ which makes them alkaline.

11. ___*___ is green-yellow contains no enzymes. It contains salts which convert large globules of fat into an ___*___ of millions of microscopic droplets. Though the fat in not changed chemically, its ___*___ ___*___ is greatly increased.

12. ___*___ juice contains a number of enzymes, including lipase, which splits fats into ___*___ ___*___ and ___*___. Other enzymes convert maltose into ___*___, and polypeptides into amino acids.

13. The area of the lining of the ___*___ intestine is increased by thousands of finger-shaped ___*___, and the surface of the cells lining the intestine is further increased by millions of submicroscopic ___*___.

14. Each ___*___ contains a network of microscopic blood vessels called ___*___, which join up to form the ___*___ ___*___ vein which takes the blood to the ___*___.

15. Each villus also contains a 'twig' of the tree-like lymphatic system called a ___*___. Fatty acids and glycerol pass through the lining of the intestine and recombine to form microscopic droplets of ___*___.

16. Glucose and amino acids are absorbed into the capillaries by ___*___ transport, which requires ___*___ produced in ___*___.

17. Most of the water is absorbed in the small intestine; most of the remainder is absorbed in the ___*___.

18. The undigested food, together with millions of bacteria, is called ___*___ and is temporarily stored in the rectum.

7 Transport in Animals

Even in the simplest animals, different parts of the body have different functions. The various parts of the body therefore depend on each other. In humans, for example, the lungs supply oxygen for the rest of the body, the gut supplies glucose and other nutrients, and the kidneys remove wastes. This *division of labour* means that substances must be transported from one part of the body to another.

In many very small animals and in some larger but inactive ones, transport is solely by *diffusion* (Fig. 7.1). As explained more fully in Chapter 21, diffusion is only rapid over very short distances and is too slow to meet the needs of larger animals.

Most larger animals have some kind of *transport system*, in which materials are carried around the body by *flow*, in a circulating fluid. Even in these animals, however, substances enter and leave this circulating fluid by diffusion (Fig. 7.2).

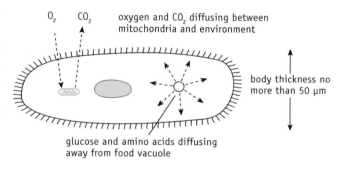

oxygen and CO_2 diffusing between mitochondria and environment

body thickness no more than 50 μm

glucose and amino acids diffusing away from food vacuole

Fig. 7.1 Some examples of substances transported in a single-celled organism such as *Paramecium*

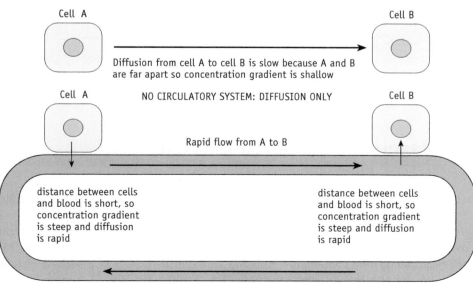

Cell A

Cell B

Diffusion from cell A to cell B is slow because A and B are far apart so concentration gradient is shallow

NO CIRCULATORY SYSTEM: DIFFUSION ONLY

Cell A

Cell B

Rapid flow from A to B

distance between cells and blood is short, so concentration gradient is steep and diffusion is rapid

distance between cells and blood is short, so concentration gradient is steep and diffusion is rapid

Fig. 7.2 How a circulatory system speeds up diffusion

CIRCULATORY SYSTEM: DIFFUSION + FLOW

This chapter describes four modes of transport:

- transport by diffusion (e.g. sea anemones and flatworms)

- transport by flow of seawater (gastrovascular system of jellyfish)

- open blood system of crayfish

- closed blood systems of earthworm, fish, mammal, and bird.

Managing without a circulatory system – planarians

Planarians belong to the phylum Platyhelminthes or flatworms. Though common in New Zealand streams and lakes, they are unfamiliar to most people because of their small size and tendency to hide under stones.

Fig. 7.3 shows a transverse section through a planarian. Though diffusion in liquids is very slow over distances of more than about 0.5 mm, they get by because of two key adaptations:

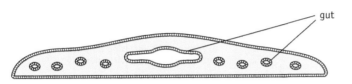

- They have a large surface compared with their volume because of their fairly small size and flat shape (Fig. 7.3).

- All the organ systems of the body are spread out like the branches of a tree. As a result no cell is more than a fraction of a millimetre from a branch of alimentary canal, or from a branch of the excretory system. Each 'twig' of one system is very close to a 'twig' of another. Fig. 7.4 shows the digestive and excretory systems, but the reproductive systems are similarly tree-like.

Fig. 7.3 A planarian in transverse section

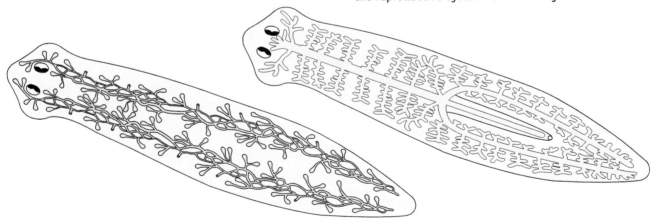

Fig. 7.4 The highly branched alimentary and excretory systems of a planarian

The gastrovascular system of jellyfish

Together with the sea anemones and corals, jellyfish belong to the phylum Cnidaria (also called Coelenterata).

Aurelia aurita is a jellyfish common in the New Zealand seas. It grows to about 25 cm across (some species reach over a metre), but unlike most large creatures, jellyfish have no blood system. Instead, they have a **gastrovascular system** containing circulating seawater. It consists of a network of tubes that are extensions of the **enteron** or digestive cavity (Fig. 7.5 and Fig. 7.6). Seawater is slowly propelled by cilia lining the channels, bringing fresh oxygen and carrying away CO_2. The currents also transport tiny semi-digested food particles from the enteron; these are taken in by the endoderm cells lining the canals and digested intracellularly (see Chapter 6). There is only one layer of cells around the canals, so oxygen and CO_2 have only a short distance to diffuse.

Transport by gastrovascular system has two major limitations:

- Oxygen is sparingly soluble in water — seawater at 15° C can dissolve only 0.7% of oxygen by volume.

- Since the water is propelled by cilia, its movement is very slow; it takes about 20 minutes for water to travel from enteron along the radial canals and back.

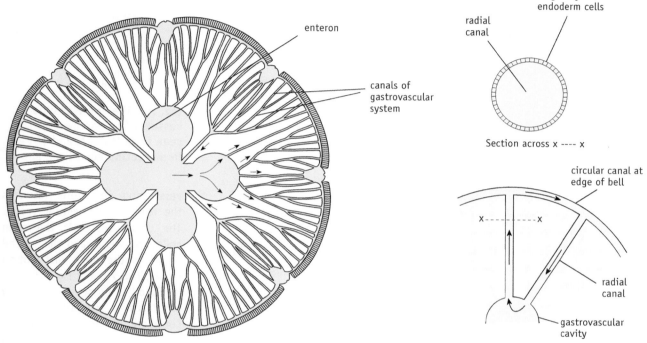

Fig. 7.5 Top view of a common jellyfish showing gastro vascular system

Fig. 7.6 Simplified view of the jellyfish gastrovascular system

Despite these limitations, jellyfish cope quite well. There are two reasons why a gastrovascular system is sufficient for their needs:

- They are relatively inactive, so their metabolic rate is fairly low.

- Most of the animal's bulk is made up of the jelly-like **mesogloea** between the outer **ectoderm** and the inner **endoderm** — only about 1% of its bulk consists of cells. The cells of the ectoderm are all in contact with the seawater. Even those of the endoderm are in contact with the water in the gastrovascular system. Though the body is large, no cells are more than a fraction of a millimetre from the circulating seawater.

A simple blood system — the earthworm

Earthworms belong to the phylum **Annelida**, or segmented worms.

Whereas the gastrovascular system of jellyfish is open to the surrounding seawater, annelid worms have a specialised *blood system* in which the circulating fluid has no connection with the outside world. The blood is confined within blood vessels so it never comes into contact with the tissues it serves. To reach the cells, substances therefore have to pass through the blood vessel walls. This is called a *closed blood system*.

There are two main blood vessels — the **dorsal vessel** running along the back, and the **ventral vessel** running along the lower side of the body. Blood flows anteriorly (towards the front) in the dorsal

External features of the earthworm

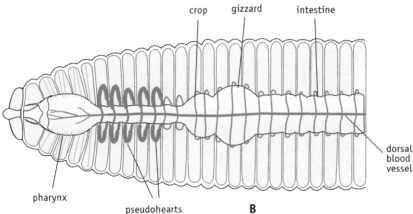

Fig. 7.7 **The anterior blood vessels of an earthworm. (A) seen from the left and (B) as in dissection seen from above.**

vessel and posteriorly (towards the rear) in the ventral vessel. The dorsal vessel is the main collecting vessel and the ventral vessel is the main distributing vessel (Fig. 7.8 and Fig. 7.9).

Blood is propelled mainly by *peristalsis* — wave-like muscular contractions travelling along the dorsal and ventral vessels. There are also five pairs of **pseudohearts** connecting the dorsal and ventral vessels (Fig. 7.9). Valves prevent backflow.

Because there is no true heart, there is no distinction between arteries and veins as there is in vertebrates. The finest vessels are the microscopic **capillaries**. Their walls consist of a single layer of extremely flattened cells. Every organ of the body has a dense network of capillaries. It is in these vessels where substances diffuse into and out of the blood. This is quite rapid because no cell is more than 30–40 µm from the nearest capillary, and the capillary walls are extremely thin.

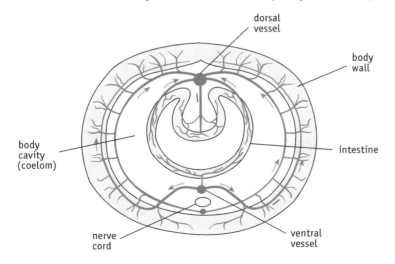

Fig. 7.8 **Transverse section through the middle of an earthworm in the region of the intestine**

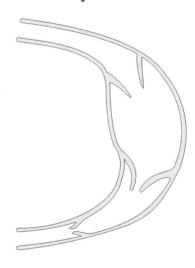

Fig. 7.9 **Longitudinal section through a pseudoheart**

ISBN: 9780170214094

The blood contains the oxygen-carrying pigment **haemoglobin**. This enables it to carry more oxygen than the seawater in the gastrovascular system of a jellyfish. The haemoglobin is dissolved in the plasma — unlike vertebrates where it is carried inside red cells.

Crayfish

The freshwater crayfish or koura (*Paranephrops*) belongs to the class Crustacea in the phylum Arthropoda.

The crayfish blood system differs in three important ways from that of the earthworm:

- As in all arthropods it is an *open* blood system, in which the blood bathes the tissues it serves. Instead of leading into microscopic capillaries, the vessels taking blood from the heart open into a series of large blood spaces collectively called the **haemocoel**. Here the blood comes into direct contact with the cells (Fig. 7.10).

- The blood contains **haemocyanin** instead of haemoglobin. This is pale blue when oxygenated and colourless when deoxygenated. Whereas haemoglobin contains iron, haemocyanin contains copper.

- There is a well-developed heart.

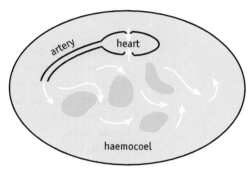

Open blood system

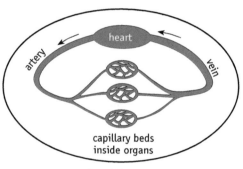

Closed blood system

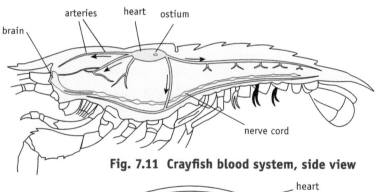

Fig. 7.11 **Crayfish blood system, side view**

Fig. 7.10 Open and closed blood systems. In an open blood system the organs (shown as grey islands) float in a 'lake' of blood which flows suggishly around them. In a closed blood system blood is separated from the tissues by the walls of capillaries.

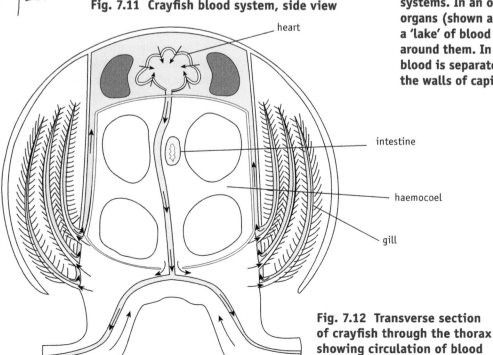

Fig. 7.12 **Transverse section of crayfish through the thorax showing circulation of blood**

Blood leaves the heart via several **arteries** (Fig. 7.11), and from the open ends of these it enters the haemocoel (Fig. 7.12). Blood flow in the haemocoel is sluggish, but is evidently sufficient to meet the animal's metabolic needs. On the way back to the heart it passes through the gills, where it picks up oxygen and changes from colourless to pale blue. Blood drains from the haemocoel into the heart via a series of openings called **ostia** (singular, *ostium*).

Single closed circulation – trout

The trout is a *bony fish*, one of the six main classes of vertebrates (backboned animals). Other vertebrates include the cartilaginous fish (sharks and rays), amphibians (frogs and salamanders), reptiles, birds, and mammals.

All vertebrates have a *closed blood system*, in which the blood is confined within blood vessels and does not touch the cells it serves. Oxygen, food and wastes therefore have to pass through blood vessel walls.

The blood consists of two parts: a liquid **plasma**, and the cells suspended in it. The blood cells are of two main kinds: red cells (erythrocytes) and white cells (leucocytes). The red cells contain the oxygen-carrying pigment **haemoglobin**. Unlike the red cells of mammals, red cells of fish, amphibians, reptiles and birds have nuclei. The white cells defend against foreign organisms such as bacteria.

A fish heart has three contractile chambers: a thin-walled *sinus venosus*, a slightly thicker-walled *atrium,* and a muscular, thick-walled *ventricle* (Fig. 7.13).

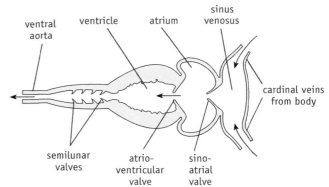

Fig. 7.13 Heart of a bony fish

Blood leaves the ventricle and enters the *ventral aorta*. This is an **artery** because it takes blood away from the heart. Branches of the ventral aorta divide repeatedly, eventually giving rise to millions of microscopic **capillaries** in the gills. Here oxygen diffuses into the blood and CO_2 diffuses out. At the same time the blood changes from dark red to scarlet. Blood leaving the gills enters the *dorsal aorta*, whose branches supply capillary networks in the other organs of the body (Fig. 7.14).

The smaller arteries, and especially the tiniest ones, the *arterioles*, contain considerable amounts of smooth ('involuntary') muscle. This enables the diameter of an artery to be varied so that blood can be diverted to parts where it is most needed — for instance active muscle.

Thus by having each organ supplied by a separate capillary network, blood flow can be regulated according to the needs of individual tissues. This is not possible in an open blood system in which organs are floating in 'shared' blood in a haemocoel.

Blood is returned to the heart by a system of **veins**. The heart pumps only deoxygenated blood. In any complete circulation, the blood thus passes once through the heart. This is called a **single circulation** (Fig. 7.15).

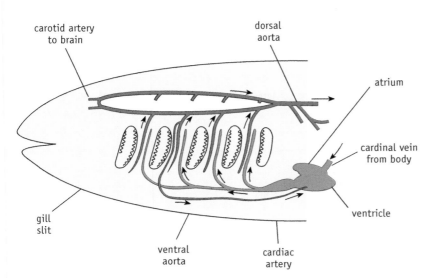

Fig. 7.14 Gill circulation in a bony fish

ISBN: 9780170214094

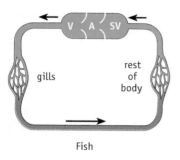
Fish

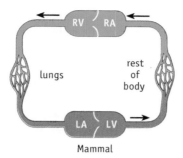
Mammal

Fig. 7.15 Diagrammatic view of the single circulation of a fish (left) and the double circulation of a mammal. In the mammal the two sides of the heart are actually together.

Double closed circulation – mammal

Like other mammals, humans have a **double circulation**, in which the blood passes *twice* through the heart in any complete circulation. The right side of the heart pumps blood through the **pulmonary circuit** (lungs) and the left side pumps blood through the **systemic circuit** (rest of body). Thus in any complete circuit the blood passes through *two* capillary networks. Although the two sides of the heart are in the same muscle mass, they are actually in *series*, so they pump *equal* volumes of blood. The left side pumps oxygenated blood and the right side pumps deoxygenated blood.

In the mammal there are two atria and two ventricles. The sinus venosus is almost absent, being represented by the **S-A (sino-atrial) node**. This is a tiny patch of tissue in the right atrium, and is the 'pacemaker', setting the rhythm for the rest of the heart. The human heart is shown in detail in Fig. 7.16.

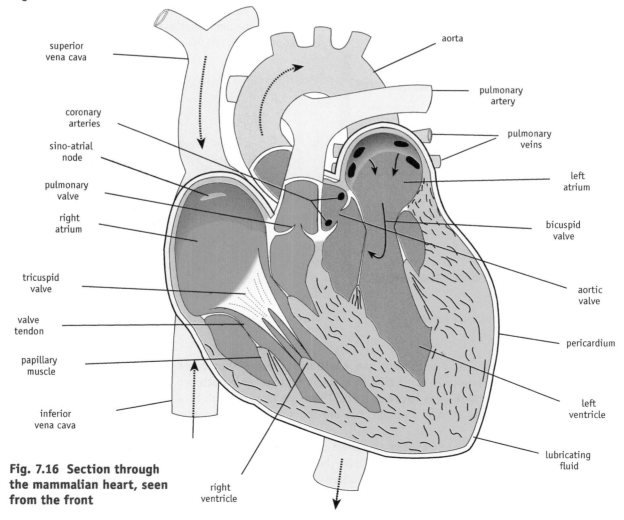

Fig. 7.16 Section through the mammalian heart, seen from the front

Oxygenated blood leaves the left ventricle via the **aorta**, the branches of which distribute blood to the various organs. Fig. 7.17 shows some of the main vessels.

Systemic arteries carry blood under high pressure and have thick walls to withstand this (Fig. 7.18). The largest arteries have much elastic tissue in their walls. Its function is to absorb the surge of the pulse. By the time the blood reaches the tiniest arteries — the **arterioles** — the pulse has all but disappeared, the pressure is steady and the flow continuous.

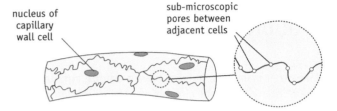

Fig. 7.18 Side view of capillary, showing tiny pores through which glucose and amino acids leave capillaries

Though the arterioles contain little elastic tissue they have much smooth muscle. This enables their diameter to be varied, so blood can be diverted to where it is most needed. For example during exercise the arterioles in the gut become narrower, reducing the flow of blood. The arterioles in the limb muscles get wider, greatly increasing the blood supply.

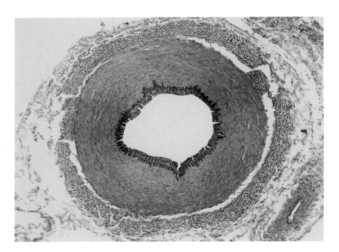

Fig. 7.17 Main blood vessels in a mammal

Transverse section through muscular artery

The arterioles branch into the **capillaries**, where substances enter and leave the blood. Capillaries have walls consisting of a single layer of very flattened cells. Adjacent capillary wall cells have minute gaps between them. These are large enough to allow glucose, amino acids and other nutrients to pass through, but are too small to allow plasma proteins to leave the blood (Fig. 7.19).

The capillaries join to form tiny veins or **venules**, and these join to form **veins**, which eventually lead into the right side of the heart. By the time the blood reaches the veins the pressure has fallen to very low levels. Most veins have valves that prevent backflow during body movements.

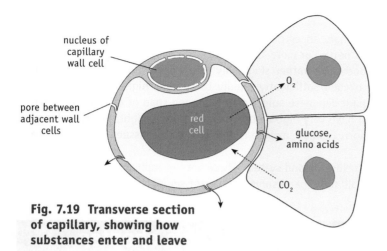

Fig. 7.19 Transverse section of capillary, showing how substances enter and leave

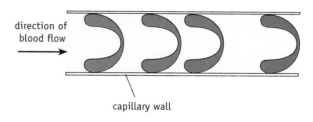

Fig. 7.20 Red cells squeezing along a capillary

Why are red cells such a peculiar shape?

Unlike other vertebrates, mammals have biconcave red cells without nuclei. In the narrowest capillaries the red cells are squeezed into a bell-shape. As a result most of the haemoglobin is displaced towards the periphery, so oxygen has a shorter distance to diffuse (Fig. 7.20).

Advantage of a closed blood system

In a closed blood system the various organs can have their blood supplies adjusted to suit ever-changing demands. In an open blood system all the organs 'share' the same blood supply.

Advantages of a double circulation

In a fish, blood pressure drops about 30% as it passes through the gills, so it is already quite low by the time it reaches the capillaries in the rest of the body. In the double circulation, the pulmonary and systemic capillaries are supplied by different pumps — the right and left ventricles respectively. As a result the blood must pass through two sets of capillaries in any complete circuit. This arrangement allows the organs supplied by the systemic circuit to receive blood at much higher pressure than would be possible with a single circulation.

Though the left and right sides of the heart pump equal quantities of blood, the left ventricle exerts about five times higher pressure than the right ventricle and is correspondingly much more muscular. There are three reasons for this:

- The right side *must* exert a low pressure to avoid forming tissue fluid in the alveoli of the lungs, which would interfere with gas exchange.

- During vigorous exercise the blood supply to the muscles increases greatly. This is partly due to the diversion of blood from regions that are not needed, such as the gut. If there were not adequate 'spare' pressure, there would be a fall in pressure to the brain (think what happens to the water pressure in the shower when someone diverts water to the kitchen sink).

- Mammalian kidneys need a high blood pressure for filtration of the blood.

Notice an important point: the fact that the left side pumps blood over longer distances than the right is *not*, as often stated, the reason for its greater muscularity. The first 99% of the journey from the left ventricle to the tissues it supplies occurs with very little drop in pressure. Most of the resistance is in the arterioles and capillaries, which account for only the last two to three millimetres of the journey from the heart.

Open versus closed blood systems

Open blood systems are characteristic of the two largest phyla, the arthropods and most molluscs. Closed blood systems are characteristic of annelids (segmented worms) and vertebrates. All are large and diverse groups, and all have evolutionary ancestries going back over 500 million years, so it is not particularly sensible to ask which is 'better' or more 'advanced'.

However, it is true to say that blood flows rather sluggishly in an open blood system, which would therefore be unable to deliver oxygen sufficiently rapidly to support high activity. Though some crustaceans are very large (the giant Japanese spider crab has a leg span of over 3.5 metres), they are fairly slow moving. Insects can have very high metabolic rates during flight, but their blood system is not involved in oxygen transport. Instead, oxygen is delivered to the tissues directly by the tracheal system (Chapter 8).

There is one group of molluscs with a closed blood system. These are the cephalopods — octopus, squid and cuttlefish. These animals are all active, and some (e.g. giant squid) are very large as well.

Summary of key facts and ideas in this chapter

○ In most very small animals, substances move around the body by *diffusion*.

○ In most larger animals, particularly the more active ones, materials are transported around the body by some form of transport system.

○ Planarians manage without a transport system because they have a very thin shape and their digestive and other systems are finely branched. As a result diffusion distances are short.

○ Jellyfish have a simple system for transporting oxygen and CO_2 between cells and environment. This consists of channels into the body in which seawater circulates.

○ Annelids such as the earthworm have a *blood system* in which liquid (blood) circulates in closed vessels. Oxygen is carried by the red pigment *haemoglobin* which is dissolved in the blood. In the earthworm there is no true heart, blood being pumped along by peristaltic contraction of the larger blood vessels. The finest blood vessels are microscopic and are called *capillaries*; it is here that substances enter and leave the blood. Because the blood never comes into direct contact with the cells it serves, it is called a *closed blood system*.

○ The crayfish and other arthropods have an *open blood system*, in which there are no capillaries; instead the blood bathes the cells directly. In the crayfish the blood contains the oxygen-carrying pigment *haemocyanin*, and blood is pumped round the body by a well-developed heart.

○ Vertebrates have a *closed blood system*. The trout and other fish have a *single circulation*, in which blood passes once through the heart in each complete circuit.

○ Mammals have a *double circulation*, in which blood passes twice through the heart in each complete circuit. The right side of the heart pumps deoxygenated blood through the lungs and the left side of the heart pumps oxygenated blood through the rest of the body.

○ Blood is carried from the mammalian heart by *arteries* and is returned to the heart by *veins*. Connecting the arteries and veins are the microscopic *capillaries*, in which substances enter and leave the blood.

ISBN: 9780170214094

Test your basics

Copy and complete the following sentences. In some cases the first letter of a missing word is provided.

1. In most very small animals, transport around the body is by ___*___.

2. Planarians and other ___*___ manage without a blood system because their thin body shape means that all the cells are close to the exterior, so diffusion distances are ___*___.

3. Arthropods (such as crayfish) have an ___*___ blood system, in which the tissues are bathed in a huge blood space called a ___*___. In the crayfish, oxygen is transported by ___*___, which is pale ___*___ when oxygenated and ___*___ when deoxygenated.

4. Earthworms and vertebrates have a ___*___ blood system, in which blood does not come into direct contact with the tissues it serves. Instead, substances have to pass through the walls of microscopic ___*___ to enter and leave the blood.

5. Fish have a ___*___ circulation, in which blood passes ___*___ through the heart in each complete circuit.

6. The blood of mammals and other vertabrates consists of a liquid ___*___ in which are suspended cells of two main kinds: ___*___ cells which carry the pigment ___*___, and ___*___ cells which defend the body against ___*___.

7. The red cells in mammals are unusual in that they have no ___*___ when mature.

8. In vertebrates, blood is carried away from the heart by ___*___s, and returned by ___*___s. Connecting these are microscopic vessels called ___*___s, where substances enter and leave the blood. Because the blood never comes into direct contact with the tissues it serves, vertebrates have a ___*___ blood system.

9. Mammals have a ___*___ circulation because the blood passes ___*___ through the heart in each complete circulation. The heart is really two pumps in one; the left side pumps ___*___ blood and the right side pumps ___*___ blood. Each side has a thin-walled ___*___ which receives blood, and a thicker-walled ___*___ which pumps blood out of the heart.

10. Oxygenated blood leaves the ___*___ ___*___ by the ___*___, and deoxygenated blood leaves the ___*___ ___*___ by the ___*___ artery.

11. Between each atrium and ventricle is a large ___*___ which prevents backflow. There is also a valve at the base of the ___*___ and ___*___ artery.

12. Arteries have much ___*___er walls than veins. The walls of the larger arteries have lots of ___*___ tissue, which helps to absorb the surge of the pulse, so by the time the blood reaches the tiny arteries or ___*___s, the flow is ___*___. These tiny arteries have lots of ___*___ muscle in their walls. This enables their diameter to be varied, thus regulating the ___*___ of blood to the various organs according to their activity. Thus during exercise the blood flow through the skeletal muscles is ___*___ and the flow through the gut is ___*___.

8 Gas Exchange in Animals

General principles

Apart from some parasites, all animals use oxygen and produce CO_2 in the process of *respiration*. As a result oxygen is less concentrated inside the body than outside, and CO_2 is more concentrated inside than outside. Because of these concentration differences, oxygen diffuses into the body and CO_2 diffuses out (Fig. 8.1). This two-way diffusion of gases across a surface is called *gas exchange*.

In multicellular creatures oxygen has to travel from the body surface to the cells, and CO_2 travels in the opposite direction. Gases must therefore be transported from body surface to cell surface. This is achieved in one of two ways:

- In very small animals and in those with very thin bodies, diffusion alone is fast enough because all the cells are near to the body surface.

- A transport system carries oxygen to within a very short distance of the cells, only the last part of the journey being by diffusion. CO_2 is transported in the reverse direction.

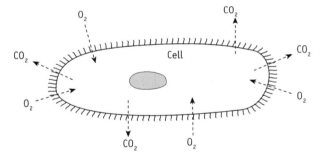

Fig. 8.1 Gas exchange in a single-celled organism such as *Paramecium* occurs over the entire body surface

Animals show a great variety of gas exchange mechanisms, but they are all affected by two things:

- The size of the body.
- Whether the environment is water or air.

The effect of body size and shape

In very small organisms such as *Paramecium*, gas exchange occurs over the whole body surface. If an animal were to get bigger without any other changes, it would become harder to obtain oxygen and get rid of CO_2. There are two reasons for this.

ISBN: 9780170214094

1. Surface : volume ratio. Since oxygen has to enter the body via its *surface*, the *supply* of oxygen is limited by (among other things) the surface area of the body. The *demand* for oxygen is affected by (among other things) the *volume*. If the body size increases but shape remains unchanged, volume (and therefore oxygen supply) increases faster than surface area (and therefore oxygen supply). For example if the body doubles in length, the surface area increases four times, but the volume increases *eight* times.
This is true of any shape, but is most easily seen with a simple shape like a cube (Fig. 8.2).

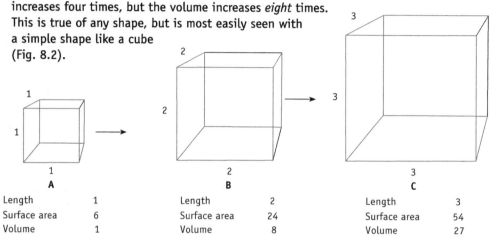

	A		B		C
Length	1	Length	2	Length	3
Surface area	6	Surface area	24	Surface area	54
Volume	1	Volume	8	Volume	27

Fig. 8.2 Showing how, as size increases, volume increases faster than surface

2. Diffusion distances. With increasing body size, the distance between the body surface and the innermost parts of the body increases. As a result the concentration gradients for oxygen and CO_2 become less and less steep, so diffusion becomes slower. If the body gets thinner but does not change its volume, the surface area increases (Fig. 8.3). For a given concentration difference, halving the length of the diffusion path *doubles* the rate of diffusion.

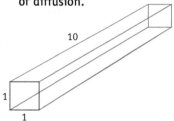
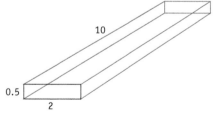

Surface area = 4 (1 x 10) + 2 x (1 x 1) = 42 units
Volume = 1 x 1 x 10 = 10 units
Surface/volume = 4.2

Surface area = 2 (2 x 10) + 2 (0.5 x 10) + 2 (0.5 x 2) = 52 units
Volume = 2 x 0.5 x 10 = 10 units
Surface/volume = 5.2

Fig. 8.3 How a change in shape (in this case, flattening) can increase body surface area

Planarians

Planarians are small animals living in lakes and streams. They have no blood system but their flat shape means that no part of the body is more than a fraction of a millimetre from the surface (Fig. 8.4). Over these short distances diffusion — which occurs over the whole body surface — is rapid enough.

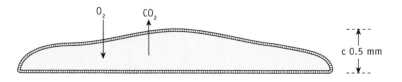

Fig. 8.4 A planarian in transverse section, showing short diffusion distances of oxygen and CO_2

ISBN: 9780170214094

Earthworm

Earthworms are much larger than planarians, so transport distances are greater. Also, their larger size and cylindrical shape means that their surface area to volume ratio is smaller. These effects of size and shape are overcome by the possession of a *blood system*.

As in planarians, gas exchange occurs by diffusion over the entire body surface, which is adapted for gas exchange in the following ways:

* It is *densely supplied with capillaries* (Fig. 8.5).

* It is very *thin*, so the distance through which gases have to diffuse is *short*, favouring a steep concentration gradient.

* It is kept moist by secretion of *mucus* and by a watery fluid released through pores in the body wall. Moist skin is necessary because oxygen cannot diffuse through a dry surface.

Though the area of the skin is not particularly large in comparison to the animal's volume, the oxygen demands of a sluggish animal are not great.

Blood low in oxygen and rich in CO_2 is brought close to the skin. By reducing the diffusion distance the concentration gradients of oxygen and CO_2 are steepened.

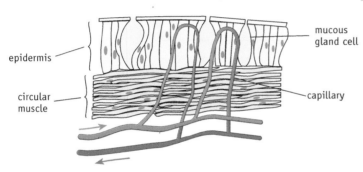

Fig. 8.5 Blood vessels in skin of an earthworm

epidermis

circular muscle

mucous gland cell

capillary

Oxygen and CO_2 readily diffuse between the worm and the soil surface via the air in the burrow and also interconnecting air spaces between the soil particles. However after flooding, the oxygen supply is greatly reduced. This is because it diffuses thousands of times more slowly in water than through air. Worms respond by coming up to the surface.

As in all terrestrial (land-living) animals, the gas exchange surface is moist and therefore a potential site of water loss. However, in the damp environment of the burrow, evaporation is unlikely to be significant.

Weta

In insects and some other terrestrial arthropods the blood is colourless and plays no part in oxygen or CO_2 transport (perhaps for this reason, insect blood is often called 'haemolymph', though many insect physiologists feel comfortable with the term 'blood'). Instead these gases are transported by a system of branching, air-filled tubes called **tracheae**, which open at the sides of the body by holes called **spiracles**. In the weta there are two pairs in the thorax and eight in the abdomen (Fig. 8.6). Each spiracle is protected by two lip-like structures that can be closed when metabolic rate is low.

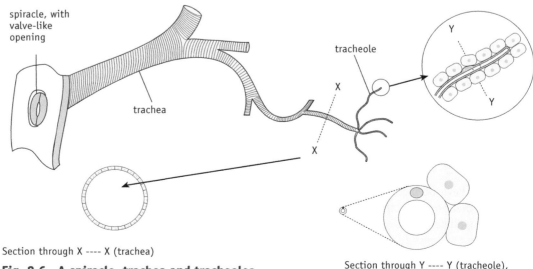

spiracle, with valve-like opening

trachea

tracheole

X

X

Y

Y

Section through X ---- X (trachea)

Fig. 8.6   A spiracle, trachea and tracheoles

Section through Y ---- Y (tracheole), and two neighbouring cells

ISBN: 9780170214094

Each trachea is prevented from collapse by a spiral lining of chitin, which is actually an intucking of the cuticle. The tracheae end in microscopic **tracheoles**. Unlike tracheae, these are *intracellular* (lie *within* individual cells). No cell is more than a fraction of a millimetre from a tracheole, so diffusion distances are short. The number of tracheoles is huge — a silkworm has about 1.5 million.

Since oxygen is being used, its concentration is lower in the cells than in the atmosphere outside, so it diffuses along the tracheae towards the cells. Similarly, the cells are producing CO_2, so it diffuses in the opposite direction (Fig. 8.7). The final stage of the diffusion pathway is in the liquid of the walls of the tracheoles and the surrounding cells. Although diffusion in solution is 10 000 times slower than through gas, the distance is at most a few hundred micrometres.

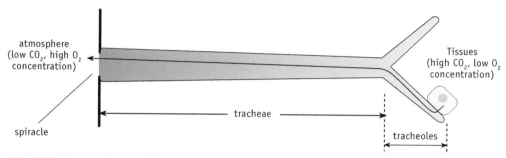

Fig. 8.7 Diffusion pathway of CO_2 and O_2. The tracheoles are relatively much smaller than shown.

In large insects such as weta, the distance between cells and spiracles may be several millimetres. When active, they speed up gas exchange by actively ventilating their tracheal systems by muscular movements of their abdomens. In weta, the volume of air exhaled and inhaled (the *tidal volume*) is increased by the presence of large air sacs (Fig. 8.8). These are thin-walled and easily compressed by the action of muscles in the body wall.

As with lungs (see below), the lining of the tracheal system is moist, and is thus a potential site of water loss. The rate of ventilation of the tracheae is regulated by the output of CO_2. A rise in CO_2 causes the spiracles to open. As a result the rate of water loss is no higher than is needed to meet the animal's respiratory needs.

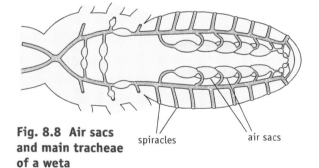

Fig. 8.8 Air sacs and main tracheae of a weta

Tracheal system and body size

Diffusion is only rapid over short distances, and it is often stated that this is the reason why no insects are very large. This is unlikely to be the true explanation, for two reasons:

- Large insects ventilate their tracheal systems by active breathing movements, especially during periods of activity.

- The last stage of oxygen transport is diffusion in solution from the tracheoles. Though oxygen diffuses much more slowly in solution, the distance of the liquid stage is very short — 50 m or so. It is this final stage of diffusion that is likely to be the rate-limiting stage, and this distance is similar for all insects.

A much more likely factor limiting insect size is the weakness of the new cuticle immediately after the old one has been shed; a soft cuticle is unlikely to be able to support a large insect.

Mammal

The gas exchange organs of a mammal are the **lungs** (Fig. 8.9). Like the tracheal system of insects they are deep, air-filled intuckings into the body. There is, however, an important difference. Whereas the tracheal system transports oxygen all over the body, in a mammal the respiratory gases are transported in the blood.

The lungs are spongy organs on either side of the heart. Air is delivered to the gas exchange surface by a system of tubes called the **bronchial tree**. The 'trunk' of the tree is the trachea ('windpipe').

This branches into two main **bronchi**, one to each lung. Each main bronchus branches repeatedly into smaller and smaller bronchi. Like the trachea, the walls of the bronchi are supported by cartilage. The smallest bronchi lead into **bronchioles**, which are not supported by cartilage. The bronchioles lead into clusters of air sacs or alveoli, where gas exchange occurs.

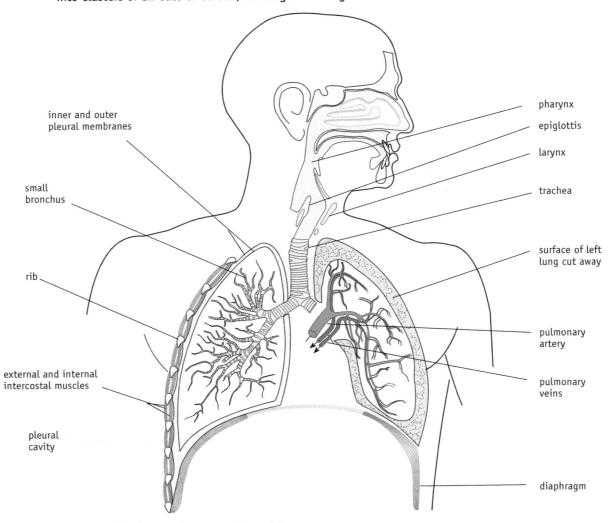

Fig. 8.9 The human lungs and breathing structures

As a gas exchange surface the alveoli have the following important features:

- Because they are so small, they have a *large surface/volume ratio*. The human lungs have about 300 million alveoli, with a total area estimated to be about 140 m².

- They have a *rich supply of capillaries*, with a total area estimated to be about 125 m² (Fig. 8.10).

- Their walls, together with those of the capillaries, are extremely *thin*, with the result that air and blood are separated by a distance of only about 0.5 μm. This is important because diffusion is only rapid over short distances i.e. if the concentration gradient is steep.

- Their lining is *moist* — gases cannot easily diffuse through a solid surface.

Fig. 8.10 shows a network of capillaries around a cluster of alveoli. In reality, each capillary lies *between* adjacent alveoli, so it is just as true to say that capillaries are surrounded by alveoli (Fig. 8.11).

The lungs are ventilated by two sets of muscles; the *intercostal muscles* between the ribs, and the *diaphragm*.

During quiet breathing, only about 10% of the alveolar air is changed in each breath. Each inhaled breath therefore simply 'tops up' a much larger volume of stagnant air already present. Moreover, the

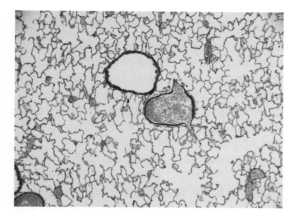

Section through lung of mammal

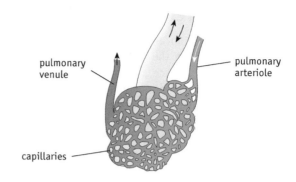

Fig. 8.10 A cluster of alveoli and surrounding capillary network

first part of the air to enter the alveoli is air that had emerged from the alveoli during the previous exhalation but had remained in the bronchi and trachea.

The air in the bronchi and trachea makes up the **dead space**, and is about 150 cm³ in an average adult human. This means that if, during quiet breathing, one inhales 500 cm³ of air, only 350 cm³ is 'fresh' air. With each 500 cm³ breath, 150 cm³ of 'stale', dead space air is shunted back and forth between the alveoli and bronchial tree.

The result of all this is that alveolar air contains only about two thirds as much oxygen as atmospheric air. The air next to the alveolar walls is stationary, so oxygen and CO_2 travel the 200 μm or so across the alveoli by diffusion. This is reasonably quick since it occurs in a gas.

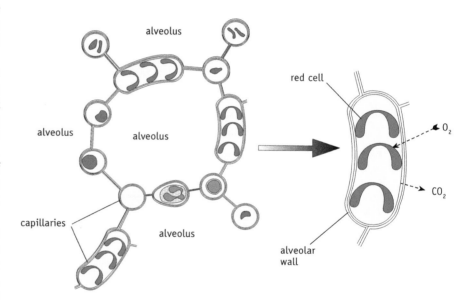

Fig. 8.11 Section through an alveolus and surrounding capillaries

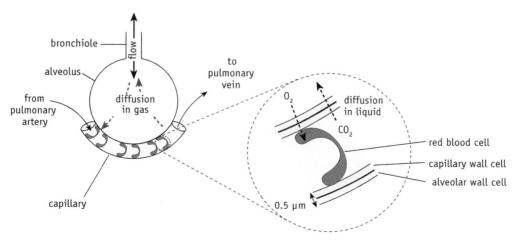

Fig. 8.12 Diagrammatic view of an alveolus and an adjacent capillary

ISBN: 9780170214094

Air-breathing and water loss

Whether an animal breathes by lungs or by tracheae, air-breathing results in the evaporation of water from the gas exchange surface. The very features that promote uptake of oxygen — such as a large surface area of thin, moist surface — inevitably lead to loss of water. In animals that live in damp environments, such as earthworms, water loss is less of a hazard. Animals living in dry environments minimise water loss by having their gas exchange surfaces tucked deep inside the body in the form of lungs or tracheal systems. The air inside the lungs or tracheae is almost saturated with water vapour, so the cells at the gas exchange surface are protected from water loss.

Nevertheless, every time air is breathed in, it becomes humidified on its way down the breathing passages, and this water is lost on breathing out.

There is a price to pay for water conservation. The air next to the gas exchange surface is not only humid, it contains less oxygen (13.8%) and more CO_2 (5.5%) than the air outside. The haemoglobin in the red cells must therefore be able to become saturated with oxygen at this lower concentration.

The rate and depth of breathing are regulated by the *breathing centre* in the brain. By means of nervous connections, this controls the activity of the breathing muscles. If the concentration of CO_2 in the blood rises, the breathing centre stimulates the breathing muscles to work harder. This raises the rate at which CO_2 is excreted, tending to bring it back to normal.

In this way, the rate of breathing is regulated so that it is no higher than is necessary to keep the blood CO_2 level constant. As a result the rate of water loss is no greater than necessary.

Water-breathing in fishes

In fishes, the parts of the body surface specialised for gas exchange are the **gills**. Like lungs, gills are highly folded, greatly increasing the surface area of the body while adding little to its volume. Like lungs, gills have a very rich blood supply separated from the environment by a very thin surface membrane.

There is however, an important difference. Gills are *outpushings* of the body into the environment, whereas lungs are *intuckings* of the environment into the body (Fig. 8.13). This is related to the fact that water is much denser than air. Its buoyancy supports the gill filaments, enabling water to flow between them. In air they would stick together, greatly reducing their effective surface (this is why fish die on land).

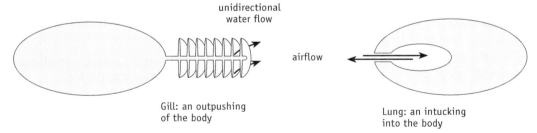

unidirectional water flow

airflow

Gill: an outpushing of the body

Lung: an intucking into the body

Fig. 8.13 Difference between a gill and a lung

Trout

Trout are bony fish, one of the two main classes of vertebrates loosely called 'fish'. Water has two major disadvantages for gas exchange:

- Even the well-aerated water of streams contains much less oxygen than air does. At 15°C, 1000 g of freshwater can dissolve up to 0.01 g of oxygen. Thus to bring a given mass of oxygen to the gills a fish must pump 100 000 times as much water.

- Much more energy is needed to propel water because of its greater density (about 1000 times greater than air) and viscosity (100 times greater than air). Over 10% of a resting fish's energy expenditure is accounted for in breathing, compared with 1–2% in a resting mammal.

These disadvantages are to some extent counteracted by two adaptations, described in more detail later:

1. Water flow over the gills is *unidirectional* and almost *continuous*, so there is no dead space as there is in a lung.

2. The blood flows in the opposite direction to that of the water, enabling most of the oxygen to be extracted from the water.

As in other vertebrates the mouth leads into a cavity that is used for both feeding and breathing. This cavity is called the **pharynx**. In fishes water passes from the pharynx to the exterior by a series of **gill slits**. In bony fish these are hidden from view by a gill cover or **operculum** (Fig. 8.14).

Between successive gill slits are the gills. Each consists of a bony arch from which extend many thin **gill filaments** (Fig. 8.15A, B, and C). The surface area of each gill filament is further increased by many thin **lamellae** projecting from it (Fig. 8.15D). The lamellae are richly supplied with capillaries (Fig. 8.15E).

Because the lamellae and capillary walls are so thin, the blood comes to within a few micrometres of the water. The blood flows in the opposite direction to that of the water. As a result of this *counterflow*, the most oxygenated blood is next to the most oxygenated water, and *vice versa*. A concentration difference is thus maintained along the whole length of each lamella, with the result that about 80% of the oxygen is extracted from the water (Fig. 8.16). The oxygen concentration in the blood leaving the gills is actually higher than that the oxygen concentration in the exhaled water.

Oxygen diffuses into the blood where it combines with haemoglobin in the red blood cells, and CO_2 diffuses in the opposite direction. These are larger than those of mammals, and have nuclei (Fig. 8.17).

Trout blood can carry 14% oxygen by volume — only about two thirds as much as human blood.

Breathing can be divided into two phases (Fig. 8.18):

1. The mouth opens and the floor of the pharynx is lowered, drawing water into the pharynx. At the same time the operculum expands, *sucking* water over the gills from the pharynx.

2. The mouth closes and the floor of the pharynx is raised, *pushing* water over the gills. The operculum moves inwards, pushing water out of the opercular cavity.

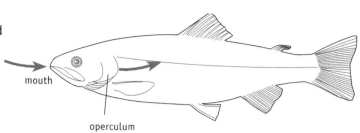

Fig. 8.14 The trout in side view showing inhalent and exhalent water flow

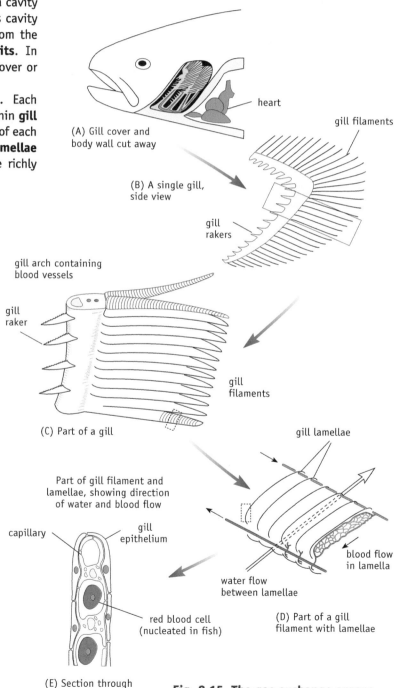

(A) Gill cover and body wall cut away

(B) A single gill, side view

(C) Part of a gill

(D) Part of a gill filament with lamellae

(E) Section through part of a gill lamella

Fig. 8.15 The gas exchange organs of a trout at successively higher magnifications

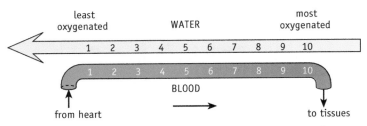

Fig. 8.16 Diagram showing how the most oxygenated water is opposite the most oxygenated blood, and vice versa

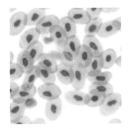

Fig. 8.17 Red cells of a trout

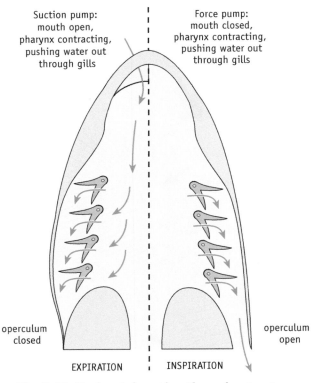

Fig. 8.18 Horizontal section through a trout pharynx showing how two pumps maintain a continuous flow of water over the gills

In pushing water over the gills, the floor of the pharynx acts as a *force pump*. By sucking water through the gill slits, the operculum acts as a *suction pump*. By working together these two pumps create an almost *continuous* current of fresh water over the gas exchange surface.

Besides the disadvantages mentioned earlier, water has one more drawback as a gas exchange medium. The solubility of oxygen in water decreases with temperature. This means that warm water saturated with oxygen contains less oxygen than colder water saturated with oxygen. Moreover, a rise in water temperature raises the rate of metabolism, and hence oxygen demand. To obtain more oxygen, breathing becomes more vigorous, further raising energy expenditure.

The cost of aquatic gas exchange

Just as gas exchange on land results in water loss, aquatic gas exchange also has costs. In freshwater fish, the large surface area of a gas exchange surface leads to osmotic uptake of water. In marine bony fish, the water concentration of the blood is higher than that of seawater and so water tends to leave the body by osmosis. Even when the blood has the same water concentration as the seawater (as in most marine worms, molluscs and crustaceans), the proportions of the various ions are different. This leads to a slow leakage of ions into and out of the body.

In all of these cases, the gain or loss of water or of ions must be counteracted by the expenditure of energy in *active transport*.

Extension: Gas exchange in birds

On 29 November 1973, a Rüppell's vulture was sucked into an engine of a plane flying at an altitude of 11.3 km — 2445 metres higher than the summit of Everest. At this altitude the oxygen concentration is only 20% of that at sea level. This is not an isolated example; many long distance migrants are known to fly at over 6000 metres — an altitude at which unacclimatised humans begin to suffer from altitude sickness. In an experiment, mice and sparrows were subjected to an atmosphere similar to that at 6000 metres; whilst the mice could barely move, the sparrows could still fly. How do they do it?

This table gives a clue; it shows that air exhaled by a bird contains less oxygen than air exhaled by a human. This means that the bird can extract a higher proportion of oxygen from the air than a human can.

Percentage oxygen in exhaled air	
Goose	Human
13.2	15.5

A bird's gas exchange system is entirely different from that of a mammal, in a number of significant ways:

- The lungs are relatively *inelastic* and *small* — about half the size of the lungs of a mammal of similar size, and they change little (less than 2%) in volume during breathing. As explained below, however, their internal surface area is relatively larger than those of a mammal.

- In a mammal, gas exchange occurs in blind-ending sacs (alveoli), but in a bird gas exchange occurs in air-filled tubes called *air capillaries* (Fig. 8.19 and Fig. 8.21).

- In a mammal the gas exchange surface is ventilated *tidally* (back and forth), but in a bird the air flows past the gas exchange surface and continues *in the same direction* even during expiration. In this respect the ventilation of the gas exchange surface resembles that of a fish (see above).

- Unidirectional airflow is made possible by a system of nine large **air sacs**, some of which have extensions that penetrate the bones (Fig. 8.22). The walls of the air sacs have a poor blood supply and take no direct part in gas exchange. Instead they act as bellows in propelling air through the lungs. Of the total volume of the breathing system, the gas exchange tissue of a bird accounts for only about 15%, compared with 95% in a mammal.

- Birds have no diaphragm, breathing movements being carried out by muscles in the abdominal wall and ribcage.

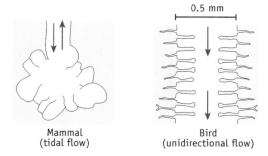

Fig. 8.19 Comparison of air flow in the finest units of the lungs of a mammal and a bird (After Schmidt-Nielsen)

The gas exchange system of a bird is (much!) more complicated than in a mammal. Inside the lung, each main bronchus or *primary* bronchus passes through the lower side of each lung and enters the large posterior air sacs in the abdomen. At its anterior end, each primary bronchus connects with the anterior air sacs.

Inside each lung, each primary bronchus divides into four to six *secondary* dorsobronchi, each of which divides into many parallel *tertiary* bronchi or **parabronchi**. These are tiny parallel tubes, less than 0.5 mm diameter. Unlike the alveoli of mammals, the parabronchi are open at each end. The 'downstream' ends of the parabronchi join to form secondary ventrobronchi (Fig. 8.21).

In strong-flying birds there are up to about 2000 parabronchi in each lung. Each parabronchus gives off hundreds of tiny **air capillaries** which form an interconnecting network running between blood capillaries. The air capillaries have a diameter of 8–20 µm. This is much narrower than the alveoli of mammals, giving a correspondingly larger surface for a given volume.

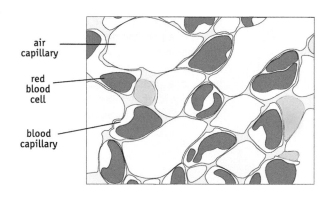

Fig. 8.20 Section through exchange tissue in the lung of a domestic chicken (drawn from an electron micrograph)

The blood and air capillaries are the site of gas exchange and are accordingly called *exchange tissue*; here the blood and air come very close together (0.09 µm in some birds – about half that in most mammals). Though the lungs of a bird have only 27% the volume of a mammal of similar mass, the gas exchange surface is about 15% greater. Per gram of lung tissue the surface area of bird lung tissue is about ten times greater than that of mammals (Fig. 8.20).

It is now known that the blood and air flow *at right angles to each other* (Fig. 8.21). This *crosscurrent* flow is not quite as efficient as the counterflow in a fish, but is superior to the situation in a mammal. This is shown by the fact that in a bird the blood leaving the lungs is richer in oxygen than is the exhaled air.

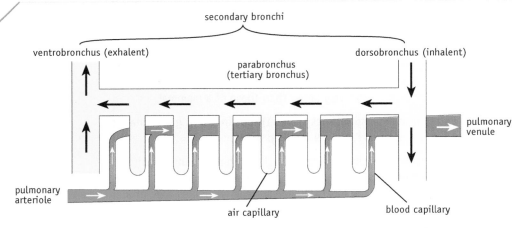

Fig. 8.21 Highly simplified diagram showing cross current flow in lung of bird. In reality, both air capillaries and blood capillaries form a branching network. (After Scheid.)

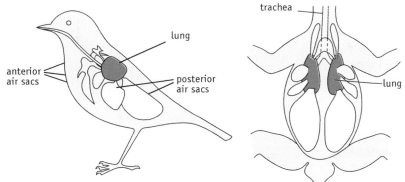

Fig. 8.22 The air sacs of a bird

The air sacs and breathing cycle

Fig. 8.22 shows the air sacs of a bird. The four posterior air sacs receive fresh (inhaled) air directly from the primary bronchi and deliver it to the gas exchange surfaces in the lungs. The five anterior air sacs receive CO_2-rich, oxygen-depleted air from the lungs and deliver it to the trachea to be exhaled.

Fig. 8.23 shows the breathing cycle in a highly simplified view of the relationship between the lungs and air sacs. The posterior air sacs are shown as a single chamber and always contain inhaled, oxygen-rich air (yellow). The anterior air sacs are also shown as a single chamber and always contain exhaled air (blue). The parabronchi are shown as straight tubes; there are course many more than shown. The important thing to note is that the airflow through them is *unidirectional* and *continuous*.

As in a fish, the breathing mechanism consists of two pumps, acting in a kind of **'push-pull' fashion**. The abdominal muscles draw inhaled air by expanding the posterior air sacs. They then compress the air sacs, forcing the air through the lungs. The rib cage and anterior air sacs expand, drawing air out of the lungs, and then contract, forcing it into the trachea.

Notice that it takes *two* breathing cycles to propel a given mass of air through the gas exchange system. Air flows through the lungs during both inspiration and expiration.

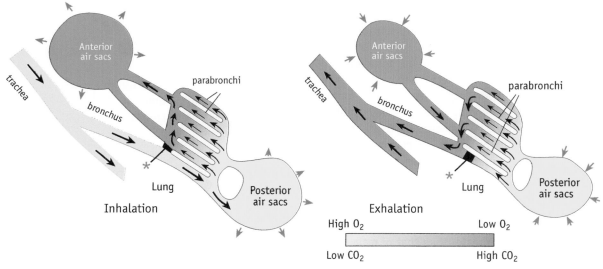

Fig. 8.23 Simplified diagram showing how the action of air sacs enables a bird to achieve continuous, one-way air flow through the lungs during both inspiration and expiration. The black part (marked with an asterisk) indicates a valve-like feature that prevents air flow during part of the breathing cycle.

Summary of key facts and ideas in this chapter

○ Gas exchange in animals is the diffusion of oxygen into a cell or organism and the diffusion of CO_2 in the reverse direction.

○ In very small organisms, gas exchange occurs over the entire body surface. This is possible because small organisms have a large surface area (which affects oxygen supply) compared with their volume (which affects oxygen demand).

○ Due to their thin shape, flatworms can obtain sufficient oxygen by simple diffusion because no cell is more than a very short distance from the surface.

○ In an earthworm the concentration gradients of CO_2 and oxygen at the body surface are steep because:
  - in the skin, blood constantly brings CO_2 to the surface and carries oxygen away from it, and
  - the skin is very *thin* so the capillaries are close to the surrounding air.

○ Insects breathe by means of branching air-filled tubes called *tracheae*. These extend throughout the body. Diffusion of gases between environment and cells is rapid.

○ The finest branches of the tracheal system are microscopic *tracheoles* which, unlike the tracheae, penetrate individual cells.

○ The tracheae open at the sides of the body by *spiracles*, whose openings can be partially or wholly closed when the animal is less active, thus reducing water loss.

○ In mammals the gas exchange organs are the *lungs*. The air at the gas exchange surface is very different from the atmosphere outside, being saturated with water vapour, with a lower oxygen concentration and a higher CO_2 concentration.

○ In fishes, the gas exchange organs are the *gills*, over which water flows continuously and unidirectionally. Because water and blood flow are in opposite directions, a high percentage of the oxygen is extracted.

Test your basics

Copy and complete the following sentences. In some cases the first letter of a missing word is provided.

1. The movement of oxygen and CO_2 into and out of the surface of an animal is called ___*___ ___*___, and occurs by the process of ___*___.

2. In very small organisms, such as *Paramecium*, gas exchange occurs over the entire body surface. This is possible because small organisms have a large ___*___ compared with their ___*___.

3. In flatworms, which are much larger than *Paramecium*, a large ___*___/___*___ ratio is achieved by having a ___*___, leaf-like body shape. In earthworms, which are larger, the rate of gas exchange is increased by transporting oxygen and CO_2 to and from the body surface by a ___*___ system.

4. In insects, the ___*___ plays no part in gas transport. Instead, oxygen and CO_2 are transported to and from the tissues by a tree-like system of ___*___-filled tubes called ___*___. These open at the body surface by holes called ___*___. They can be partially closed to reduce the loss of ___*___.

5. The finest branches of the ___*___ system are microscopic ___*___, which penetrate individual cells.

6. Even in large insects, movement of gases through the tracheal system is by diffusion, but when active (e.g. during flight), the tracheal system is v___*___ by active pumping movements of the abdomen.

7. In mammals, gas exchange occurs in the ___*___. These are the blind-endings of a branching system of tubes called the ___*___ tree. The larger tubes are supported by ___*___ and are called ___*___. The smallest ones are called ___*___ and have no ___*___ in their walls, but have a layer of ___*___ muscle, which enables their width to be controlled.

8. Between adjacent alveoli is a dense network of microscopic blood vessels called ___*___. These are supplied by branches of the ___*___ ___*___ and drained by branches of the ___*___ ___*___.

9. Gas exchange occurs by __*__, because the concentration of oxygen in the alveoli is __*__ than in the blood, and the concentration of CO_2 in the alveoli is __*__ than in the blood.

10. The distance between the blood and air is extremely short because the walls of the capillaries and alveoli are very __*__. Also, the capillaries have a large total __*__ __*__.

11. The alveoli are continually ventilated by breathing movements. These are due to two sets of muscles; the __*__ between the ribs, and the dome-shaped __*__ forming the floor of the chest.

12. Whereas the lungs of a mammal are __*__ of the outside world into the body, the gills of a fish are highly folded __*__.

13. A gill consists of two rows of gill __*__, each of which has many smaller __*__ projecting from it.

14. Whereas in a mammal ventilation of the gas exchange surface is two-way or t__*__, in a fish the water flows over the gills c__*__ and in the same __*__. This makes it possible for the flow of water and blood to be in __*__ directions. This __*__ enables a high proportion of the oxygen to be extracted from the water. This to some extent compensates for the fact that oxygen is sparingly __*__ in water, so a given volume of water contains much less oxygen that the same volume of air.

15. In a fish such as a trout the gills are ventilated by two pumps acting alternately. The floor of the pharynx p__*__ water over the gills, and the gill cover or __*__ s__*__ water over them.

ISBN: 9780170214094

9 Adaptation in Plants

The plant way of life: Light to chemical energy

From a very early age we learn that plants are very different from animals. Whereas animals typically move quickly and usually move around, plants are anchored to one spot and move only slowly, by growth. Also, they have finely divided bodies, are green, and require light to survive.

These characteristics are linked to the way they obtain their raw materials and energy. Whereas animals are **heterotrophic**, feeding on ready-made organic substances obtained from other organisms, plants are **autotrophic**. This means that they make their own organic compounds from simple inorganic raw materials — carbon dioxide and water. To do this plants use light energy from the sun in **photosynthesis**, summarised thus:

$$\text{light + carbon dioxide + water} \xrightarrow{\text{Chlorophyll}} \text{carbohydrate + oxygen}$$

This can be summarised by a simple chemical equation, in which CH_2O is chemical shorthand for carbohydrate:

$$\text{light} + CO_2 + H_2O \xrightarrow{\text{Chlorophyll}} CH_2O + O_2$$

Besides CO_2 and water, plants also require mineral ions which, like CO_2, are present in low concentrations. These are brought to the plant's surface by mass movements of air or, in the case of water in the soil, by diffusion.

The life of a flowering plant is associated with a number of adaptive features:

- Under the light microscope, plant cells have a number of distinctive features (Fig. 9.1).

- They are surrounded by a **cell wall**, consisting largely of *cellulose*, which is strong enough to prevent excessive water uptake. The cell wall also prevents plant cells from rapidly changing shape, unlike the muscle cells of animals.

- Mature plant cells typically have a large **vacuole**, filled with cell sap, so the cytoplasm is very close to the cell wall.

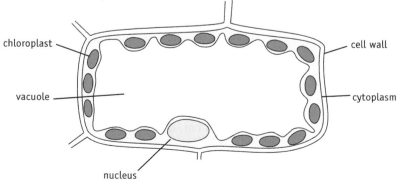

Fig. 9.1 A section through a plant cell, as seen under the light microscope

- At least some of the cells contain **chloroplasts**. Because of the vacuole, these are very close to the cell surface and the source of CO_2.

- To absorb raw materials and energy, plants need to expose a large surface area to the environment. This is achieved by having a finely divided, branching body.

- A land plant lives in a two-part environment (Fig. 9.2):

  - an above-ground *shoot system* which absorbs light and CO_2 in photosynthesis, and also carries out sexual reproduction, and

  - a below-ground *root system* which absorbs water and minerals and anchors the plant.

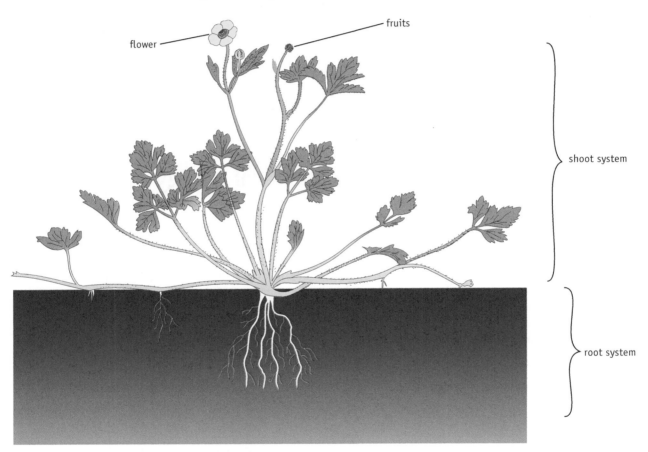

flower

fruits

shoot system

root system

Fig. 9.2   The creeping buttercup, a terrestrial plant

The root and shoot systems are *interdependent*; the shoot system needs the root system for water and mineral ions, and the root system depends on the shoot system for organic materials. Details of the ways in which root and shoot are adapted for these functions are described in Chapters 10–12.

These features are common to land plants in general. But of course terrestrial environments vary, and each species of plant is adapted to particular conditions, for example:

- Although almost all plants photosynthesise, some are adapted to living in very bright light ('sun plants'), while others are adapted to living is lower light intensities of the forest floor ('shade plants'). In addition, some have a photosynthetic mechanism that works particularly well in hot, dry conditions, as discussed in Chapter 11.

- Some plants, called *xerophytes*, are adapted to dry environments. Others, called *hydrophytes*, are adapted to living in water or very wet terrestrial (land) habitats. Examples are described in Chapter 11.

ISBN: 9780170214094

Extension: Genes or environment?

In any suitable environment, a plant develops all the features described above; such features are determined by genes. Yet some features develop differently in different environments; for example, plants growing in shade tend to have thinner leaves than plants of the same species growing in more sunny situations (see Chapter 10).

Phenotypic characteristics are determined by both genes and environment. The effects of genes and environment on physical characteristics (phenotype) were investigated by three American scientists using yarrow (*Achillea lanulosa*), a weed that has also been introduced into New Zealand (Fig. 9.3).

In California, this plant has a wide distribution, growing at altitudes ranging from sea level to the mountains of the Sierra Nevada. Conditions at high altitude are very different from those at sea level; temperatures are much lower and the growing season shorter. Not surprisingly, plants of this species are shorter in the mountains than at lower altitudes.

There are three possible explanations:

- The differences are adaptive in that they are the result of genes that have been selected for their advantages. If this is true, then the environment was acting on *previous* generations via natural selection.

- They are the direct effect of the environment on growth of the present generation.

- Some combination of the two.

The scientists tackled the problem in the following way.

Specimens were collected from sites at various altitudes along a transect extending from the coast to the high mountains (Fig. 9.4). The plants were then grown in a garden and from each, a clone of many genetically identical plants was produced. Since the plants were all grown in the same garden, their physical

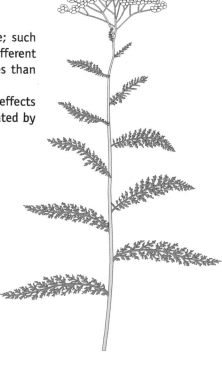

Fig. 9.3 *Achillea lanulosa* **(yarrow), a plant used in the experiment illustrated in Fig. 9.4**

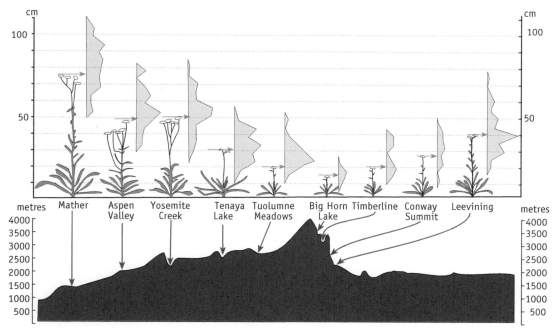

Fig. 9.4 Results of an investigation into the effect of genes and environment on phenotypic characters of *Achillea lanulosa*. Plants were collected from sites along the transect shown below, and grown in the same garden at low altitude. The mean heights to which they grew are shown by the red arrows. For each clone, the variation in height is shown by a green shape, tipped on its side.

environments were very similar. The height of each plant was measured when it was mature. The results are shown in Fig. 9.4.

Even when grown in the same environment, plants from different altitudes clearly grew to different heights. This shows that populations of plants at different localities along the transect must have been genetically different. Plants from high altitude grew lower than plants collected from lower habitats because of differences in their genetic makeup.

Summary of key facts and ideas in this chapter

○ Plants are adapted to absorb light, CO_2 and minerals by having a finely divided body, and consequent large surface area.

○ Plants also show structural adaptations for living in differing light intensities and widely varying water availability. Examples are described in later chapters.

Test your basics

Copy and complete the following sentences.

1. Plants are ___*___, meaning that they can produce their own organic compounds from inorganic raw materials in the process of ___*___.

2. In photosynthesis the raw materials are ___*___ from the air and ___*___ from the soil. The products are ___*___ and ___*___ (gas).

3. CO_2 and mineral salts are present in the environment in ___*___ concentrations. To absorb them therefore requires a finely ___*___ body, giving it a large ___*___ ___*___ compared with its ___*___.

4. Plant cells are able to limit the uptake of water by the presence of a ___*___ ___*___ made of ___*___.

5. Photosynthesis takes place in organelles called ___*___. These are normally close to the cell surface because of the presence of a large, fluid-filled space called a ___*___. As a result the diffusion path of ___*___ into the chloroplasts is ___*___.

6. The body of a typical flowering plant is divided into an above-ground ___*___ system and a below-ground ___*___ system. The ___*___ system imports organic matter from the ___*___ system, and the ___*___ system imports water and ___*___ from the ___*___ system.

7. This interdependence is associated with a two-way transport system; ___*___ carrying water and minerals, and ___*___ carrying organic matter.

8. Plants are adapted to a wide range of water availability; ___*___ are adapted to living in dry conditions, and ___*___ are adapted to living in wet conditions. Most plants are adapted to conditions between these two extremes and are called ___*___.

10 Nutrition in Plants

Nutrition in plants involves the uptake and utilisation of raw materials and energy. It therefore includes *photosynthesis* and *mineral nutrition*. There are no significant differences in photosynthesis or mineral nutrition that can be linked to particular taxonomic groups such as flowering plants, conifers and ferns, but there are some important differences as regards functional groups. This chapter deals with three such groups:

- C_3 sun plants, adapted to bright sunlight in temperate conditions.

- C_3 shade plants, adapted to dim light.

- C_4 plants, adapted to high light intensities, high temperatures and restricted water availability.

Outline of the photosynthetic process

As explained in more detail in Chapter 20, photosynthesis is the process in which plants convert CO_2 and water into sugar and oxygen:

$$\text{light} + CO_2 + \text{water} \xrightarrow{\text{Chlorophyll}} \text{sugar} + \text{oxygen}$$

The entire process occurs in organelles called **chloroplasts** (chapters 16 and 17), which are particularly abundant in leaves.

Photosynthesis and respiration

It is easy to forget that, while photosynthesis only occurs in the light, respiration occurs all the time. As a result, the organic matter that accumulates in the daytime is actually *less* than the amount the plant actually produces.

A good way to envisage this is to liken a plant to a water tank, as in Fig. 10.1.

Water in the tank represents organic matter in the plant (dry mass). Water entering the tank represents gain in organic matter by photosynthesis, and water leaving it represents loss of organic matter in respiration. The rate of each process is represented by the rate of water flow, indicated by the number of drops of water.

During the night, respiration continues and the organic matter content falls. At dawn, photosynthesis begins and, once it exceeds respiration, the organic matter content begins to rise. By midday, photosynthesis is much faster than respiration and the plant is gaining rapidly in organic matter. During the night there is no photosynthesis so the plant is living on its 'savings' until dawn.

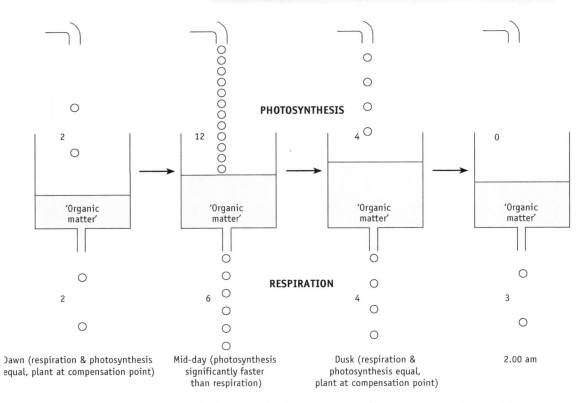

Fig. 10.1 Respiration and photosynthesis, represented by water entering and leaving a tank. Numbers represent imaginary relative rates.

Although the organic content falls, the plant uses less during the night than it gained the previous day, so over the 24-hour period it has gained dry mass. Notice that it is usually cooler at dawn than at dusk, so respiration is slower late in the night than early in the night. For the same reason, a plant loses less organic matter in a cool night than in a warm one.

Gross and net photosynthesis

Photosynthesis can be measured by finding the amount of oxygen given off per hour, or the amount of CO_2 taken in per hour. It is complicated by the fact that respiration takes place at the same time. Respiration takes place in the mitochondria and can be summarised as:

organic matter + oxygen ⟶ carbon dioxide + oxygen

Since respiration has the opposite effect to photosynthesis, true photosynthesis is always a little faster than it appears. If a leaf is making 10 units of oxygen per hour in photosynthesis and using four units of oxygen per hour in respiration, only six units per hour leave the leaf to be measured (Fig. 10.2). Thus the leaf *seems* to be producing six units per hour (the net rate of photosynthesis) but is actually making 10 units per hour (the true, or *gross* rate). To find the true rate of photosynthesis, the rate of respiration has to be measured (in the dark), and added to the net rate of photosynthesis.

true photosynthesis = net photosynthesis + respiration

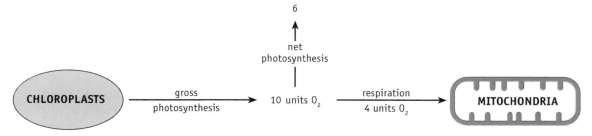

Fig. 10.2 Why a plant produces oxygen faster than it appears

How a leaf is adapted for photosynthesis

Fig. 10.3 shows a thin section (slice) through a rhubarb leaf. Its structure is highly adapted for absorbing CO_2 and light, and importing water.

Fig. 10.3 Section through a rhubarb leaf

midrib or main vein

(area enlarged right)

Section through the leaf along X ----- X

palisade cells

spongy cells

cuticle

upper epidermis

xylem

phloem

substomatal air space

stoma

Absorbing light

In dicotyledonous plants growing in sunny situations the leaves grow so they face the light. The mesophyll is divided into two distinct layers; an upper **palisade** layer rich in chloroplasts, and a lower **spongy** layer. Because the palisade cells are elongated along the light path, there are fewer cell walls to reflect and scatter the light. Most of the light passing through the palisade cells is absorbed by the chloroplasts.

Because the palisade cells also need to absorb CO_2 (see below), there are extensive air spaces between them (Fig. 10.4).

About a third of the light reaching the leaf thus passes *between* the palisade cells and is reflected down to the spongy cells. By reflection and refraction, much of this light is scattered back up again through the palisade layer, giving it a 'second chance' to be absorbed (Fig. 10.5). Reflection by the spongy layer is the main reason why the lower surface of a leaf looks paler than the upper.

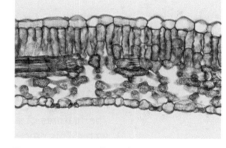

Transverse section through privet leaf

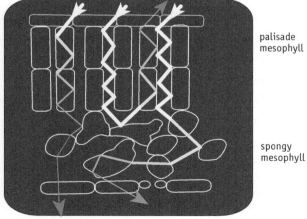

parallel rays of direct sunlight entering leaf

palisade mesophyll

spongy mesophyll

scattered, mainly green light passing through

Fig. 10.5 How the organisation of the mesophyll maximises light absorption

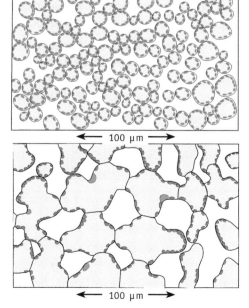

100 μm

100 μm

Fig. 10.4 Tobacco leaf: Section through palisade mesophyll (top) and spongy mesophyll (bottom), parallel to the epidermis, showing extensive air spaces

ISBN: 9780170214094

Absorbing CO$_2$

As a result of photosynthesis the concentration of CO$_2$ inside the leaf is lower than it is outside, so it diffuses into the leaf. The structure of the leaf helps this in several ways (Fig. 10.6):

- The leaf is *thin*, giving it a large surface area compared with its volume.

- Its thin shape shortens the distance between the atmosphere and mesophyll cells. Diffusion is faster over shorter distances, in which the concentration gradient is steeper.

- The extensive air spaces between the mesophyll cells extend between the palisade cells, so most of the diffusion pathway is through gas (in which diffusion is 10 000 times faster than in solution).

- The large vacuole in the mesophyll cells keeps the chloroplasts close to the cell surface, so diffusion through liquid is very short.

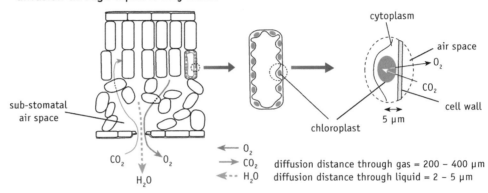

Fig. 10.6 **The diffusion path of CO$_2$**

Conserving water

The features of a leaf that promote the inward diffusion of CO$_2$ also help the outward diffusion of water. This process is called **transpiration** and is the 'price' the plant has to pay to absorb CO$_2$. A leaf is adapted to reduce transpiration in two distinct ways:

- A water-resistant, waxy **cuticle** is secreted by the epidermis. In plants living in dry environments (xerophytes), the cuticle may be very thick. The thickness of the cuticle is thus an adaptation to long-term conditions, or *climate*.

- Short-term fluctuations in water supply ('*weather*') are dealt with by tiny holes in the epidermis called **stomata** (singular, stoma). Each stoma is surrounded by two **guard cells**. In the daytime, and when water supply is good, these are tightly inflated (turgid) with water and are bent, keeping the stoma wide open (Fig. 10.7).

Plant	Water lily	Oak	Apple	Sunflower	Potato	Tomato	Oat	Cabbage	Geranium
Upper	460	0	0	207	51	12	25	141	29
Lower	0	346	294	250	161	130	23	226	179

Table 10.1 **Number of stomata per mm² on the upper and lower epidermis of various plants**

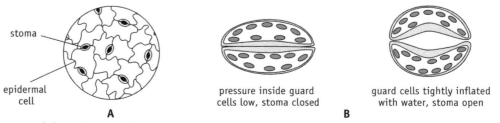

Fig. 10.7 **(A) Epidermis of a leaf in face view. (B) Enlarged view of a stoma closed and open.**

When water is scarce, and also at night, the guard cells lose water and become straight, closing the stomata. In herbaceous plants stomata are more numerous on the lower surface, and in most trees and shrubs they are absent from the upper epidermis.

There is a marked tendency for stomata to be more numerous on the lower epidermis in dicotyledons. There are two possible advantages to the plant:

- Since the lower surface is slightly cooler, transpiration would be slightly slower.

- Guard cells in the upper epidermis would reflect some light before it could enter the palisade cells.

Despite this generalisation, many plants have some stomata on the upper surface, as Table 10.1 shows.

Although stomata occupy only about 1–3% of the leaf surface, diffusion through them is much faster than you would expect. This is because diffusion occurs around the edge of each stoma into a *substomatal air space* (Fig. 10.8).

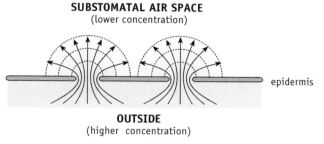

Fig. 10.8 Diffusion of CO_2 through stomata

Despite adaptations for conserving water, a rapidly photosynthesising plant transpires over a hundred times faster than it absorbs CO_2. This does not matter when the soil is moist, because water is absorbed by the roots as fast as it is lost by the leaves. It is transported by specialised cells in the **xylem** tissue.

Exporting sugar

The sugar made in photosynthesis is transported in a tissue called the **phloem**. Phloem cells lie adjacent to the xylem cells and together make up the veins or **vascular bundles**. On a sunny day sugar is produced by the leaf much faster than it can be exported, and it is temporarily converted into starch in the chloroplasts. At night the starch is reconverted to sugar. This is then transported to the other parts of the plant such as roots and developing flowers, fruits and seeds.

Most plants growing in open situations are adapted to living in bright light, and are called **sun plants**, for example rhubarb, tomato, sunflower. Because they can photosynthesise very rapidly they are 'high-earners', and can thus grow quickly.

Photosynthesis on the forest floor – shade plants

Plants of the forest floor are called **shade plants** e.g. the Mexican breadfruit plant and most other house plants. They are adapted to make the most of light that is not only dim, but the rays are scattered in all directions (Fig. 10.9).

Fig 10.9 How light changes as it passes through the canopy

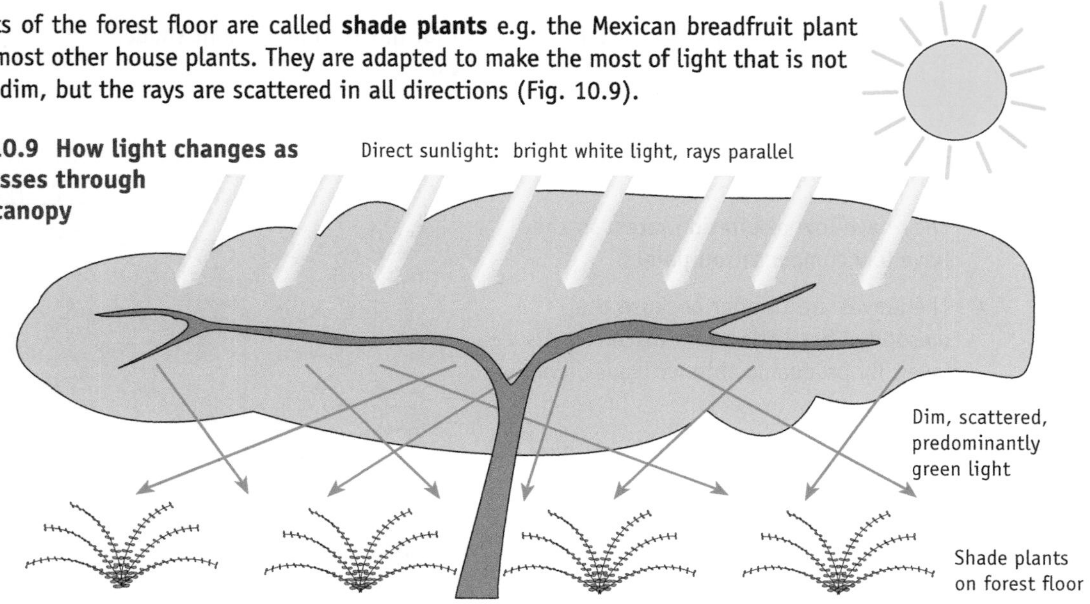

Direct sunlight: bright white light, rays parallel

Dim, scattered, predominantly green light

Shade plants on forest floor

Since shade plants have a low 'income', they grow slowly. Fig. 10.10 shows why. Look at the graph for the sun plant and notice the following:

1. In dim light (up to about light intensity 6), the rate of photosynthesis increases in proportion to the light intensity (graph is a straight line). In dimmer light, the rate of photosynthesis is said to be *limited* by light, meaning that the rate depends on the light intensity.

2. In very bright light the graph levels off, showing that the plant cannot make use of extra light. This is because some other factor is limiting the rate, such as the CO_2 supply, or temperature.

3. In very dim light (intensity 2 on the graph) the rate of photosynthesis *seems* to be zero. This is because the rate of CO_2 production in respiration *just* balances the rate of CO_2 absorption in photosynthesis. This light intensity is the *compensation point*, at which the plant is 'marking time'.

Now look at the curve for the shade plant. Notice the following:

1. In dim light (e.g. below light intensity 3) the shade plant absorbs CO_2 faster than the sun plant.

2. The graph for the shade plant levels off at much lower light intensities than the graph for the sun plant. In bright light, therefore, the shade plant is wasting more light than the sun plant.

3. The shade plant has a lower compensation point than the sun plant. The shade plant can therefore gain carbohydrate in light that a sun plant cannot.

4. To survive, a plant needs brighter light than its compensation point, because it has to make up all the organic matter it loses during the night.

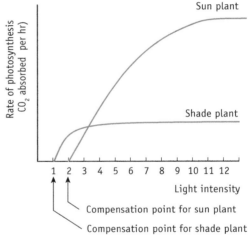

Fig. 10.10   Effect of light intensity on rate of photosynthesis in sun and shade leaves

Sun leaves and shade leaves compared

Fig. 10.11 shows a section through a leaf of a plant that grows in permanent shade.
Shade leaves differ from sun leaves in several ways:

- They have a higher concentration of chlorophyll, so a higher proportion of the light can be absorbed.

- They have low respiration rates, so they have low compensation points.

- The leaves are thinner because the mesophyll has little or no distinct palisade layer. By producing thinner leaves, energy is saved.

- Sun leaves have more stomata per mm². To use more light, the leaf must be able to absorb more CO_2, requiring more stomata. The cost is a greater rate of transpiration.

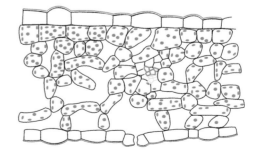

Fig. 10.11 Section through twinleaf, a shade plant

Sun and shade leaves in trees

The distinction between sun and shade leaves is not absolute. For example the upper leaves of a large tree may be in full sun, whereas those on the lower branches are in deep shade. Also, most large forest trees spend their youth in the shade of much taller trees, and develop shade leaves. If the tree survives long enough to reach bright light, it develops sun leaves on the upper branches. Lower branches, which receive less light, continue to produce shade leaves. Thus leaves with the same genetic makeup develop different adaptive characteristics. Fig. 10.12 shows sections through a sun leaf and a shade leaf taken from the same maple tree.

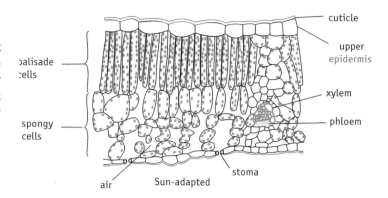

Fig. 10.12 Maple, sections through sun-adapted and shade-adapted leaves

Summary of comparison between sun and shade-adapted leaves

	Sun-adapted	**Shade-adapted**
Palisade mesophyll	Strongly developed	Less developed, or absent
Stomatal density	Higher	Lower
Respiration rate	Higher	Lower
Compensation point	Higher	Lower
Photosynthetic rate in dim light	Lower	Higher
Chlorophyll content	Lower	Higher

Extension: Photosynthesis in hot, dry environments

When, in the early 1950s, scientists discovered the chemical pathway in photosynthesis, it was assumed that the process was the same for all plants. In the 1960s two Australian scientists showed that some plants have a different process. It was called C_4 **photosynthesis** because the first chemical on the pathway is a *four*-carbon compound. Most plants have C_3 **photosynthesis** because the first stable compound on the pathway has *three* carbon atoms.

C_4 photosynthesis occurs in many tropical plants, especially grasses like maize and sugarcane. It also occurs in certain introduced grasses such as paspalum and kikuyu grass. C_4 plants have the following features:

- They use bright light more efficiently than C_3 plants because they do not become light-saturated (Fig. 10.13).

- At high temperatures photosynthesis is much faster than in C_3 plants (Fig. 10.13).

- C_4 plants use water much more efficiently than C_3 plants.

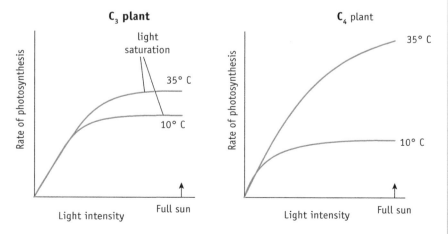

Fig. 10.13 The different responses of C_3 and C_4 plants to high light intensities and high temperatures

In C_3 photosynthesis the carbon pathway consists of the light-independent reactions of the **Calvin cycle** (Ch. 20). The first step in the cycle is catalysed by an enzyme called **rubisco** (short for ribulose bisphosphate carboxylase). Rubisco has one major weakness – it not only catalyses the uptake of CO_2, it also catalyses a process called **photorespiration**. In this process one of the key chemicals in the Calvin cycle (a 5-carbon sugar) is oxidised to CO_2:

$$\text{five-carbon sugar + oxygen} \longrightarrow \text{organic compounds} + CO_2$$

The process resembles respiration only in so far as organic matter is oxidised to CO_2; rather than making ATP, photorespiration uses it. Because it counteracts photosynthesis, photorespiration is disadvantageous to the plant.

Thus rubisco has two alternative substrates, CO_2 and oxygen, which 'compete' with each other for the active site of the enzyme. At lower temperatures, or at higher CO_2 concentrations, CO_2 'wins', and photorespiration is insignificant. At higher temperatures or at lower CO_2 concentrations, the reverse is true.

This explains why C_3 plants become 'light-saturated' in full sunlight. The brighter the light, the lower the CO_2 concentration inside the leaf, so the more photorespiration counteracts photosynthesis.

C_4 plants avoid this problem by a process in which the CO_2 concentration in the chloroplasts is raised to a level at which photorespiration is insignificant. In these plants the photosynthetic tissue is of two kinds:

- The mesophyll, which does not contain rubisco and the Calvin cycle does not occur.

- The **bundle sheath**, which consists of tightly packed cells round the vascular bundles (Fig. 10.14).

What happens is that CO_2 is in effect 'pumped uphill' from a low concentration in the mesophyll cells to a higher concentration in the bundle sheath cells, where rubisco is present.

In the mesophyll cells CO_2 combines with a three-carbon compound to form a four-carbon compound (hence 'C_4'). This compound is then transported via plasmodesmata into the bundle sheath cells where it breaks down into the three-carbon compound again, releasing CO_2 (Fig. 10.15).

This process of concentrating CO_2 in the bundle sheath costs energy in the form of ATP. However, at higher temperatures and in bright light this is more than offset by the benefit of avoiding photorespiration. In cooler climates C_3 photosynthesis is more efficient.

There is another, highly significant benefit of C_4 photosynthesis; water is used far more efficiently. To

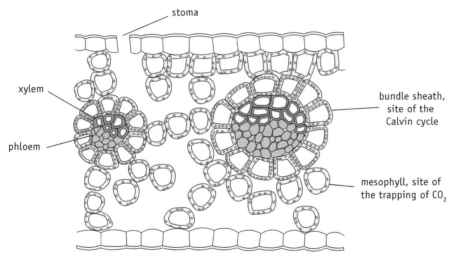

Fig. 10.14 The characteristic leaf anatomy of a C_4 plant

photosynthesise, plants must absorb CO_2 from a very low atmospheric concentration – about 390 parts per million. To do so they must expose a large area of photosynthetic tissue to the air. This inevitably results in the evaporation of water from the leaves. Transpiration is thus the price a plant has to pay for absorbing CO_2. Of the water absorbed by a plant, less than 1% is used in photosynthesis; most is lost in transpiration. If water is plentiful this does not matter, but in dry conditions a plant must restrict water loss by at least partly closing its stomata. This slows photosynthesis. Thus the plant has to compromise between the demands of photosynthesis and water conservation.

When a C_3 plant is moderately short of water, the stomata partly close. This reduces the CO_2 concentration in the leaf, and photorespiration becomes significant. In a C_3 plant therefore, water shortage severely reduces photosynthesis.

ISBN: 9780170214094

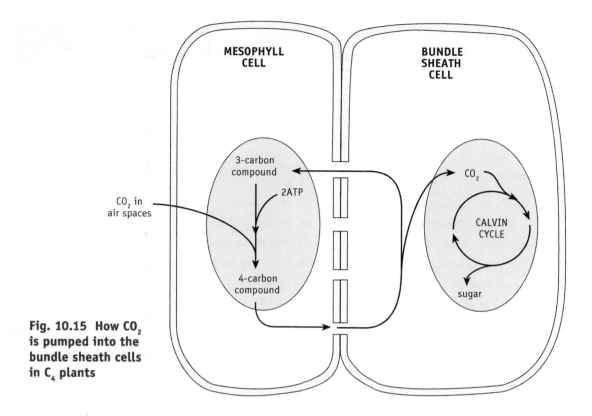

Fig. 10.15 How CO$_2$ is pumped into the bundle sheath cells in C$_4$ plants

C$_4$ plants are far less affected. Because mesophyll cells don't undergo photorespiration, a C$_4$ plant can continue to fix CO$_2$ when its stomata are almost closed, saving water. At 30 °C a C$_3$ grass loses over 800 water molecules for every CO$_2$ molecule fixed, a C$_4$ grass under the same conditions loses fewer than 300 water molecules. C$_4$ plants can thus continue to photosynthesise in drier conditions than a C$_3$ plant can.

Ecological significance of C$_3$ and C$_4$ photosynthesis

C$_4$ plants grow faster in hot, dry conditions and C$_3$ plants grow better in cooler, wetter environments. You can see evidence of this on many lawns in northern New Zealand. In summer, C$_4$ grasses such as paspalum and kikuyu thrive and begin to displace C$_3$ plants such as clover and most lawn grasses. By the autumn, the balance of advantage begins to shift back to the C$_3$ plants. Over a year as a whole, neither type has the advantage; there is an ever-changing shift in competitive advantage. One result of global warming may be the gradual replacement of many C$_3$ plants on New Zealand lawns.

CAM photosynthesis

Pineapple, cacti, and many other succulents have a kind of photosynthesis that saves water even more effectively than C$_4$ photosynthesis does. This is also true of many epiphytes (plants living on the branches and trunks of trees, gaining anchorage but not raw materials). This mechanism is called CAM photosynthesis. 'CAM' stands for 'Crassulacean Acid Metabolism' and the term was coined because it was first discovered in plants of the Crassulacae (stonecrop family).

Most plants close their stomata at night and open them in the daytime, but CAM plants close their stomata in the daytime and open them at night, when water loss is much slower. The CO$_2$ taken in at night is temporarily stored by combining it with an organic acid. In the daytime the CO$_2$ is released and used in the Calvin cycle. As a result these plants lose very little water — in fact they lose only about 50 water molecules for every CO$_2$ molecule taken in.

Summary of key facts and ideas in this chapter

○ The true rate of photosynthesis is always higher than it appears because of respiration. Apparent or net photosynthesis is equal to the true rate of photosynthesis *minus* the rate of respiration.

○ A leaf is an *organ* in which various tissues cooperate to carry out photosynthesis.

○ The absorption of direct sunlight is promoted by the division of the mesophyll into palisade and spongy layers.

○ The absorption of CO_2 is promoted by the:
- thin shape of a leaf,
- presence of extensive air spaces, and
- short distance between chloroplasts and cell surface.

○ The absorption of CO_2 is accompanied by an inevitable evaporation of water, or *transpiration*.

○ There is thus a conflict between the need to absorb CO_2 and the need to conserve water.

○ The function of the stomata is the regulation of water loss.

○ The supply of water and minerals to the leaf is the function of the *xylem*; the export of sugar is the function of the *phloem*.

○ In a certain low light intensity (the *compensation point*), the rate of photosynthesis appears to be zero because photosynthesis (which produces oxygen and carbohydrate) is balanced by the rate of *respiration* (which uses oxygen and carbohydrate).

○ Shade plants have a low compensation point because they have a low rate of respiration, so they can grow in dimmer light than sun-adapted plants.

○ Sun-adapted plants can 'afford' to have a higher respiration rate because they have a higher 'income' from photosynthesis.

○ Certain plants are adapted to living in hot, dry conditions with intense sunlight. In these C_4 *plants* carbohydrate is produced in the chloroplasts of the *bundle sheath* cells. In these cells the concentration of CO_2 is raised by an active process involving the expenditure of ATP.

○ C_4 plants also use water much more efficiently than C_3 plants.

○ Many succulents (such as cacti) have a different kind of photosynthesis. In these CAM plants, the stomata open at night and CO_2 is absorbed and stored. In the daytime, with the stomata closed, the CO_2 is released and used in photosynthesis.

Test your basics

Copy and complete the following sentences. In some cases the first or last letters of a missing word are provided.

1. Whereas photosynthesis only occurs in the light, ___*___ occurs all the time. This means that the actual or ___*___ rate of photosynthesis is always ___*___er than the apparent or ___* photosynthetic rate.

2. A plant can only grow if, over a 24 hour period, it makes more carbohydrate in photosynthesis than it uses in respiration. The light intensity at which these processes are equal is the ___*___ point, at which the plant is 'marking time'. During the day, a plant must make more sugar than it uses in the entire 24 hour period.

3. In a dicotyledonous leaf the photosynthetic tissue or ___*___ is divided into two layers. Nearest the light is the ___*___ mesophyll, consisting of cells elongated more or less ___*___ to the incident light. As a result light that passes between these cells undergoes little scattering and reaches the ___*___ cells below, where much of it is reflected back into the ___*___ layer.

4. The absorption of CO_2 is facilitated ('helped') by the ___*___ shape of the leaf. As a result a leaf has a large ___*___ area compared with its ___*___, and the distance CO_2 has to diffuse is relatively ___*___.

ISBN: 9780170214094

5. The diffusion of CO_2 is facilitated by the extensive system of ___*___ spaces between the cells, because diffusion is much faster in gas than in ___*___.

6. The last stage of diffusion of CO_2 is in solution through the cytoplasm. This stage is extremely short because the large ___*___ displaces the chloroplasts and the rest of the cytoplasm near to the air spaces between the cells.

7. The uptake of CO_2 is inevitably accompanied by ___*___ or the evaporation of water. Water loss is reduced by a waxy ___*___ secreted by the ___*___.

8. Whereas the thickness of the ___*___ is related to climate, plants can respond to temporary water shortage by partially or wholly closing their ___*___. In doing so, ___*___ uptake and (therefore photosynthesis) is reduced.

9. In most trees stomata are absent from the upper ___*___, but in many herbs they may also be present (though in smaller numbers) in the upper epidermis.

10. On a warm, sunny day a leaf makes sugar much faster than it can be exported by the ___*___ tissue, so the excess is temporarily stored in the ___*___ as ___*___. At night this is reconverted to sugar and exported to other parts of the plant.

11. Plants adapted to shade are able to survive because they have a low ___*___ point because of a low rate of ___*___. This means that they can make a 'profit' in light that is too dim for a sun-adapted plant to survive.

12. Sun-adapted leaves have ___*___er stomatal densities (number of stomata per unit ___*___) than shade-adapted leaves. This enables them to absorb CO_2 more ___*___ than shade leaves. Sun-adapted leaves also have well-developed ___*___ layers in the mesophyll, whereas in shade leaves there is little or no distinct ___*___ layer.

ISBN: 9780170214094

11 Plants and Water

What is transpiration and why is it inevitable?

Transpiration is the evaporation of water from the shoots of a plant. In order to absorb CO_2, the photosynthesising cells must be in communication with the air outside. Although this enables CO_2 to diffuse in to the leaf, it is also allows water vapour to diffuse out. On a warm sunny day the difference in water vapour concentration may be 100 times greater than the difference in CO_2 concentration, so water diffuses out much faster than CO_2 enters. Transpiration is thus the inevitable 'price' a land plant has to pay for photosynthesis.

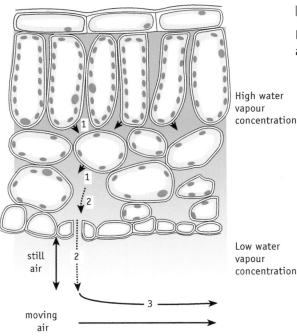

High water vapour concentration

Low water vapour concentration

still air

moving air

1 = evaporation from cell walls
2 = diffusion through air spaces, stomata and boundary layer
3 = carried away by wind

Fig. 11.1 Pathway of water in transpiration

How does transpiration occur?

Fig. 11.1 shows the route water takes during transpiration. There are essentially three stages:

1. Evaporation from the surface of the mesophyll cells.

2. Diffusion through the still air in the mesophyll, through the stomata, and through the very thin layer of stationary air next to the leaf surface.

3. Mass movement by wind.

The rate of transpiration is affected by the following factors:

- Humidity. The drier the air, the faster is transpiration.

- Wind. Transpiration is faster in windy conditions.

- Temperature. Higher temperatures speed up transpiration.

- Altitude. Transpiration is faster in lower atmospheric pressures.

- Light. In most plants, stomata close at night, so transpiration is faster in the daytime.

- Soil water content. In drought conditions, stomata close even in daytime, reducing transpiration.

- Leaf structure. Plants that are adapted to dry conditions have a leaf structure that reduces transpiration rate.

ISBN: 9780170214094

How weather affects transpiration

Humidity, wind and temperature are different factors, but they all affect the steepness of the concentration gradient. The idea of a humidity gradient is illustrated in Fig. 11.2, using figures that are common on a cool day and on a warm day.

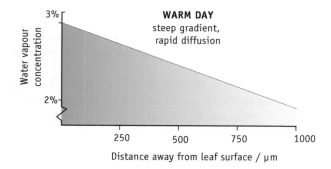

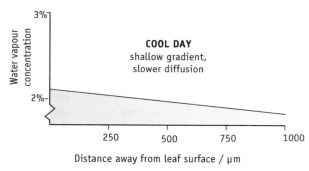

Fig. 11.2 Humidity gradients at the surface of a leaf

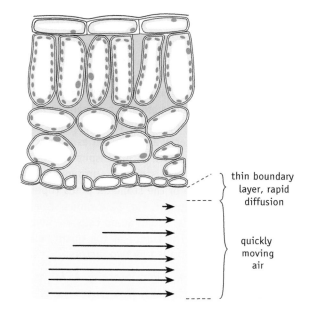

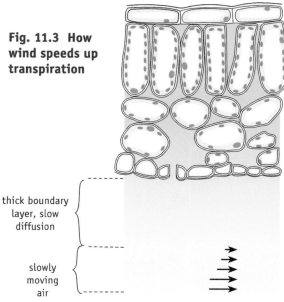

Fig. 11.3 How wind speeds up transpiration

The steepness of the water vapour gradient depends on two things:

1. The difference in concentration of water vapour inside and outside the leaf. This in turn can be influenced by two things:

    - Atmospheric humidity affects the water concentration *outside* the leaf.

    - Temperature affects the water vapour concentration *inside* the leaf by affecting the rate of evaporation from cell surfaces in the leaf.

2. The length of the path along which water is diffusing. The air immediately next to any surface is always still, *even in windy conditions*. This layer of still air is called the **boundary layer** (a similar still layer exists next to the lining of a blood vessel and the surface of a fish's gill). The effect of wind is to reduce the thickness of the boundary layer, thus steepening the concentration gradient. On a windy day it may be a fraction of a millimetre, and several centimetres on a still day (Fig. 11.3).

ISBN: 9780170214094

Advantages of transpiration

Because transpiration is inevitable, it cannot be said to have a *function*. It can however have two *advantages*:

- As explained it Chapter 12, the upward flow of xylem sap from roots to leaves is caused mainly by transpiration. The xylem sap contains dissolved mineral salts. The transport of minerals from roots to leaves is therefore helped by transpiration.

- The evaporation of water uses up heat. In hot weather therefore, transpiration helps to keep a plant cool.

The role of stomata

The stomata (singular, stoma) are tiny pores present in the epidermis of leaves and young stems. Their function is to regulate the balance between photosynthesis and water conservation in changing environmental conditions.

As previously explained, the 'cost' of photosynthesis is the loss of water. In most plants there has to be a compromise between the need to take up CO_2 and the need to prevent excessive water loss. At night, when photosynthesis is impossible, water conservation is the only factor that matters. In the daytime, water availability varies with the weather. Provided there is adequate soil moisture, water uptake can keep pace with transpiration, but in dry conditions water conservation take priority.

Each stoma is surrounded by two **guard cells**. By changing their shape the guard cells vary the size of the stomatal pores (openings). The changes of shape result from changes in internal pressure or **turgor**. Fig. 11.4 shows stomata of a dicotyledonous leaf. Due to the uneven thickening of the cell walls, turgid guard cells are bent, opening the pore. When they are limp or flaccid, the stoma closes.

The turgor of the guard cells (and therefore the degree of opening of the stomatal opening) is controlled by two factors:

- Light intensity. As CO_2 is used up in photosynthesis the CO_2 concentration in the leaf cells falls. This causes the guard cells to absorb water and thus become more turgid (see Extension section below for explanation).

- The water content of the leaf. When water loss exceeds water uptake the mesophyll cells lose turgor pressure and the guard cells close.

In dry sunny weather these two factors oppose each other. The actual stomatal aperture is a 'compromise' between them.

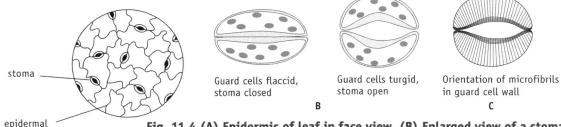

stoma

epidermal cell

A

Guard cells flaccid, stoma closed

Guard cells turgid, stoma open

B

Orientation of microfibrils in guard cell wall

C

Fig. 11.4 (A) Epidermis of leaf in face view. (B) Enlarged view of a stoma, closed and open. (C) Orientation of microfibrils in guard cell wall.

Extension: How do guard cells work?

Changes in turgor result from changes in solute concentration of the guard cells. In the daytime the guard cells become more turgid and the stoma opens. At night the reverse happens. If the plant is short of water even in the daytime, the stomata close.

Why should guard cells be more turgid in the daytime? In the light, the CO_2 concentration in the leaf falls because it is being used up in photosynthesis. The fall in CO_2 stimulates the guard cells to absorb potassium ions from the surrounding cells. To maintain electrical neutrality, negatively charged organic ions are produced within the guard cells. This may explain why, unlike other epidermal cells, guard cells have chloroplasts and can photosynthesise. The increase in solute concentration causes water to enter by osmosis from the surrounding cells and the stoma opens (Fig. 11.4). At night the reverse happens. This pumping of potassium ions is helped by the small size of the guard cells, so they have a large surface relative to their volume.

Because guard cells close as a result of loss of turgor, it is tempting to think that wilting of a leaf automatically results in stomatal closure. In fact it turns out to be more complicated. Stomatal closure during water shortage results from the action of a plant growth substance called **abscisic acid** (ABA). This is synthesised in leaves in response to water shortage, causing the guard cells to lose potassium ions and hence to close.

How do changes in turgor cause the guard cells to change their shape? The answer lies in the way the cell walls are thickened. Even under the light microscope it can be seen that their inner walls are noticeably thicker than the outer walls, but the key lies in the organisation of the cellulose microfibrils. Most are laid down around the width of the cells, like hoops on a barrel (Fig. 11.4c). Along the inner walls there are many additional microfibrils running *lengthwise*. As a result, when the guard cells absorb water the outer walls stretch more lengthwise than the inner walls, causing the guard cells to curve outwards. When they lose water the reverse happens.

The effect of leaf structure

Stomata enable a plant to respond continually to short-term environmental changes (weather). *Climate*, on the other hand, involves factors that operate over a much longer timescale. Many plants have evolved anatomical and physiological features that enable them to survive where there is prolonged water scarcity or water excess. Plants adapted to withstand water shortage are called **xerophyte**s, while those adapted to experience water excess are called **hydrophytes**. Those in between are called **mesophytes**. There are of course no sharp divisions between these categories.

Xerophytes

Water shortage can reduce photosynthesis in two ways:

- A wilting leaf cannot maintain the stiff shape needed to intercept light.
- The closure of stomata of a wilting plant reduces its supply of CO_2.

Incidentally, water is never in short supply as a *raw material* for photosynthesis — most plants die from dehydration before they have lost half their water.

Not all xerophytes live in dry climates. Epiphytes (plants that live on other plants but do not obtain any raw materials from them), rely on what little water the roots can scavenge from crevices in the bark of the tree on which they are growing. Examples are *Astelia* (perching lily) and some ferns (Fig. 11.5).

Xerophytes have evolved a number of features that enable them to balance water loss and uptake:

- Reduced transpiration.
- Storage of water after rare periods of heavy rain (e.g. cacti and other succulents).
- Some have CAM photosynthesis, which uses water much more efficiently (see Chapter 10).
- Increased lignification.
- Deep roots e.g. some *Eucalyptus* species have roots 20 m deep.

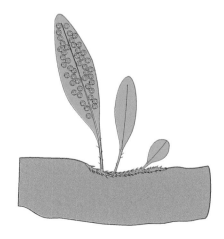

Fig. 11.5 *Pyrrosia eleagnifolia, an epiphytic fern*

Reducing transpiration rate

Transpiration rate is expressed as the amount of water lost per unit time per unit area of leaf ($mg\ hr^{-1}\ cm^{-2}$). Rates not only vary from one species to another, they can vary even in the same plant under the same conditions. In trees, for example, sun-adapted leaves (adapted for rapid CO_2 uptake) transpire faster than shade-adapted leaves (Chapter 10).

Leaves have a variety of anatomical features that reduce transpiration:

- Low stomatal density. Plants that are adapted to living in dry soils may have fewer than 150 stomata per mm^2 of leaf, while those living in moist climates may have over 500 per mm^2.

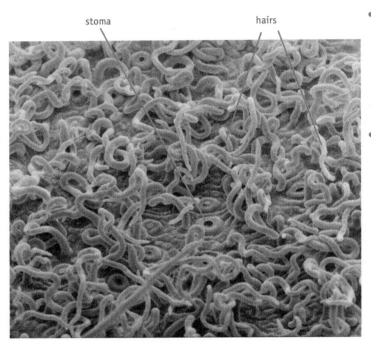

stoma　　　　　hairs

Fig. 11.6 Scanning electron micrograph of lower epidermis of pohutukawa leaf. (Courtesy of Doug Hopcroft, Hort Research.)

- Thick cuticle. Since the cuticle is not completely impermeable to water, the rate of cuticular transpiration depends on its thickness. In leaves with a thick cuticle, it may be as low as 1% of transpiration with stomata open. In leaves with a thin cuticle, it may account for 40% of total transpiration.

- Plants can decrease the humidity gradient between the still air in the mesophyll and the moving air outside in various ways such as:

  - Hairy epidermis, e.g. pohutukawa (Fig. 11.6).

  - Leaves that can roll up in dry conditions, e.g. grasses such as the sand dune grass *Spinifex* (Fig. 11.7). In this and other grasses, large 'hinge cells' are responsible for the rolling up movement. When water loss is excessive the hinge cells shrink. This causes the leaf to roll up, enclosing a tube of still, humid air. Perhaps surprisingly, *Spinifex* leaves do have some stomata on the outside of the leaf as well as the inside.

  - Sunken stomata, e.g. pine (Fig. 11.8).

- Reduced leaf area. Many xerophytes have small or needle-shaped leaves, such as pines. Though this does not in itself reduce the rate of water loss per cm_2, it does reduce total loss by the shoot system as a whole. In cacti the leaves are reduced to spines, the fleshy stems taking over the function of photosynthesis (Fig. 11.9).

Notice that reducing the steepness of the concentration gradient of water vapour also reduces the CO_2 concentration gradient, and thus the rate of photosynthesis.

Fig. 11.7 Transverse section of a *Spinifex* leaf

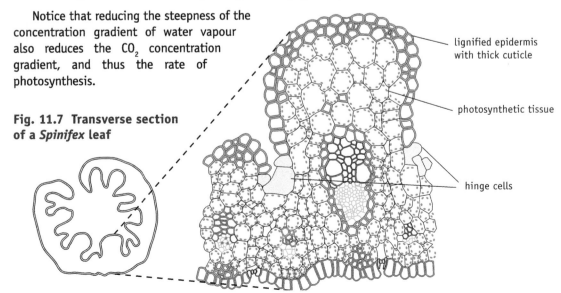

lignified epidermis with thick cuticle

photosynthetic tissue

hinge cells

ISBN: 9780170214094

Storage of water

Cacti and other succulents have large amounts of water storage parenchyma tissue. During a long drought (in some areas this may last several years) the plant gradually shrinks (Fig. 11.9).

CAM photosynthesis

Cacti and many other succulents, and also some epiphytes, have a kind of photosynthesis that uses water much more efficiently than other plants. They open their stomata at *night*, when it is much cooler and transpiration is slower. During the night they absorb CO_2 and store it by combining it with organic compounds. In the daytime (with stomata closed) the CO_2 is released and used in photosynthesis (see Chapter 10).

Increased lignification

Most xerophytes have a thick cuticle, and their leaves are also resistant to wilting by having large amounts of *lignin*, which helps prevent cell walls collapsing when cells lose water.

'Resurrection' plants

Many mosses that live on bare rock (e.g. *Tortula muralis*, common on walls) cannot reduce their water loss but are able to survive complete drying of their tissues. The plant appears to be dead, yet within minutes of being rehydrated, it becomes active again. It survives by being able to take advantage of brief periods of wetting.

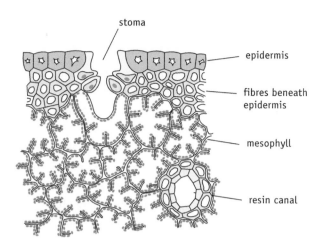

Fig. 11.8 Section through outer layers of pine leaf, showing sunken stoma

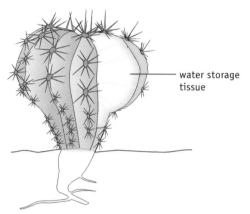

Fig. 11.9 A cactus (*Echinocactus*) with part cut away to show water storage tissue

Hydrophytes

Though water is essential to all living things, too much brings problems as well as advantages. The main one is that gases diffuse very slowly in solution. Hydrophytes usually have extensive internal air space systems, enabling oxygen and CO_2 to diffuse rapidly. In water lilies the leaves float on top of the water, so the upper epidermis is open to the air. The stomata are on the upper surface of the leaves (Fig. 11.10). Xylem is weakly developed (in some hydrophytes it is absent altogether), and there are huge air spaces in the petioles (leaf stalks).

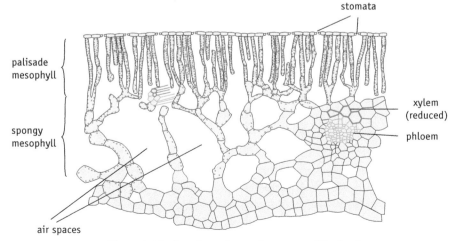

Fig. 11.10 Section through a water lily leaf

ISBN: 9780170214094

Summary of key facts and ideas in this chapter

○ Transpiration is the evaporation of water from the leaves and stems of a plant, and is the inevitable 'price' the plant has to pay to absorb CO_2 in photosynthesis.

○ The rate of transpiration is increased by environmental factors such as wind, a rise in temperature, and low atmospheric humidity.

○ The rate of transpiration can be regulated by varying the aperture (degree of opening) of the *stomata*.

○ In most plants, stomata close at night and open during the day. They also close in drought conditions; this is under the control of abscisic acid (ABA) produced in the leaves in response to water shortage.

○ Plants that are adapted to grow in prolonged dry soils are called *xerophytes*.

○ Plants that are adapted to conditions of water excess are called *hydrophytes*.

Test your basics

Copy and complete the following sentences. In some cases the first letter of a missing word is provided.

1. Transpiration is the ___*___ of ___*___ from the aerial parts of a plant.

2. On a sunny day, the loss of ___*___ is many times faster than the uptake of ___*___.

3. The rate of transpiration depends on the difference in ___*___ of water vapour inside and immediately outside the surface of the leaf. The water concentration in the air spaces inside the leaf depends on the ___*___. The water concentration in the air immediately outside the leaf depends on the atmospheric ___*___, and also on the ___*___ speed.

4. Even in darkness, when in most plants the ___*___ are closed, some transpiration occurs through the ___*___.

5. The rate of transpiration through the stomata is controlled by the ___*___ cells. In darkness there is no photosynthesis and the concentration of ___*___ rises. This causes the ___*___ to lose water and become less bent, closing the stoma. In the light, photosynthesis causes the concentration of CO_2 in the leaf to fall. The guard cells absorb water and become more ___*___, opening the stoma.

6. Plants adapted to living in dry soils are called ___*___. They are able to reduce transpiration rates in various ways, such as low ___*___ densities, hairy ___*___, ___*___ stomata, or leaves that can roll up, e.g. ___*___.

7. Plants adapted to living in conditions of water excess are called ___*___. For a plant, the main disadvantage living in water is that CO_2 and oxygen diffuse ___*___. Mud at the bottom of a pond can become ___*___, and the roots of many hydrophytes obtain their oxygen by an extensive system of ___*___ spaces running from leaves to roots.

Excellence in Biology Level 2

ISBN: 9780170214094

12 Transport in Plants

In a floating plant, all parts have access to water, CO_2 and light, but in a terrestrial (land) plant, water comes from below and light from above. Land plants thus live in a two-part environment, and their organisation reflects this. In most plants the upper part is concerned with photosynthesis and the lower part with anchorage and the absorption of water and minerals. The upper, light-harvesting part is the **shoot system** and consists of a stem and leaves. The below ground part is the **root system**.

Root and shoot systems are interdependent (Fig. 12.1). The leaves supply the roots with energy in the form of organic matter, and the roots supply the leaves with water and minerals. There is thus a two-way transport of materials between shoot and root.

Except in liverworts and mosses, transport is carried out by specialised *vascular tissues*. This is why ferns and seed plants are called **vascular plants**. These have two kinds of tissue specialised for conduction:

- **Xylem**, which transports water and minerals.

- **Phloem**, which transports sugar and amino acids.

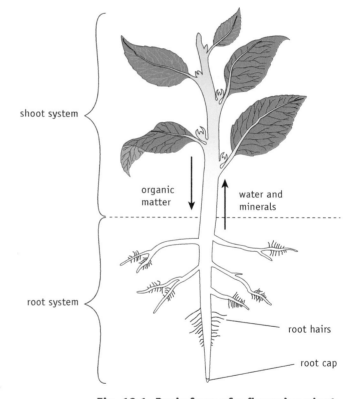

Fig. 12.1 Basic form of a flowering plant

The path of water

Before reading further you will need to read the sections on diffusion, osmosis and active transport in Chapter 21.

The path of water from soil to atmosphere can be divided into four stages:

- From soil to root xylem.

- From root xylem to leaf xylem.

- From the leaf xylem to the surface of the leaf mesophyll cells.

- Through the leaf air spaces to the atmosphere.

The absorption of water

Most water is absorbed by young roots, in the region a few millimetres behind the tip. The surface area of the root is greatly increased by outgrowths of the epidermal cells called **root hairs** (Fig. 12.2). A single maize root has about 10 000 root hairs, each of which is about 7–8 mm long. The root system of a single rye plant growing in less than a tenth of a cubic metre of soil was estimated to have a total surface area of about 640 m², or about 130 times the area of the shoot system!

Inside the epidermis is the **cortex**, the innermost layer of which is the **endodermis** (Fig. 12.3). The endodermis plays an essential part in regulating the entry of ions into the root (see below). Inside the endodermis are the xylem and phloem. These are separated from the endodermis by the **pericycle**, a tissue from which lateral roots develop.

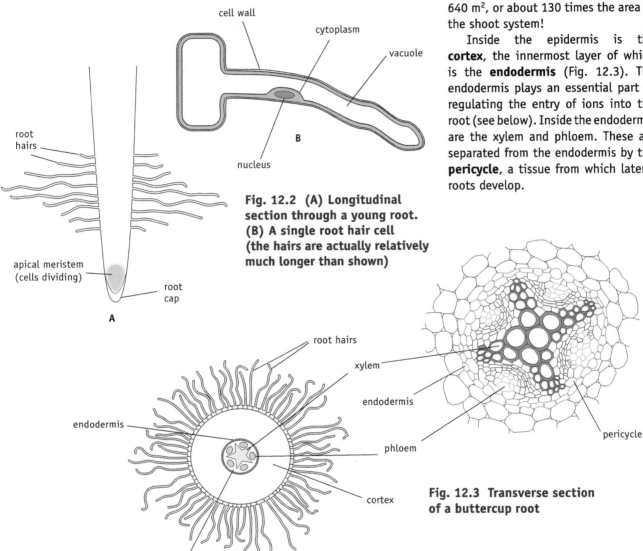

Fig. 12.2 (A) Longitudinal section through a young root. (B) A single root hair cell (the hairs are actually relatively much longer than shown)

Fig. 12.3 Transverse section of a buttercup root

Apoplast and symplast

A plant cell wall consists of a network of interwoven *microfibrils* of cellulose, together with proteins and other substances (Chapter 17). The spaces between the cellulose microfibrils are occupied by water. Since cells are in contact with each other, the water in the cell walls forms a continuous network called the **apoplast**. The conducting cells of the xylem (vessel members and tracheids) have no cytoplasm so the water in their cavities is continuous with the water in the walls. The cavities of the vessels and tracheids are thus part of the apoplast.

The cytoplasm of adjacent plant cells is also interconnected via fine threads called **plasmodesmata**. Water and other materials can thus pass from cell to cell through the plasmodesmata without passing through the cellulose of the cell wall. The cytoplasmic parts of plant cells thus form a second continuous system, called the **symplast**.

It follows that water can take three possible routes through a plant tissue (Fig. 12.4):

- The apoplastic pathway, which avoids any cell membranes. In moving through the apoplast, water moves as it does through filter paper, by capillary flow.

- The symplastic pathway, passing from the cytoplasm of adjacent cells via the plasmodesmata.

- The transcellular pathway, moving through the cytoplasm and vacuoles of adjacent cells.

The path of least resistance is the apoplast, so most water moves through this route. Relatively little water flows through the vacuoles.

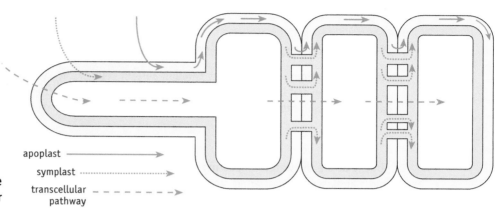

apoplast
symplast
transcellular pathway

Fig. 12.4 The three routes water can take in moving through a plant tissue

The role of the endodermis

As water moves through the apoplast across the root cortex, it reaches a barrier in the form of the **endodermis** (Fig. 12.3 and Fig. 12.5). In this layer, the cell walls that are at right angles to the epidermis are thickened by a band of waterproof **suberin** called the **Casparian strip** (suberin is chemically similar to the cutin that makes up the cuticle). The walls that are parallel to the epidermis do not have a Casparian strip, so water can pass through them (Fig. 12.5).

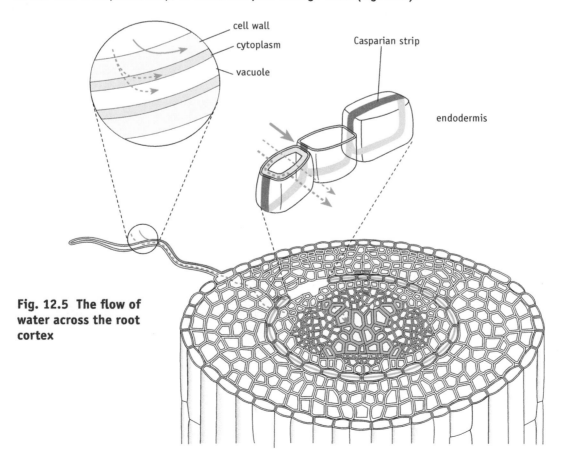

cell wall
cytoplasm
vacuole

Casparian strip

endodermis

Fig. 12.5 The flow of water across the root cortex

To cross the endodermis, water has to pass through the partially permeable membranes of the cytoplasm. The endodermis thus divides the apoplast of the root into an outer and an inner compartment. Were it not for the endodermis, water and minerals could travel from soil to leaves without passing through a single cell membrane, so the plant would have no control over the mineral ions reaching the leaf.

From root to leaf — the xylem

Water moves from roots to leaves through certain cells of the xylem. This is a complex tissue in that it consists of several kinds of cell (Fig. 12.6). In flowering plants these are *tracheids*, *vessel members*, *fibres* and *parenchyma cells*. All but the last of these are dead when mature (the parenchyma cells store starch).

Tracheids

Tracheids are the water-conducting cells of gymnosperms (e.g. conifers) and ferns, but are also present in most angiosperms (flowering plants). Tracheids are adapted for water conduction in three ways:

* They have no living contents, so resistance to flow is reduced.

* Their walls are impregnated with a complex organic substance called **lignin**. This makes the walls stiff and prevents them caving in because the xylem sap is usually under tension.

* Although lignification makes the walls less permeable to water, there are many small thin areas called **pits**. Pits of adjacent tracheids are opposite each other and provide low-resistance pathways.

Vessel members

Vessel members are the chief water-conducting cells of almost all flowering plants. They are more efficient than tracheids in two ways:

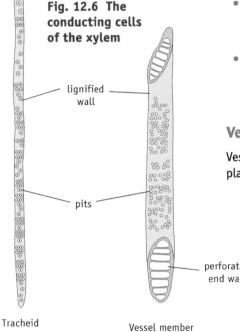

Fig. 12.6 The conducting cells of the xylem

lignified wall

pits

perforated end wall

Tracheid

Vessel member

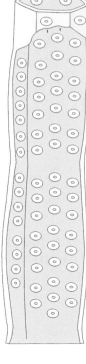

Fig. 12.7 How a vessel develops

| Cells cease dividing | Cells develop vacuoles | Walls become pitted | Walls become lignified | End walls break down |

- Their end walls are perforated and they are joined, drainpipe fashion, into continuous tubes called **vessels** (Fig. 12.7). These vary in length from a few centimetres to several metres.

- They are usually much wider than tracheids, which reduces resistance to flow.

- In young stems, in which elongation is still occurring, the walls of tracheids and vessels are lignified in the form of rings or spirals. This allows the unlignified regions between to stretch as the stem grows in length (Fig. 12.8).

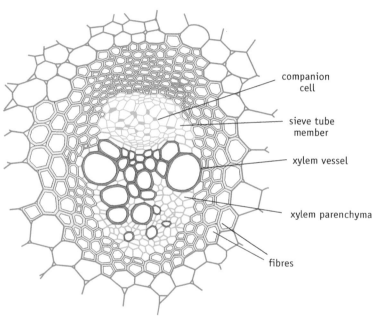

Fig. 12.8 Longitudinal section of xylem of a young stem

lignified cellulose

unlignified cellulose

Fig. 12.9 Transverse section of a buttercup stem

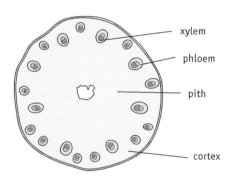

xylem

phloem

pith

cortex

companion cell

sieve tube member

xylem vessel

xylem parenchyma

fibres

Xylem also functions in support

The walls of vessel members and tracheids are lignified, which prevents them collapsing under the tension of transpiration pull (see below). As a result, xylem also plays an important role in support. Stems have to withstand the bending forces due to gravity and wind. The most economical way to distribute strengthening tissue is in the form of a hollow cylinder (as in long bones and in scaffolding).

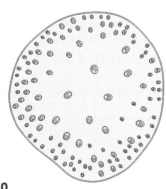

Fig. 12.10 Transverse section of a maize stem

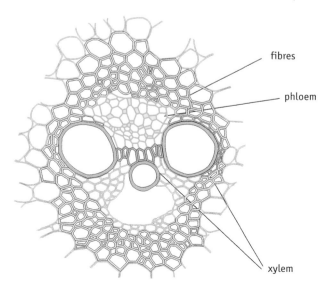

fibres

phloem

xylem

Excellence in Biology Level 2

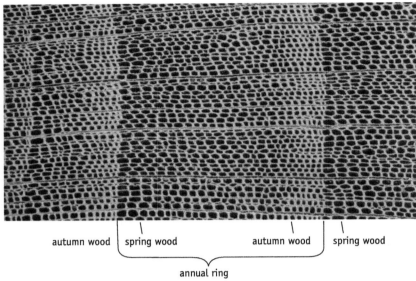

autumn wood | spring wood | autumn wood | spring wood

annual ring

Fig. 12.11 Annual rings in kahikatea (white pine).
(Courtesy of Dr Brian Butterfield, University of Canterbury.)

What propels the water?

Water transport involves two forces:

- A pull from above, called **transpiration pull**, powered by heat energy from the sun. This is by far the more important, and so far as the plant is concerned, is a passive process.

- A push from below called **root pressure**, powered by energy from respiration. Root pressure is a minor force and some plants do not develop it.

Fig. 12.9 shows the distribution of vascular tissues in a dicotyledonous stem (buttercup).

In most other dicotyledons the bundles become joined by a continuous cylinder of *secondary vascular tissue* as the stem gets older. In woody plants more secondary vascular tissue is produced every year (Fig. 12.10). Each season's secondary xylem forms an *annual ring*, and by counting these it is possible to determine the age of the tree.

In monocotyledon stems the vascular bundles are more scattered. Even so, they tend to be concentrated towards the outside (Fig. 12.10).

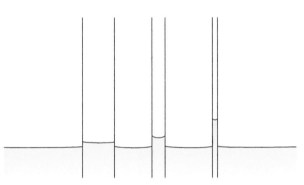

Fig. 12.12. Capillarity in glass tubes of different diameters

Transpiration pull

Transpiration pull develops in the leaves. It is due to *capillary forces* set up as a result of evaporation from the leaf cells. Fig. 12.12 shows capillarity in glass tubes. The narrower the tube, the higher the water rises.

Fig. 12.12 shows how capillarity can pull water up a glass tube. The pull results from the attraction of water molecules for the glass (adhesion) and for each other (cohesion). In creeping up the glass, the surface of the water is stretched, creating a

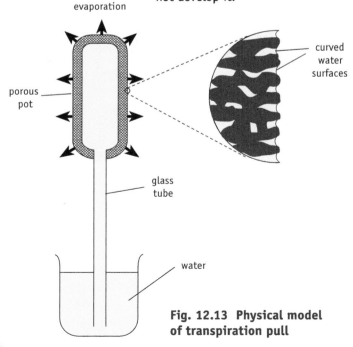

evaporation

curved water surfaces

porous pot

glass tube

water

Fig. 12.13 Physical model of transpiration pull

Diameter of tube	Height to which water rises
1 mm	3 cm
10 m	300 cm
100 nm	300 m
10 nm	3 km

Table 12.1 The heights to which a column of water can rise in capillary tubes of different diameters

ISBN: 9780170214094

curved surface or **meniscus** (plural, *menisci*). As a result of attraction of water molecules for each other, the meniscus is in a state of *tension*, like a miniature trampoline. The effect of this is to pull the water up. The narrower the tube, the greater the curvature of the meniscus and the harder it pulls on the water beneath.

Fig. 12.13 shows a physical model of transpiration pull. The tiny spaces in the porous pot are filled with water. They act as a network of interconnected microscopic capillary tubes. As water evaporates, the water retreats into the spaces, creating curved surfaces which set up a tension in the water.

The tension in the menisci is transmitted down the glass tube. The narrower the spaces in the pot, the greater is the curvature of the menisci at its surface and the greater the tension developed.

Each mesophyll cell is like a tiny, elastic porous pot (Fig. 12.14). The walls of the mesophyll cells consist of a fine network of cellulose microfibrils, with water between them. As water evaporates from the cell wall it retreats into the spaces between the microfibrils. This creates millions of tiny menisci, setting up a tension (Fig. 12.15). The spaces between the microfibrils are about 10 nm wide. As table 12.1 shows, this is enough to support a column of water three kilometres high!

No mesophyll cell is more than a fraction of a millimetre from the nearest vein ending (Fig. 12.16). Since the mesophyll cells are in contact with each other and with the xylem, the tension in the water in the cell walls is transmitted from cell to cell all the way down to the roots. In a rapidly transpiring plant therefore, the pressure in the xylem sap is *negative*.

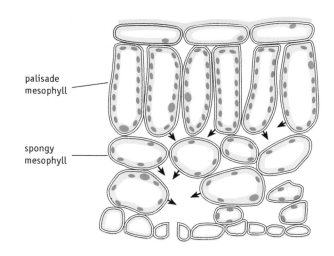

water surfaces between microfibrils are flat

palisade mesophyll

spongy mesophyll

Fig. 12.14 Section through a dicot leaf showing the mesophyll

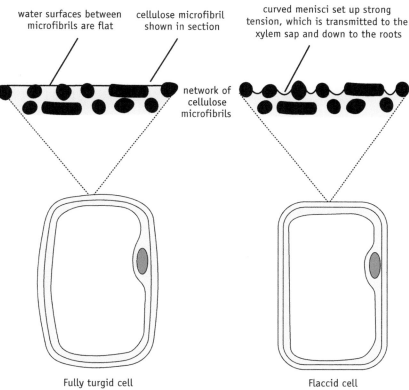

water surfaces between microfibrils are flat

cellulose microfibril shown in section

curved menisci set up strong tension, which is transmitted to the xylem sap and down to the roots

network of cellulose microfibrils

Fully turgid cell

Flaccid cell

Fig. 12.15 How water tension develops within the cell wall

From root cortex to root xylem

In the roots, water crosses the cortex and passes through the cytoplasm of the endodermal cells and enters the xylem. The sap in the xylem vessels is a very dilute solution, whereas the vacuoles of the pericycle cells contain a fairly concentrated solution. Despite this, water moves from the pericycle into the xylem.

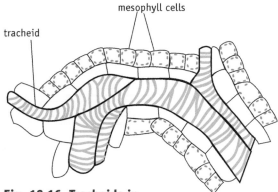

mesophyll cells

tracheid

Fig. 12.16 Tracheids in a vein-ending

ISBN: 9780170214094

To understand how water can move from a concentrated solution to a more dilute solution, we need to understand the idea of *water potential*, explained in more detail in Chapter 21. For the moment, it is sufficient to say that the water potential of a solution is the *combined* effect of its water concentration and its *pressure*, and that water always tends to move from a higher to a lower water potential.

Two factors combine to ensure that the water potential of the pericycle cells is higher than that of the xylem cells:

- The pericycle cells are under a fairly high turgor pressure.

- In a rapidly transpiring plant the xylem sap is under *tension*, which is *negative* pressure.

The difference in water potential of the pericycle and root xylem is part of a gradient of water potential that extends all the way from soil to the atmosphere. The movement of water through the plant is simply the movement down this gradient.

Root pressure

When transpiration is prevented because the air is very humid, drops of liquid may form on the leaves of grasses and some other plants. Most of this 'dew' is not condensation, but is liquid that seeps out of the xylem under pressure from below. This is called *root pressure*. It can also be seen when sap 'bleeds' from the severed stump of a plant.

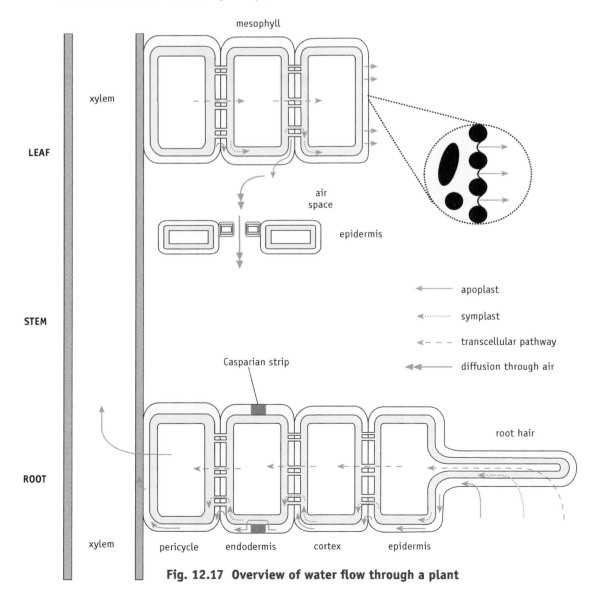

Fig. 12.17 Overview of water flow through a plant

As in the case of transpiration pull, water enters the xylem because its water potential is lower than that of the adjacent pericycle cells. In this case, the xylem sap has a lower water potential because dissolved salts are passing into the xylem vessels from the pericycle cells, reducing its water concentration. For this to continue, there must be continued active uptake of mineral ions from the soil. This process requires energy from respiration.

How fast does sap move in the xylem?

Rates of flow vary greatly. Sap moves fastest in warm, windy, sunny conditions, in which transpiration is fastest. Rates also vary with the kind of plant — in some dicotyledonous trees it may by 50 metres per hour, and higher values have been obtained for other plants.

Transport of organic matter

The transport of organic matter in the phloem is called **translocation**. It differs in several ways from sap flow in the xylem:

- Whereas vessels and tracheids are dead, sieve tube members are living and carry out metabolic processes. Killing the phloem with steam stops phloem transport immediately and permanently.

- Phloem sap is under high pressure (about 150 times human arterial blood pressure), whereas xylem sap is usually under tension. Whereas xylem sap is pulled, phloem sap is *pushed*.

- Phloem sap moves more slowly than xylem sap, about 50–150 cm hr^{-1}.

- Whereas water always moves from roots to shoots, the pattern of movement of organic materials is more variable, and frequently changes over time. Underground parts normally import carbohydrate, but storage tissues export carbohydrate to the young shoots in the spring. Very young leaves import sugar and amino acids, but mature leaves export them.

Sugar is translocated from *sources*, or sites of sugar production, to *sinks*, or sites of sugar consumption. Sources include rapidly photosynthesising leaves, and also storage tissues in which sugar is being mobilised, such as the cotyledons and endosperm in germinating seeds. Sinks include most roots, young leaves, and developing seeds and fruits.

The structure of phloem

Like xylem, phloem is a complex tissue containing several kinds of cell. In flowering plants the conducting tubes are called **sieve tubes**, and are formed by end-to-end joining of individual cells called **sieve tube members**. The end walls of the sieve tube members form **sieve plates**, perforated by many small holes (Fig. 12.18). Sieve tube members are lined by a thin layer of cytoplasm. Unique amongst plants, the nucleus degenerates during early development. Next to each sieve tube member are one or more **companion cells**. Their function is to transport sugar into or out of the sieve tubes.

An ingenious method of directly analysing phloem sap uses the fact that aphids have stylets so fine that they pierce a single sieve tube. By anaesthetising a feeding aphid the animal can be cut away from its mouthparts, after which sap continues to exude from the cut surface for long enough for it to be analysed. In most plants phloem sap contains about 20–30% sucrose, with smaller amounts of amino acids.

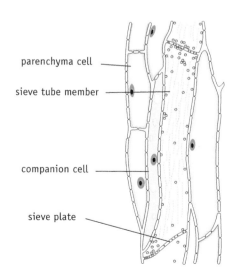

parenchyma cell

sieve tube member

companion cell

sieve plate

Fig. 12.18 Cells of the phloem, in longitudinal section

ISBN: 9780170214094

The mechanism of translocation

The generally accepted mechanism for phloem transport is the **pressure-flow hypothesis**. According to this view phloem sap is pushed down a pressure gradient, from a high pressure in a source to a lower pressure in a sink. A physical model is shown in Fig. 12.19. The two compartments A and B are each separated from the surrounding water in C by a partially permeable membrane. A contains concentrated sugar solution and B contains a more dilute sugar solution. Water passes from C to A by osmosis, raising the pressure inside A. Since they are connected, the pressure also rises in B. The pressure rise in B is sufficient to reverse any tendency for water to enter B from C, so water leaves B.

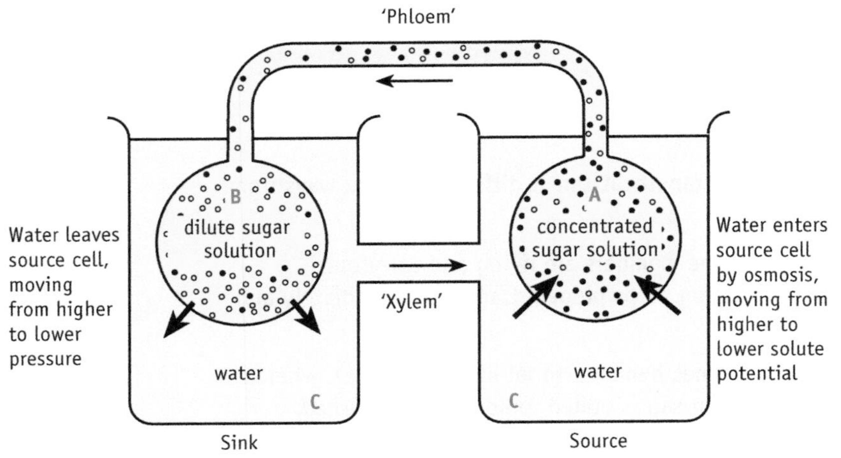

Notice that the external solution does not have to be pure water; it does not even have to be less concentrated than in B. The only necessary condition is that the sugar is more concentrated in A than in B or C, so water has a greater tendency to enter A than it has to enter B.

With continued movement of sugar from A to B, the laboratory model will eventually 'run down' as the difference in sugar concentrations decreases. In the living plant, the concentration difference is maintained by active transport. In the source, represented by compartment A, sucrose is actively transported from the companion cells into the sieve tubes. Water follows by osmosis, raising the pressure in the source. In the sink, represented by B, sucrose is removed, either by respiration or in the synthesis of other molecules, and water follows.

Notice that although sucrose moves from source to sink, water *circulates*, moving back from sink to source via the xylem, as shown in Fig. 12.20.

The transport of mineral ions

Although mineral ions are mainly transported in the xylem, some also move in the phloem. For example, before leaf fall nitrogen, phosphorus and potassium are all removed from the leaves via the phloem and are thus retained by the plant. Others, such as boron, iron and calcium are immobile and are lost at leaf fall.

This difference in mobility shows in the effects of deficiency. Shortage of nitrogen, potassium and phosphorus is shown in the oldest leaves because the young shoots are kept supplied by translocation from older leaves. Deficiency of calcium, on the other hand, shows first in the younger leaves.

Fig. 12.19 Physical model of the pressure-flow mechanism

Water leaves source cell, moving from higher to lower pressure

'Phloem'

B dilute sugar solution

A concentrated sugar solution

Water enters source cell by osmosis, moving from higher to lower solute potential

'Xylem'

water

water

C

C

Sink

Source

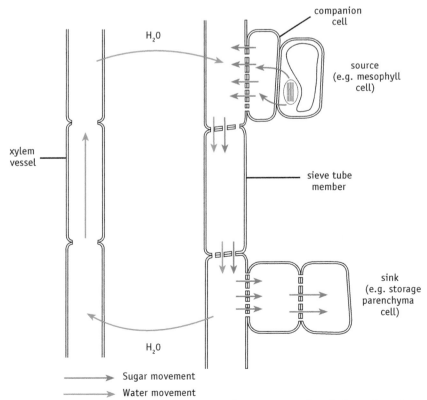

xylem vessel

H_2O

companion cell

source (e.g. mesophyll cell)

sieve tube member

sink (e.g. storage parenchyma cell)

H_2O

→ Sugar movement

→ Water movement

Fig. 12.20 Role of the xylem in phloem translocation

ISBN: 9780170214094

Transport in non-flowering vascular plants

Except for tree ferns, the stem of a fern is an underground, horizontally growing structure called a **rhizome**. The vascular tissue is organised into clusters, each consisting of a central core of xylem surrounded by phloem, extending down into the roots and up the leaves. The xylem in ferns is similar to that of flowering plants except that it is simpler. With a few exceptions, the conducting cells are tracheids only, and there are no fibres.

Transport in mosses

Bryophytes (mosses and liverworts) are the simplest plants. The upper part of a moss consists of a stem bearing light-harvesting leaves and reproductive organs. This produces threadlike structures called **rhizoids** that anchor the plant. These are single-celled structures that are superficially root-like.

In most mosses there are no specialised transport tissues. Fig. 12.21 shows a section across the stem of *Funaria*, a common moss. Though there is a central strand of elongated cells, there is no evidence that they have a transport function. Water travels mainly by capillary flow (as in a wick) up the outside surfaces of the shoots, especially when they are clustered together. The rhizoids serve for anchorage rather than water absorption. Organic materials move from cell to cell through plasmodesmata.

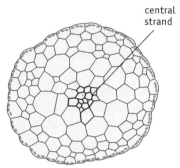

central strand

Fig. 12.21 Transverse section through a stem of a moss

In a few mosses, especially the larger ones, the central strand contains elongated cells that are thought to function as primitive transport cells. The cells are of two kinds:

- **Hydroids**, that carry water. These have very thin end walls that are perforated with pores, and are not lignified.

- **Leptoids**, that carry organic matter. Studies with radioactively labelled sugars show that the rate of transport may be as high as 50 cm per hour, which is much faster than could be accounted for by diffusion.

Summary of key facts and ideas in this chapter

○ The division of plants into two interdependent parts, an underground root system and an above-ground shoot system, has resulted in the evolution of two transport systems:
  - Xylem, which transports water and minerals from root to shoot, and
  - Phloem, which transports the products of photosynthesis, predominantly from shoot to root systems.

○ Water can move through a plant tissue along three routes:
  - The *apoplastic* route, through cell walls only, skirting around the cytoplasm and thus avoiding crossing cell membranes. Most water passes through this route.
  - The *symplastic* route, through the cytoplasm via the plasmodesmata (cytoplasmic connections between cells).
  - The *transcellular* pathway, through cytoplasm and vacuoles.

○ The *endodermis* is the innermost layer of the root cortex. Its cell walls that are at right angles to the epidermis have a layer of fatty material that is impermeable to water. As a result water is forced to pass through the cytoplasm of the endodermal cells. In this way the endodermis enables the plant to control the entry of ions.

○ The conducting cells of the xylem are of two kinds: *tracheids* and *vessel members*. Both reduce resistance to water flow because they have no cell contents when mature, and vessel members have perforated end walls.

○ Tracheids and vessel members resist collapse because the walls are *lignified*, so xylem also provides *support*.

○ As a result of transpiration, the water potential in the leaf is lower than it is in the soil, and water moves from soil to leaf down this gradient of water potential.

○ Organic matter is transported along the *sieve tubes* of the phloem, from *sources* to *sinks*. In a source, pressure is raised due to the active accumulation of sugar, which causes the osmotic entry of water. In a sink, pressure is lowered by the active removal of sugar, resulting in the osmotic loss of water.

Test your basics

Copy and complete the following sentences.

1. Ferns, conifers and flowering plants have two transport or ___*___ tissues: ___*___, which transports water and minerals, and ___*___, which transports sugar and other organic compounds.

2. The ability of a young root to absorb water and minerals is increased by the presence of thousands of root ___*___, each of which is a thread-like extension of a single ___*___ cell.

3. Plant cell walls consist of a network of cellulose ___*___, with water in the spaces between them. These spaces form a continuous system called the ___*___, through which water can move freely, as it does through filter paper.

4. The cytoplasm of adjacent plant cells is also connected by fine cytoplasmic threads called ___*___. Via these threads, water can also move from cell to cell without passing through cellulose. The cytoplasmic parts of plant cells thus form a continuous system called the ___*___.

5. The innermost layer of the cortex in a root is the ___*___. The endodermal cell walls that are at right angles to the epidermis are thickened with a fatty, waterproof material called ___*___. This prevents water from crossing the endodermis except through the partially ___*___ membranes of the cytoplasm. Without the endodermis the plant would have no control over the ions travelling from root to leaf.

6. In ferns and conifers the water-conducting cells in the xylem are called ___*___. These elongated cells have no living contents, which ___*___ the resistance to water flow. Their walls are stiffened by ___*___, which prevents them collapsing as a result of the ___*___ in the xylem sap. Water moves from cell to cell through small thin areas called ___*___.

7. In flowering plants the most important water-conducting channels are called ___*___. Each is a continuous tube like a drainpipe formed by the joining together of individual cells called ___*___ ___*___. At maturity, each ___*___ ___*___ is like a drainpipe segment, with no cell contents and perforated end walls.

8. Because lignin stiffens cell walls, xylem plays an important part in ___*___; in bulk, xylem forms wood.

9. There are two forces propelling water through a plant. More important is a *pull* from above called ___*___ pull and generated by heat from the ___*___. Less important is a *push* from below called ___*___ ___*___, which is developed by energy from respiration.

10. Transpiration pull develops in the *leaves*. When water evaporates from the surface of mesophyll cells, water retreats into the spaces between the cellulose ___*___, setting up curved surfaces (called menisci). The surface tension in these menisci is transmitted all the way down to the roots as transpiration pull.

11. The movement of material in the phloem is called ___*___, and depends on ___*___ expended by living phloem cells.

12. Sugar is transported from sites of sugar production, called ___*___, to sites of sugar consumption, called ___*___.

13. The accepted mechanism of phloem transport is the ___*___ ___*___ hypothesis. Phloem sap is propelled from high ___*___ in a ___*___, to lower pressure in a ___*___. Sucrose is pumped by ___*___ transport into sieve tubes in the source, and water follows by ___*___, raising the pressure. In the sink, sucrose is transported out of the sieve tubes and water follows, lowering the pressure.

ISBN: 9780170214094

13 Sexual Reproduction in Plants

The sexual cycle

As in animals and fungi, sexual reproduction in plants involves two essential events: *meiosis*, in which there is a reduction from the diploid to the haploid number of chromosomes, and *fertilisation*, in which two haploid gametes join to form a diploid zygote. Between them these two processes bring about new combinations of genes, and thus genetic variation.

The life cycle of plants is quite different from that of animals. There are two multicellular generations that alternate (Fig. 13.1):

- A *diploid* **sporophyte**, each cell having two sets of chromosomes. The sporophyte produces spores by *meiosis*. When a spore germinates it develops into:

- A *haploid* **gametophyte**, each cell having one set of chromosomes. The gametophyte produces gametes by *mitosis* (in animals, gametes are produced by *meiosis*). After fertilisation the zygote develops into a new sporophyte.

All plants have life cycles with this **alternation of generations**, but the emphasis on the two generations differs markedly from one group to another.

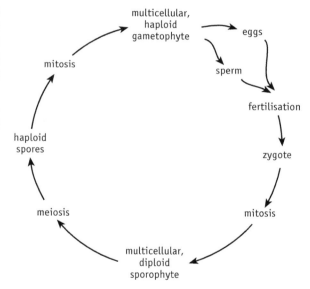

Fig. 13.1 Key events in the plant life cycle

The life cycle of *Funaria*, a moss

Mosses and liverworts belong to the phylum **Bryophyta**. The familiar moss plant is the haploid gametophyte. It consists of a stem with leaves that photosynthesise, and thread-like **rhizoids** that anchor the stem to the ground (Fig. 13.2). Moss rhizoids are quite different from roots, as each consists of only a few cells. The gametophyte is fully independent and persists from year to year.

Funaria is a moss common in glasshouses and on the sites of recent fires. The gamete-producing organs are at the tips of the shoots. The plant is *monoecious*, male and female organs being situated on different shoots on the *same* plant (some mosses are *dioecious*, male and female organs being

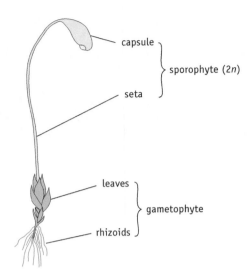

Fig. 13.2 A moss gametophyte and sporophyte

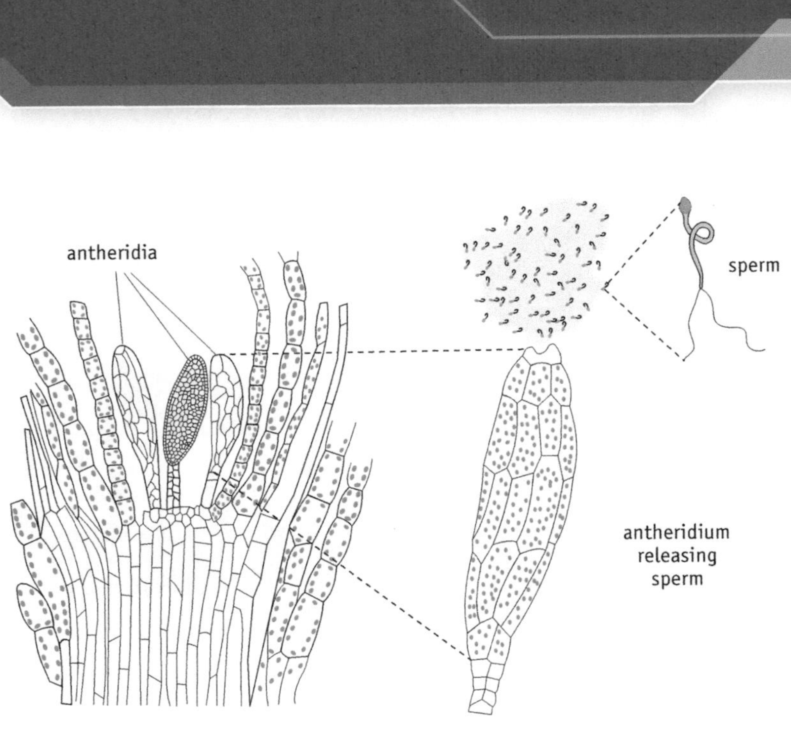

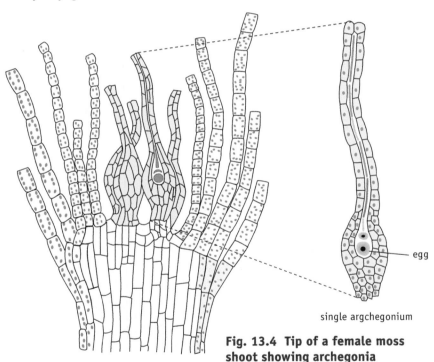

Fig. 13.3 Tip of a male moss shoot showing antheridia

Fig. 13.4 Tip of a female moss shoot showing archegonia

on *different* plants). The male organs or **antheridia** produce male gametes or sperm (sperm is both singular and plural). Since it is already haploid, it produces gametes by *mitosis*, so all the gametes produced by a given plant are genetically identical. In wet conditions each antheridium ruptures, releasing the sperm into the thin film of surrounding water (Fig. 13.3).

The sperm swim using flagella, and some may reach the bottle-shaped female organs or **archegonia** (Fig. 13.4). The base of each archegonium contains a female gamete or egg, which releases a chemical that attracts sperm. The first sperm to swim down the neck of the archegonium fertilises the egg to produce a diploid zygote.

By mitotic divisions the zygote grows into a multicellular sporophyte. This looks quite different from the parent gametophyte, consisting of a **capsule** on a long stalk that remains attached to the parent gametophyte (Fig. 13.5). For a while the capsule may be covered by a 'hat' formed from the remains of the archegonium.

The sporophyte has chlorophyll and can photosynthesise, though it is dependent on the gametophyte for water and minerals. The epidermis of the unripe capsule has a waxy, water-resistant cuticle and it has stomata that enable it to regulate water loss.

Inside the capsule, diploid cells called **sporocytes** undergo meiosis to produce haploid **spores** that are genetically different from each other. In dry weather, the teeth guarding the opening to the capsule bend outwards, releasing the spores to the wind. The sporophyte then dies.

A spore is the first stage of the gametophyte generation. If a spore lands in a suitably damp place it germinates, developing into a branching, thread-like *protonema*, which produces buds that grow into moss plants (Fig. 13.6).

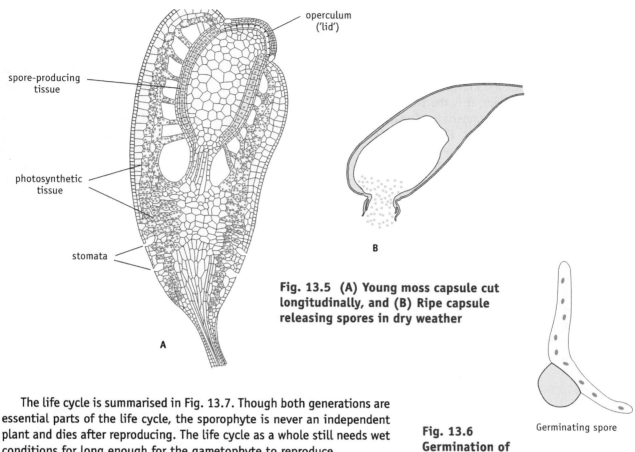

spore-producing tissue

photosynthetic tissue

stomata

operculum ('lid')

A

B

Fig. 13.5 (A) Young moss capsule cut longitudinally, and (B) Ripe capsule releasing spores in dry weather

The life cycle is summarised in Fig. 13.7. Though both generations are essential parts of the life cycle, the sporophyte is never an independent plant and dies after reproducing. The life cycle as a whole still needs wet conditions for long enough for the gametophyte to reproduce.

The gametophyte generation in plants is often referred to as the 'sexual' generation. Sexual reproduction involves both meiosis and fertilisation. Since neither can occur without the other, the two processes are complementary, and neither is more 'sexual' than the other. Rather than being confined to the gametophyte, the sexual process encompasses both generations, meiosis occurring in the sporophyte, and gamete production in the gametophyte.

Fig. 13.6 Germination of a moss spore

Germinating spore

buds that grow into moss plants

Protonema

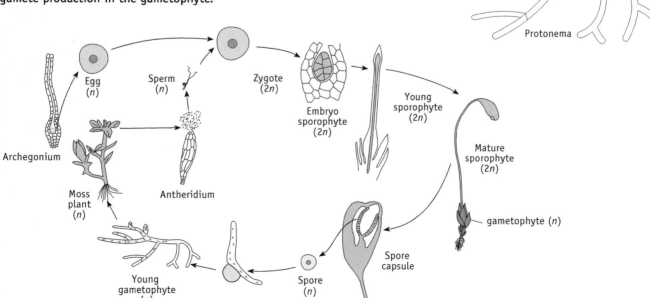

Egg (n)

Sperm (n)

Zygote (2n)

Embryo sporophyte (2n)

Young sporophyte (2n)

Mature sporophyte (2n)

gametophyte (n)

Archegonium

Moss plant (n)

Antheridium

Spore capsule

Spore (n)

Young gametophyte (n)

Fig. 13.7 Summary of moss life cycle

ISBN: 9780170214094

The life cycle of a fern

The life cycle of a fern differs from that of a moss in that the familiar fern plant is the sporophyte. This is fully independent and persists from year to year (Fig. 13.8). Though the gametophyte is also independent, it is short-lived, and is so tiny that it is seldom noticed.

On the underside of leaves or *fronds* are clusters of small sacs called **sporangia** (Fig. 13.9 and Fig. 13.10). Each cluster is called a *sorus*. Inside each sporangium, diploid sporocytes undergo meiosis to produce haploid spores. In most ferns each sorus is protected by a thin layer of tissue called an **indusium**, which shrivels as the sporangia ripen. The rim of each sporangium is called the **annulus**, and plays an important part in the release of the spores.

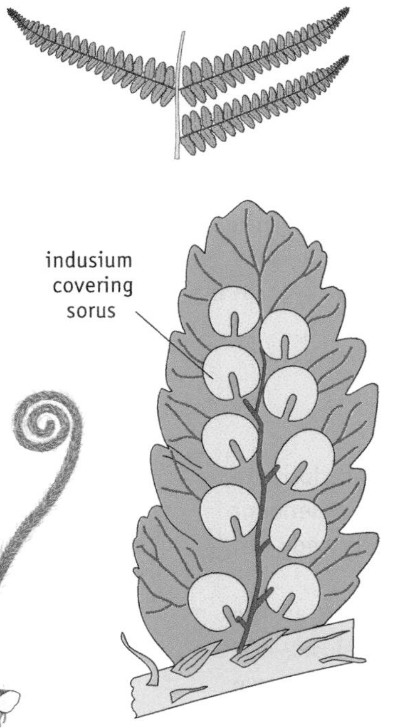

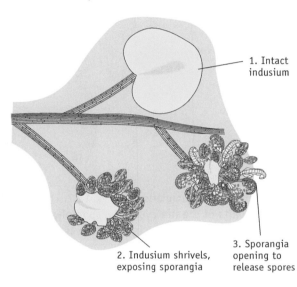

1. Intact indusium

2. Indusium shrivels, exposing sporangia

3. Sporangia opening to release spores

Fig. 13.10 Stages in the ripening of a sorus

indusium covering sorus

Fig. 13.8 A fern sporophyte, with parts of a frond enlarged

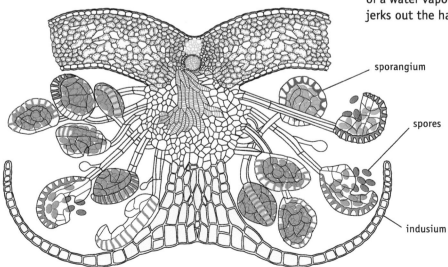

sporangium

spores

indusium

Fig. 13.9 Section through part of a fern frond showing sporangia

As the ripe sporangium dries out, tension develops in the cells of the annulus, causing it to bend back. There is a sudden release of tension caused by the formation of a water vapour-filled cavity in each annulus cell. This jerks out the haploid spores into the wind (Fig. 13.11).

If a spore lands in a damp place with enough light it germinates, developing into a small, thin, heart-shaped gametophyte (Fig. 13.12). It has neither cuticle nor stomata to restrict water loss, and is anchored by rhizoids rather than roots. Since it cannot survive drying, it is less well adapted to terrestrial conditions than the gametophyte of some mosses. Ferns thus require damp conditions for long enough for the gametophyte generation to be completed.

The antheridia and archegonia are produced on the underside of the gametophyte, where it is wettest (Fig. 13.13 and Fig. 13.14).

ISBN: 9780170214094

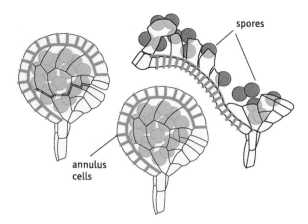

Fig. 13.11 Sporangium releasing spores

spores

annulus cells

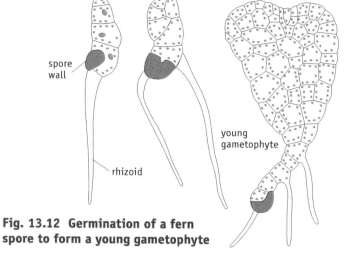

Fig. 13.12 Germination of a fern spore to form a young gametophyte

spore wall

rhizoid

young gametophyte

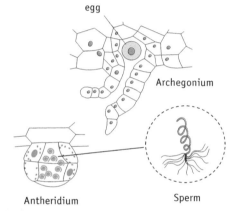

egg

Archegonium

Antheridium

Sperm

Fig. 13.14 Archegonium and antheridium of a fern

After being released from the antheridia, the sperm swim in the surface film of water to an archegonium, where the egg is fertilised. In most species the archegonia and antheridia ripen at different times so cross-fertilisation usually occurs, producing new combinations of genes. The zygote then undergoes repeated mitotic divisions, developing into a new sporophyte (Fig. 13.15).

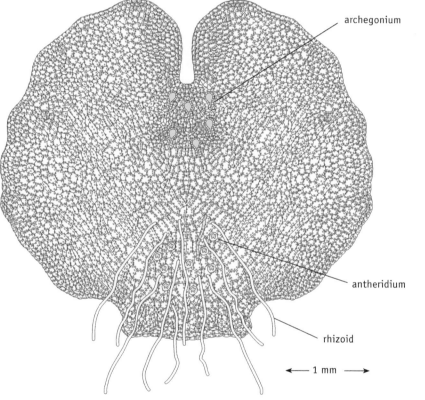

archegonium

antheridium

rhizoid

⟵ 1 mm ⟶

Fig. 13.13 Underside of a fern gametophyte

gametophyte

zygote in archegonium

embryo sporophyte

rhizoid

first root

first leaf

Fig. 13.15 Sections through fern gametophyte and sporophyte at early stages of development

Though it is at first nutritionally dependent on the gametophyte, the young sporophyte develops its own roots and leaves and becomes independent. The gametophyte then dies. The life cycle is summarised in Fig. 13.16.

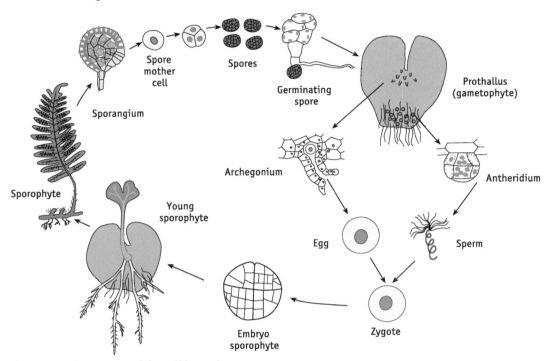

Fig. 13.16 Summary of fern life cycle

Ferns still need external water for the growth of the gametophyte and for fertilisation, but this dependence is limited to the few weeks needed to complete the gametophyte phase of the life cycle. The sporophyte can survive for year after year in conditions too dry for the gametophyte.

Extension: *Selaginella* — a link between ferns and seed plants

The colonisation of the land by plants involved an increase in the importance of the sporophyte relative to that of the gametophyte. In ferns the gametophyte, though small and short-lived, is still an independent plant. In conifers and flowering plants it is microscopic and never makes its own living. The change from fern-type life cycle to that of a conifer is so great that it helps if we take a brief look at a small plant called *Selaginella*, which bridges the gap (Figs 13.17–13.21).

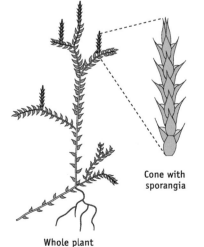

Cone with
sporangia

Whole plant

Microsporophyll
with
microsporangium

Megasporophyll
with
megasporangium

Fig. 13.17 Sporophyte of *Selaginella*

ISBN: 9780170214094

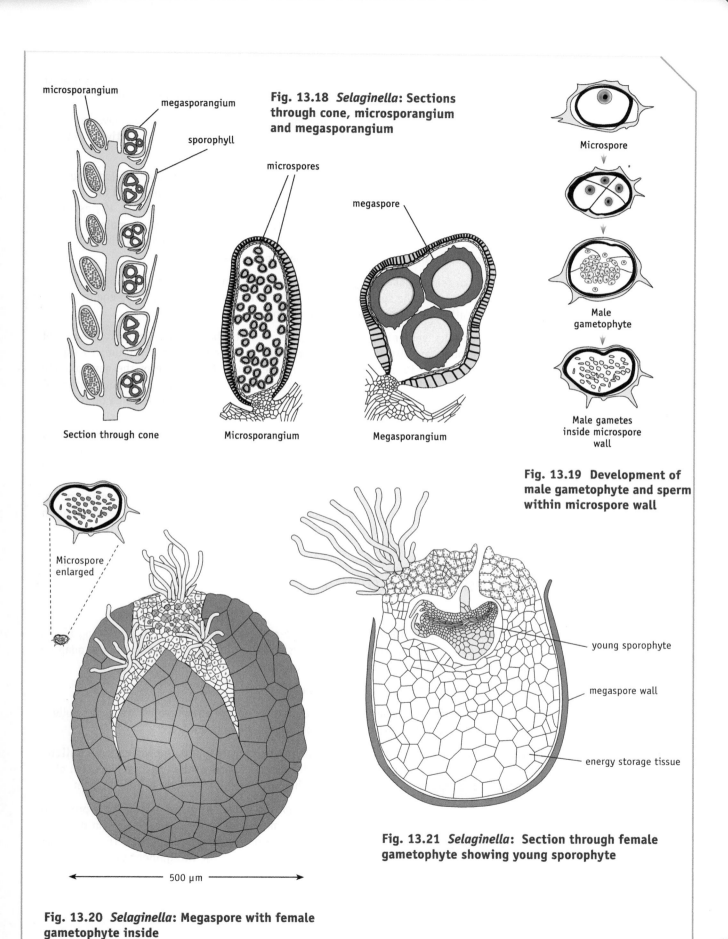

microsporangium

megasporangium

sporophyll

Fig. 13.18 *Selaginella*: **Sections through cone, microsporangium and megasporangium**

microspores

megaspore

Section through cone

Microsporangium

Megasporangium

Microspore

Male gametophyte

Male gametes inside microspore wall

Fig. 13.19 Development of male gametophyte and sperm within microspore wall

Microspore enlarged

young sporophyte

megaspore wall

energy storage tissue

Fig. 13.21 *Selaginella*: **Section through female gametophyte showing young sporophyte**

← 500 µm →

Fig. 13.20 *Selaginella*: **Megaspore with female gametophyte inside**

Selaginella is an introduced plant, widespread in New Zealand. It is a kind of 'living fossil'; its nearest relatives became extinct over 300 million years ago. It differs in three important ways from ferns:

- The leaves are of two kinds: ordinary photosynthetic leaves, and reproductive leaves or **sporophylls** that bear sporangia. The sporophylls are aggregated into reproductive shoots called **cones**.

- Whereas almost all ferns are **homosporous** (produce only one kind of spore), *Selaginella* produces two kinds and is thus **heterosporous**. Small **microspores** are produced in **microsporangia** and much larger **megaspores** are produced in **megasporangia**. Microspores develop into tiny **male gametophytes** and megaspores develop into larger **female gametophytes**. In most species of *Selaginella* there is only one megasporocyte in each megasporangium, so after meiosis only four megaspores are produced.

- Neither male nor female gametophyte is a fully independent plant. Both develop and produce gametes inside the spore wall while still within the sporangium, using energy reserves bequeathed to it from the parent sporophyte.

After release from the sporangia, the microspores fall to the ground. In most species of *Selaginella* the megaspores are also released. The wall of the megaspore partly ruptures to expose the female gametophyte with its archegonia. Although cells near the surface of the gametophyte may develop a few chloroplasts, the amount of photosynthesis is insignificant. If there is sufficient water, the microspore wall ruptures to release the sperm. These swim to the archegonia on a nearby female gametophyte. Usually only one zygote develops, using the energy reserves stored in the gametophyte. These reserves were received from the sporophyte parent of the female gametophyte, so the gametophyte has no nutritional role. Because the female gametophyte does not need to build up its own energy reserves by photosynthesis, it only needs wet conditions for the short period during which the sperm are swimming.

In some species of *Selaginella* the megasporangium opens without releasing the megaspores. For fertilisation to occur, microspores must land on or in the open megasporangium. Fertilisation in these species occurs on the parent sporophyte, as in seed plants.

The life cycle of a conifer

Whereas mosses and ferns produce microscopic, single-celled spores, conifers and flowering plants reproduce by producing much larger, multicellular *seeds*. The life cycle of seed plants is better adapted to life on land than that of ferns in three fundamental ways:

- Fertilisation does not depend on external water.

- Because it carries far more energy reserves than a spore, a seed can germinate successfully in the dark below ground, where water supplies are more reliable.

- Instead of releasing motile sperm, each pollen grain produces an outgrowth called a **pollen tube.** This grows through the nucellus and delivers the sperm to the egg. The evolution of the pollen tube thus frees seed plants from the need for external water for fertilisation.

The conifer life cycle also differs from that of *Selaginella* in that only *one* megaspore is produced from each megasporocyte (the other three products of meiosis die). The megasporangium does not open, so the gametophyte develops within the protection of the parent sporophyte. In seed plants the megasporangium is called the *nucellus*. A collar of tissue called the **integument** grows round the megasporangium, enclosing it except at a tiny hole called the **micropyle**.

The megasporangium and its enclosed female gametophyte, together with the integument, make up an **ovule**. In pines there are two ovules on each megasporophyll (Fig. 13.22).

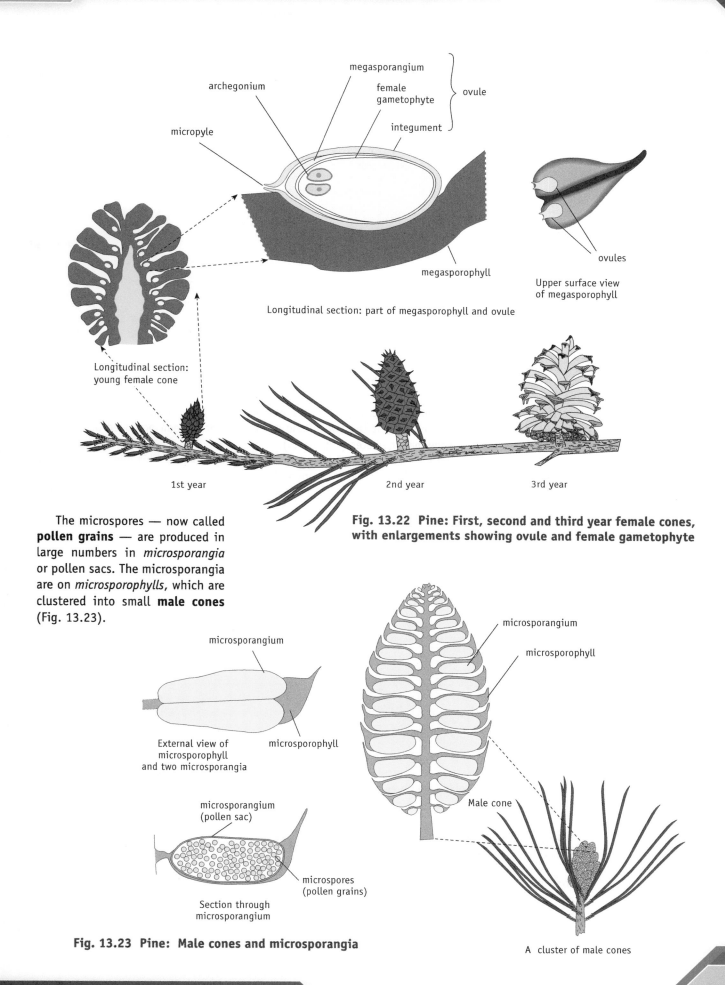

archegonium

micropyle

megasporangium

female
gametophyte

integument

ovule

Longitudinal section: part of megasporophyll and ovule

megasporophyll

ovules

Upper surface view
of megasporophyll

Longitudinal section:
young female cone

1st year

2nd year

3rd year

**Fig. 13.22 Pine: First, second and third year female cones,
with enlargements showing ovule and female gametophyte**

The microspores — now called **pollen grains** — are produced in large numbers in *microsporangia* or pollen sacs. The microsporangia are on *microsporophylls*, which are clustered into small **male cones** (Fig. 13.23).

microsporangium

External view of
microsporophyll
and two microsporangia

microsporophyll

microsporangium
(pollen sac)

microspores
(pollen grains)

Section through
microsporangium

microsporangium

microsporophyll

Male cone

A cluster of male cones

Fig. 13.23 Pine: Male cones and microsporangia

ISBN: 9780170214094

After the pollen is shed it is blown by the wind to the female cones. The male gametophyte is reduced to a few cells inside the pollen grain. Before release from the microsporangium, the nucleus of each microspore divides mitotically several times. The resulting cells are the vestigial male gametophyte (Fig. 13.24).

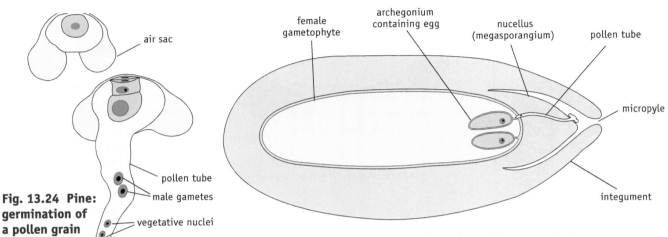

Fig. 13.24 Pine: germination of a pollen grain (microspore)

Fig. 13.25 Pine: Section through ovule at time of fertilisation

If a pollen grain lands on an ovule, it germinates by producing a *pollen tube.* This grows into the female gametophyte, carrying with it the two of the nuclei that have been produced by the male gametophyte; these are the male gametes (Fig. 13.25).

One of the male gametes joins with the female gamete, forming a zygote. This divides repeatedly to form an embryo plant that is the young sporophyte. It grows using organic compounds it imports from the parent sporophyte. The young sporophyte is still surrounded by the integuments, which form the seed coat or **testa**. During the development of the seed, it develops a wing-like outgrowth that helps dispersal by wind (Fig. 13.26).

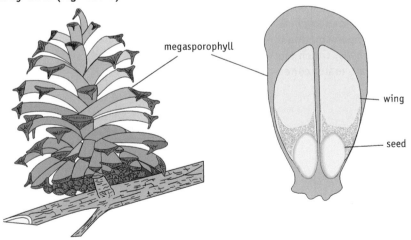

Fig. 13.26 Pine Left: Female cone at time of dispersal. Right: Ripe seeds on megasporophyll.

The life cycle of the flowering plant

The flowering plant life cycle differs from that of a conifer in that the megasporophylls surround and protect the ovules, forming an **ovary**. This ripens to form a **fruit**. In flowering plants, each megasporophyll is called a *carpel* (Fig. 13.27). The **microsporophylls** are called *stamens*. In addition, other leaves have become part of the reproductive shoot as **petals** and **sepals**. The entire reproductive shoot is called a **flower**.

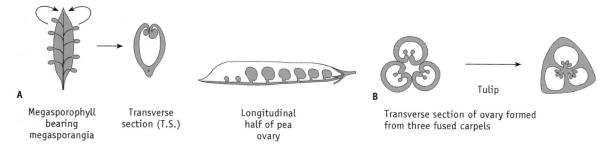

A   Megasporophyll bearing megasporangia   Transverse section (T.S.)   Longitudinal half of pea ovary   Tulip   B   Transverse section of ovary formed from three fused carpels

Fig. 13.27 Two ways in which an ovary has been formed in evolution: (A) from one carpel, as in the pea, or (B) by fusion of three carpels, as in the tulip. A carpel is actually a reproductive leaf (megasporophyll).

Pollination

As in conifers, fertilisation is preceded by pollination. In flowering plants this is the transfer of pollen from the male part of a flower to the female part of a flower on the same plant or a different plant (Fig. 13.28). In cross-pollination the pollen is transferred between different plants, and leads to *outbreeding*. In self-pollination pollen transfer is to a stigma on the same plant (though often to a different flower), and leads to *inbreeding*. Each kind of pollination has costs and benefits, and the balance of advantage depends on ecological circumstances.

In New Zealand, pollination may be by animals (mainly insects), wind or, in a few cases, by water.

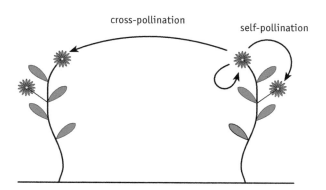

cross-pollination   self-pollination

Fig. 13.28 Self- and cross-pollination

Insect-pollination

Successful insect-pollination requires three key features:

- Insects must be 'rewarded' with some form of food. This is usually *nectar*, a solution containing various combinations of sucrose, glucose and fructose. Some flowers provide food in the form of pollen, which is rich in protein. By far the most effective pollinators are bees, because they have to collect food for their larvae as well as themselves. Some bee-pollinated flowers make no nectar, relying entirely on pollen as a lure — for example broom, gorse, and poppies.

- The flowers must advertise their presence, usually by being conspicuous and/or by producing some kind of scent.

- The pollen must be sticky so as to adhere to the insect's body. This is quite accidental so far as the insect is concerned.

Some flowers have quite complex structures, and it is better to begin with a simple example, the buttercup.

The buttercup

The tip of the flower stalk is the **receptacle**, to which the floral parts are attached (Fig. 13.29).

Outermost are the *sepals*, which protect the inner parts in the bud stage, and collectively form the **calyx**. When the flower opens these are bent back to reveal the five large *petals*, collectively called the **corolla**. To humans they look a uniform yellow, but there are lines radiating from the base of each petal that reflect ultraviolet light, which most insects can see. These lines are called *nectar guides*. They help to guide the insect to the nectary at the base of each petal. This is a small flap of glandular tissue that

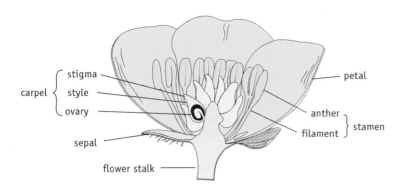

Fig. 13.29 Vertical half of a buttercup flower

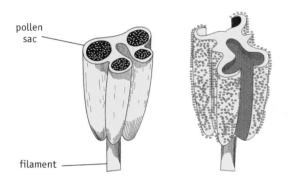

Fig. 13.30 Buttercup: an anther cut transversely, before and after shedding pollen

secretes nectar. The nectar is easily accessible to insects with a short proboscis (plural, *proboscides*) such as small beetles and flies.

Inside the corolla is the male part of the flower, the **androecium**, which consists of many **stamens**. Each stamen consists of an **anther**, held up by a long stalk or **filament** (Fig. 13.30). The anther contains four **pollen sacs** (microsporangia), in which pollen grains are produced by meiosis. When the anther is ripe it splits down the middle to release the pollen, some of which sticks to the bodies of insects feeding on the nectar.

The centre of the flower forms the **gynoecium**, or female part of the flower. It consists of many **carpels**, which are the female equivalent of stamens. Each consists of the three parts: *stigma*, *style*, and *ovary* (Fig. 13.31). The **stigma** not only receives pollen adhering to the body of insects, but also secretes sucrose, which stimulates pollen grains to germinate. The **style** is a short stalk holding the stigma up. Each **ovary** houses a single **ovule**, which develops into a seed.

The buttercup is often said to be a 'typical' flower, but in reality there is no such thing, as flowers vary enormously in almost all their parts. In dicotyledons the floral parts are usually in multiples of four or five, while in monocotyledons they are usually in multiples of three (Fig. 13.32).

In the tulip and many other monocotyledons there appear to be six petals in two layers or 'whorls' of three. Only the inner whorl is the equivalent of the petals of the buttercup. The outer

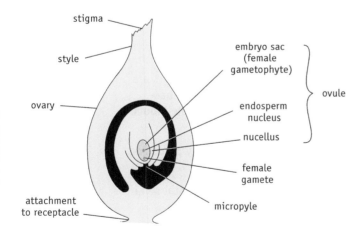

Fig. 13.31 Buttercup: Vertical section through a carpel

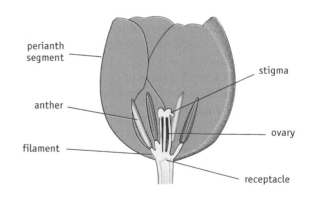

Fig. 13.32 Vertical half of a flower of tulip, a monocotyledon

whorl is the equivalent of the sepals, but they resemble petals. These two whorls together make up the *perianth*. In most dicotyledons the two whorls are different and are given different names (calyx and corolla). In tulips and many other monocotyledons the two whorls are similar and function as petals.

Flower parts may be separate (as in buttercup) or various parts may be joined together. Some flowers are clustered into **inflorescences**, e.g. dandelion (Fig. 13.33). In this plant each 'petal' is actually the corolla of a tiny flower or *floret*, the five petals being fused into a single strap-shaped structure. The sepals are highly modified as tiny hairs that function as a parachute to disperse the fruit.

ISBN: 9780170214094

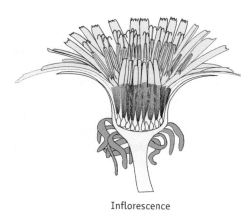

Fig. 13.33 Vertical half of a dandelion inflorescence, with one floret enlarged

Inflorescence

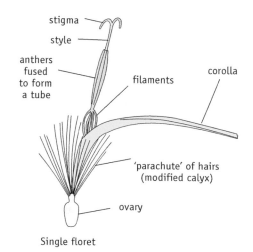

Single floret

Concealing the reward

Many flowers have their nectaries situated at the end of long tubes. As a result only insects with a long proboscis or 'tongue' (e.g. bees, butterflies and moths) can reach it. Two examples are shown in Fig. 13.34. It pays bees to concentrate on flowers with concealed nectar, since they have an advantage over insects with short tongues. It is also advantageous to the plant since bees visiting them are likely to carry pollen of the same species. This is an example of *co-evolution*, in which two species influence each other's evolution.

Some flowers restrict the range of insect visitors even more. The sweet pea and many other members of the legume family secrete their nectar into a trough formed from the filaments of nine of the ten stamens (Fig. 13.35). To reach the nectar an insect needs more than a long proboscis. Having landed on the wing petals, it must be heavy enough to depress the keel petals to which the wing petals are attached. Only bumblebees can do this.

The flowers of the sweet pea and other legumes are bilaterally symmetrical or **zygomorphic**. This means that they can be cut into two equal halves in only one plane. It forces a visiting insect to take up the same position relative to the flower. As a result the same part of the insect's body touches the stigma and stamens, so less pollen is wasted.

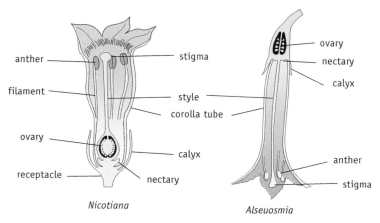

Nicotiana

Alseuosmia

Fig. 13.34 Vertical half of a flower of *Nicotiana*, a common garden flower, and toropapa (*Alseuosmia*), a native shrub. Both are pollinated by moths, which have a very long proboscis.

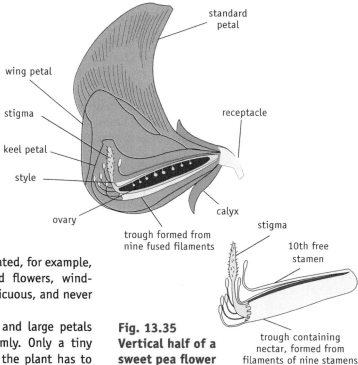

Fig. 13.35 Vertical half of a sweet pea flower

Wind-pollination

Like all conifers, many flowering plants are wind-pollinated, for example, grasses. In contrast to the showy animal-pollinated flowers, wind-pollinated flowers are usually small, green and inconspicuous, and never produce nectar or scent.

The energy saved by not producing nectar, scent and large petals is offset by the fact that wind carries pollen randomly. Only a tiny proportion of the pollen grains find their target, and the plant has to produce large quantities (a single ryegrass anther produces about 20 000 grains). Since the pollen grains are very small they have a high surface to mass ratio, so they fall slowly and are easily caught by the wind. (Conifer pollen has air sacs that add extra surface with very little extra mass.) They are also non-sticky so they do not adhere to leaves, branches and other obstacles. Many wind-pollinated trees produce their flowers before the leaves open, thus helping free air movement.

Fig. 13.36 shows a flower of ryegrass, a wind-pollinated flower. The inflorescences are produced on long stems, well above the leaves, so the wind is able to catch them. Each flowering stalk bears many miniature inflorescences or **spikelets**, each consisting of several tiny flowers. A grass flower has three stamens, each with its filament attached midway along the anther so that it rocks in the wind, shaking out the pollen. The stigmas are feathery, exposing a large surface to the wind. Even so, the chances of a stigma intercepting more than one pollen grain are not high, and like most wind-pollinated flowers, grasses have only one ovule in each ovary. Other examples of wind-pollinated flowers are plantains (Fig. 13.38) and *Coprosma* spp. (Fig. 13.39).

The differences between wind- and insect-pollinated flowers are summarised in Table 13.1.

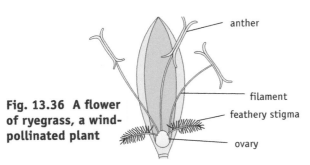

Fig. 13.36 A flower of ryegrass, a wind-pollinated plant

anther

filament

feathery stigma

ovary

Typical insect-pollinated flower	Typical wind-pollinated flower
Flowers large and conspicuous	Flowers small and green
Usually produce nectar and scent	Never produce nectar or scent
Anthers do not pivot on filaments	Anthers usually pivot on filaments
Pollen sticky	Pollen non-sticky
Pollen grains relatively large	Pollen grains very small
Smaller amounts of pollen	Larger amounts of pollen
Stigmas not usually feathery	Stigmas feathery or hairy

Table 13.1 Comparison between insect-pollinated and wind-pollinated flowers

Water-pollination

Some flowering plants live permanently submerged, and of these, a few are pollinated by water. An example is eelgrass (*Zostera novazelandica*), a flowering plant (though not a member of the grass family). It grows on sandy and muddy shores below low tide mark, where it may form extensive 'meadows' (Fig. 13.37).

The greatly reduced male and female flowers are produced in clusters on the same shoot, partly concealed by the leaf bases. Each female flower consists of an ovary and two stigmas, and each male flower consists of an anther. Female flowers ripen a few days before the male flowers. When ripe, each

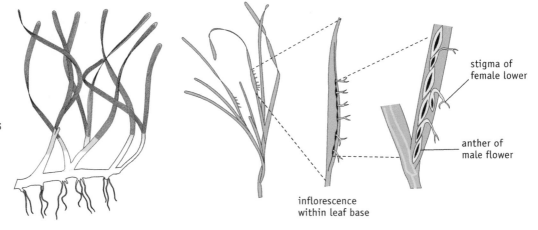

Fig. 13.37 Eelgrass (*Zostera*), a water-pollinated plant

stigma of female lower

anther of male flower

inflorescence within leaf base

ISBN: 9780170214094

anther bursts to release the pollen into the water. Each pollen grain is thread-like, about 2 mm long and 3 m wide. Since the pollen threads have the same density as water, they float freely. On coming into contact with a stigma, a pollen grain curls around it and germinates.

Preventing inbreeding

Darwin's statement that 'nature abhors self-fertilisation' was based on the fact that many plants have adaptations that reduce the likelihood of self-pollination. As explained in Chapter 15, inbreeding reduces genetic variability, producing a high proportion of homozygotes. Plants have several of ways of reducing the chances of self-pollination.

Many flowers are **dichogamous**, male and female parts maturing at different times. Most flowers are **protandrous**, the anthers shedding their pollen before the stigma is ripe. Some, such as plantains and eelgrass, are **protogynous**, the gynoecium ripening first (Fig. 13.38).

Although this makes cross-pollination more likely, it does not usually ensure it, for two reasons. First, dichogamy in most plants is incomplete, with a period of overlap when the flower is functionally hermaphrodite. Second, most plants have flowers of differing ages, enabling pollen to be transferred from flowers that are functionally male to flowers that are functionally female.

Many plants are **monoecious**, with unisexual flowers of both types on the same plant, as in beech, nikau palm and sweetcorn. Others are **dioecious**, each plant producing flowers of only one sex, for example kiwifruit, pepper tree and *Coprosma* (Fig. 13.39). In these species, only female plants produce berries.

Some plants are **self-incompatible**, pollen growing more slowly, or not at all, on a stigma of the same plant, e.g. clover, apples and pears.

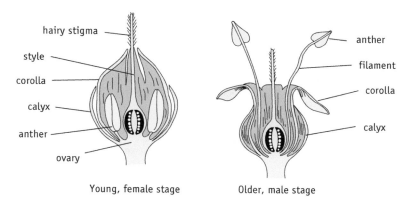

Fig. 13.38 Younger and older flower of plantain, a wind-pollinated flower

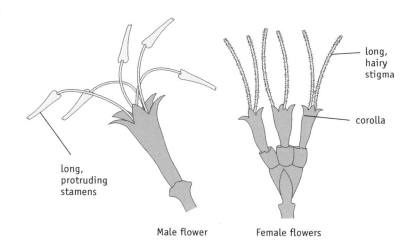

Fig. 13.39 Male and female flower of *Coprosma robusta*, a wind-pollinated, native shrub

The formation of pollen

Pollen grains (microspores) are produced within the four pollen sacs (microsporangia) of each anther. Each pollen sac of a young stamen contains many diploid pollen mother cells or **microsporocytes**. These undergo meiosis to produce the pollen grains or microspores.

A wall of a mature pollen grain is two-layered. The inner **intine** consists of a continuous and unmodified cellulose wall. The outer **exine** is waterproof and very resistant to decay. It is perforated by a number of pores and is sculpted in a manner characteristic of the species.

The wall of the anther contains a fibrous layer in which the cells have unevenly thickened walls. As the anther ripens, tensions are set up as the fibrous layer dries, causing it to split longitudinally along two lines, releasing the pollen.

As the anther ripens, the haploid nucleus of the pollen grain divides mitotically into a **generative nucleus** (so-called because it later gives rise to the male gametes), and a **vegetative nucleus**, which is the vestigial male gametophyte.

The ovule and egg

An ovule is a potential seed, and it develops on a part of the ovary wall called the **placenta**. It originates as a swelling called the **nucellus**. This becomes partially surrounded by two collars of tissue called **integuments**, with a pore or **micropyle** (Fig. 13.40).

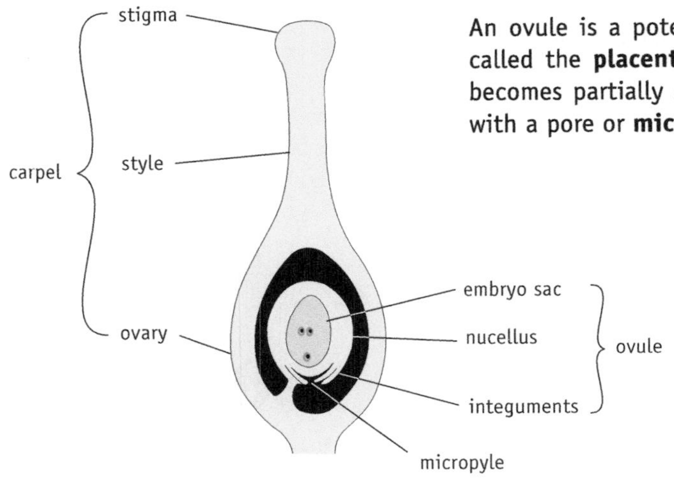

The nucellus is actually a **megasporangium**. Inside it develops a megaspore mother cell or **megasporocyte**, which undergoes meiosis to produce four potential **megaspores**.

One of the four haploid products of meiosis becomes a megaspore; the other three die. The megaspore then undergoes three consecutive mitotic divisions to produce eight haploid nuclei. These, and the surrounding cytoplasm, form the **embryo sac**. This is the greatly reduced **female gametophyte**.

Two of the eight nuclei then move to the centre of the embryo sac and are called **polar nuclei**. Of the other nuclei, the one nearest the micropyle is the **egg**. The others appear to have no function.

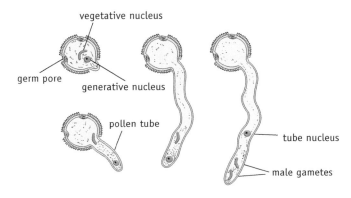

Fig. 13.40 An ovary and ovule in a flowering plant (only three of the eight nuclei in the embryo sac are shown)

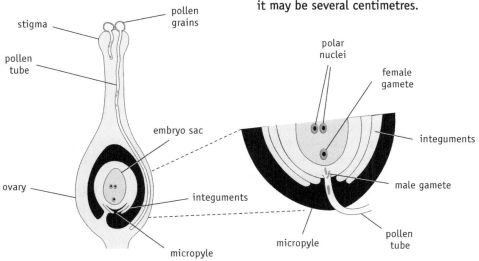

Fig. 13.41 Germination of a pollen grain

Fertilisation

Pollination merely brings the pollen to a stigma — the egg is still some distance away inside an ovule. The next stage is the growth of the **pollen tube**, which brings the sperm to the egg — a process that takes several days.

Stimulated by sugar secreted by the stigma, the pollen tube grows out through one of the pores in the outer layer of the pollen grain (Fig. 13.41). The tube grows down into the stigma and style, secreting digestive enzymes as it goes. The simple products of digestion (such as sugars and amino acids) are absorbed by the pollen tube and used for growth. The length to which the pollen tube grows depends on the length of the style. In a buttercup the stigma is little more than a millimetre away from the ovule, but in some species it may be several centimetres.

Fig. 13.42 Fertilisation in a flowering plant

ISBN: 9780170214094

During its growth down the style, the generative nucleus divides mitotically into two *male gametes*, each of which retains a thin envelope of cytoplasm.

When the pollen tube reaches the ovary, it grows towards an ovule, attracted by chemicals produced by the egg. The pollen tube usually grows through the micropyle. On reaching the egg the tip breaks down, releasing the two sperm (Fig. 13.42). One joins with the egg to form a **zygote**, and the other fuses with the two polar nuclei to form a triploid **endosperm nucleus**. This *double fertilisation* is unique to flowering plants. In most plants the ovary contains many ovules, so many pollen grains will be needed if they are all to develop into seeds.

The formation of fruit and seed

After fertilisation the primary endosperm nucleus divides many times to form a spongy storage tissue called the **endosperm**. Though all seeds have an endosperm to begin with, in *non-endospermic* seeds it disappears early in development, to be replaced by energy and raw materials stored in the embryo itself. In *endospermic* seeds, such as cereals (e.g. rice, corn) the endosperm continues to develop and forms a large part of the mature seed.

Meanwhile the zygote has been dividing mitotically and develops into the **embryo**. This has three parts: the **radicle** or embryonic root, the two embryonic leaves or **cotyledons** (one in monocotyledons) and the embryonic shoot or **plumule**.

Pollination and fertilisation also trigger a number of other changes:

1. The integuments continue to grow and become the seed coat or **testa**.

2. The ovary grows into the **fruit**, and the ovary wall becomes the fruit wall or **pericarp**.

3. In most flowers the sepals, petals and stamens fall off.

The developing seed lays down large amounts of nutrient in the form of protein, and either fat or starch. In most seeds the final stage of development is marked by a drastic decrease in water content to about 5–20% and the cessation of metabolic activity. These changes are a necessary preparation for dispersal, in which the seed has to travel through conditions hostile to growth.

Fruits

A fruit is usually defined as a ripened ovary, and in many plants it plays an important part in the dispersal of the seeds. According to the way the pericarp develops, fruits fall into two categories:

1. In **succulent fruits** the pericarp becomes soft and juicy and is eaten by animals.

2. In **dry fruits** it becomes quite tough and is not eaten.

Dry fruits containing more than one seed are usually **dehiscent** — that is, they open to release the seeds. In these fruits the seeds are dispersed individually. **Indehiscent fruits**, which do not open, usually contain only one seed. Some fruits are formed from parts of the flower other than the ovary and are called 'false fruits'. A strawberry, for example, is a greatly swollen receptacle, the actual fruits being the little seed-like structures on the surface. An apple is formed by the ripening of a deeply cup-shaped receptacle, the actual ovary being the 'core'.

Dispersal of fruits and seeds

In most plants the survival chances of seeds are increased by *dispersal*. This always involves a considerable element of chance, and many do not reach a suitable habitat. The successful ones do not have to compete with each other or with the parent for resources, and there is always a chance that some seeds may reach a better environment than that of the parent.

Dispersal by wind

Wind-dispersed fruits and seeds have one essential feature in common: they have a large surface/mass ratio (Fig. 13.43). In most fruits and seeds this is achieved by a complex shape in the form of wings (e.g. sycamore) or feathery outgrowths (e.g. dandelion). Although these outgrowths may be superficially similar, in some cases they have evolved by modification of different structures. In willowherb, for example, the parachute is an extension of the testa, and in dandelion it is a highly modified calyx called a **pappus**.

In orchids a high surface/mass ratio results from the minute size of the seeds, which are blown about like dust. With masses as small as 0.005 mg, orchid seeds do not carry enough stored energy to get established by themselves. Instead they obtain energy provided by a saprobic fungus, with which the orchid lives in mutualistic relationship. Pohutukawa and manuka seeds are not as small as orchid seeds, but their elongated shape further increases their surface area/mass ratio.

Wind-dispersal of seeds and fruits is obviously wasteful, but this is offset by the fact that because the seeds or fruits have to be light, larger numbers can be produced without great cost.

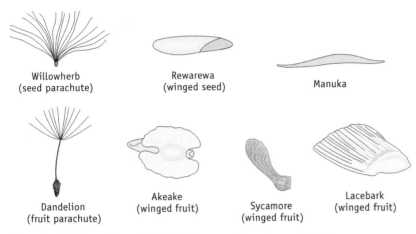

Fig. 13.43 **Examples of wind-dispersed seeds and fruits**

Dispersal by attachment

Dispersal by attachment to animals is somewhat less wasteful than wind-dispersal because animals tend to frequent areas supporting plant growth. In pre-human New Zealand flightless birds such as kiwi, weka and moa were probably important in the dispersal of a number of kinds of fruits seeds. Hooks can be developed in various ways. In hook 'grass' (actually a sedge) the hook is the tip of the tiny stem to which each flower is attached. Bidibid has barbs projecting from the base of the joined sepals. In Onehunga weed, the style hardens to form a spine that becomes embedded in animals' feet. (Fig. 13.44 shows examples of fruits dispersed in this way.)

Fig. 13.44 **Examples of fruits dispersed by attachment to animals**

Dispersal by being eaten

Many fruits are nutritious and are eaten by birds, but the seeds pass unharmed through the gut, to be deposited when the bird defaecates several hours later (Fig. 13.45). For example, the fruits of the puriri, *Coprosma*, fivefinger, supplejack and nikau palm are all favourite food of kereru (native pigeon). These fruits advertise themselves either by their bright colours or, in some fruits, being shiny blue-black. Once eaten the seeds are protected from the digestive juices of the animal by some form of hard coat. In berries such as cabbage tree, fuchsia, and tomato this is the seed coat. In puriri and other fruits with a 'stone', it is the inner layer of the pericarp. The food content of succulent fruits represents a cost to the plant and puts a limit on the number of fruits it can produce.

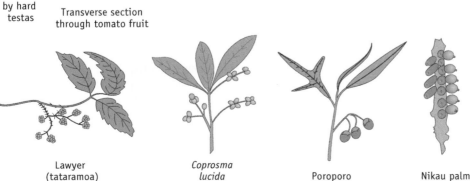

Fig. 13.45 **Examples of fruits dispersed as a result of being eaten by animals**

ISBN: 9780170214094

Water dispersal

The native mangrove is dispersed by water (Fig. 13.46). The seeds begin to germinate before they are released from the fruit, and float in the water, eventually coming to rest. In the coconut the pericarp consists of many fibres which contain air, so the fruit floats. The outer layer of the pericarp is waterproof, so the fruit does not become waterlogged and can be carried for hundreds of kilometres. (The coconut you buy has had the fibrous layer removed.)

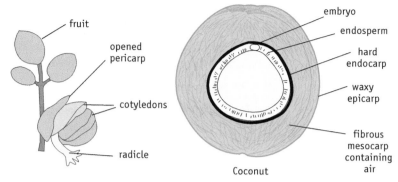

Fig. 13.46 Examples of dispersal by water

Self-dispersal

In some dehiscent fruits the pericarp is so constructed that it warps as it dries out. Tensions are set up, which are released suddenly with the violent opening of the fruit, jerking out the seeds, up to several metres in lupin (Fig. 13.47). Explosive dispersal is rare among native plants, being known only in matagouri, a spiny shrub.

Pepper-pot dispersal

Another form of dispersal is the censer, or 'pepper-pot' mechanism, in which seeds are jerked out of a fruit such as a poppy when knocked by a passing animal (Fig. 13.48). Wind is probably too gentle to jerk the stems sharply enough.

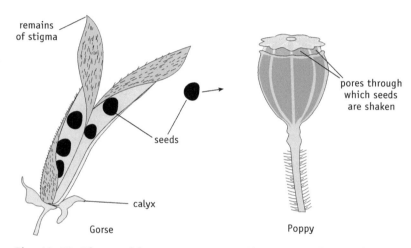

Fig. 13.47 Dispersal by explosion

Fig. 13.48 Dispersal by pepper-pot mechanism

Structure and germination of seeds

A flowering plant leaves its parent as a very young plant or embryo. In most plants the embryo ceases metabolic activity before it leaves the parent. In many seeds the water content falls to less than 15% before dispersal.

Since most seeds germinate in darkness, they require a store of organic materials. This satisfies two requirements:

- *building materials* in the form of **protein**, and

- a store of *energy*.

In some the energy is stored in the form of **starch** (cereals, beans, and peas). Many others store **fat**, for instance *Astelia* (perching lily) and sunflower. The advantage of this is that gram for gram, fat contains more than twice as much energy as carbohydrate. Storing fat thus saves weight, which is important in wind-dispersed seeds.

Fig. 13.49 Structure of a pea seed

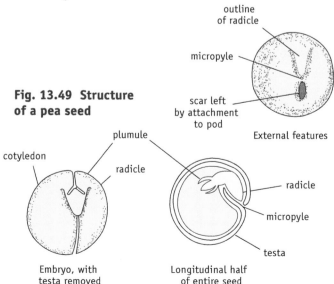

The garden pea

A pea has a thin outer coat or *testa*, and has a scar left by its attachment to the fruit (Fig. 13.49).

Just next to the scar is a tiny hole, the **micropyle**. After soaking a pea in water for 24 hours the testa becomes soft and can easily be removed, revealing the embryo inside. This consists of three parts.

- Two very large seed leaves or *cotyledons*. These do not look like leaves; they are very thick because they are swollen with stored starch and protein.

- The young root or *radicle*.

- The young shoot or *plumule*, tucked in between the cotyledons.

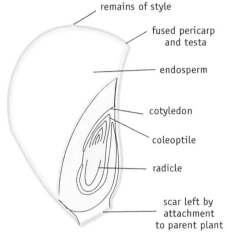

Fig. 13.50 Longitudinal section of a maize grain

Maize: a monocotyledonous 'seed'

Like all cereals, a maize grain is a one-seeded fruit, the outer layer being the pericarp and testa fused together (Fig. 13.50). Cereals store energy in the **endosperm**. This is packed with starch, and its outer layer stores protein.

Becoming independent

When a seed leaves the parent it takes with it a 'savings account' of stored energy bequeathed to it by the parent. To become independent the young plant has to use this 'capital' to produce roots and shoots to generate its own 'income'. This early phase of growth is called *germination*, and is one of the most hazardous periods in a plant's life. Because the seed has no control over where it lands, many seeds don't even reach this stage.

Different ways of germinating

Germination is the process by which the young plant emerges from the seed and becomes independent. Although the seed has an energy store, it depends on the soil for water.

The first stage of germination is the uptake of water, causing the seed to increase in mass. Once this has happened, the most urgent priority is to secure a reliable water supply, so the first visible change is the emergence of the radicle.

No matter which way the seed lands, the radicle grows downwards. This is **positive gravitropism**, or growing towards gravity (geotropism is an older term). As it reaches deeper soil where the water supply is more permanent, lateral branches grow outwards from the main root, at an angle to gravity.

Since the seed's energy supply is limited, the young plant must get its leaves up into the light as soon as possible. There are two different ways this can happen. In the garden pea (Fig. 13.51) and in some native plants such as karaka, tawa and kowhai, the plumule emerges from between the cotyledons and grows upwards. Since the cotyledons remain below the ground, this kind of germination is said to be **hypogeal** (hypo = below, geo = earth). Because the plumule grows away from gravity it is said to show **negative gravitropism**.

While it is growing through the soil the plumule is pale yellow with very small leaves. It is also hook-shaped, thus protecting the delicate leaves from being damaged by the soil. When it reaches the light the plumule straightens out and its leaves expand and become green. Once photosynthesis begins, the plant is independent. The cotyledons eventually wither and die as all the energy reserves are removed from them.

Fig. 13.51 Germination in the garden pea

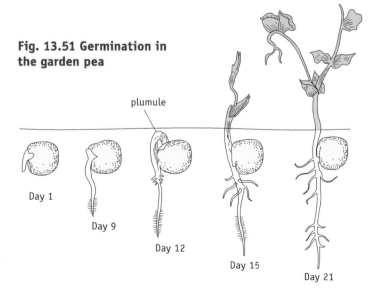

In the sunflower and in most native plants (Fig. 13.52), germination is said to be **epigeal** because the cotyledons are carried upwards above the ground (epi = above). In these seeds, the region just below the cotyledons — the **hypocotyl** — elongates, bringing not only the plumule but also the cotyledons up to the surface. The cotyledons then expand and become green, pushing off the pericarp. The plumule remains very small for some time, photosynthesis being carried out by the cotyledons.

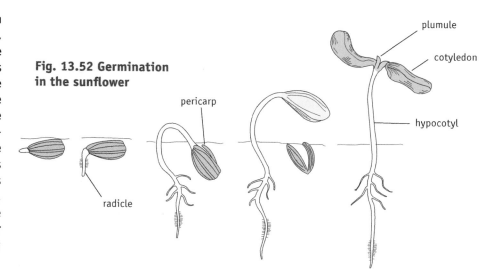

Fig. 13.52 Germination in the sunflower

plumule

cotyledon

pericarp

hypocotyl

radicle

Changes you cannot see: the physiology of germination

Some of the most important changes in germination are invisible. Soon after absorbing water, the energy reserves in the storage tissue are digested by **enzymes**. Starch is converted into *sugar*, and proteins are converted into *amino acids*. These are simpler, soluble substances that can be transported from the food store to the radicle and plumule where growth occurs. The amino acids are built up into new proteins and some of the sugar is used to make cellulose in new cell walls. The remaining sugar is used in respiration to supply energy. Fat is digested to fatty acids and glycerol, which are then converted to sugar.

Until the plumule reaches the light, the embryo depends on energy it received from its parent. Though the seedling is getting bigger, the amount of organic matter it contains (indicated by the ***dry mass***) is actually *decreasing*. The *wet mass* (live mass) continues to increase because the amount of water absorbed is greater than the amount of organic matter used in respiration.

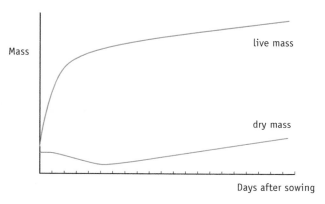

Fig. 13.53 Changes in live mass and dry mass during germination

Different kinds of life cycle

All organisms have the same 'aim' — to produce as many surviving offspring as possible — and they have different life cycle 'strategies' for achieving this. Some concentrate all their resources on rapid growth and early reproduction. Others grow more slowly and delay reproduction until a larger size has been reached.

Based on the length of their life cycles, plants can be divided into four groups:

- **Ephemerals** complete their life cycles in a few weeks, and under favourable conditions are capable of several generations a year. Ephemerals are 'opportunists', exploiting habitats that are only available for short periods of time, such a flower bed, or a desert after the heavy rain that may fall every few years. Examples of ephemerals are groundsel, shepherd's purse and many other 'weeds'.

- **Annuals** complete one life cycle per year. Like an ephemeral, an annual puts all its resources into a single reproductive effort, after which it dies.

- **Biennials**, such as cabbage and carrot, have a life cycle that spans two seasons. In the first year the plant does not flower but concentrates on building up its energy reserves. The following year it uses all its stored energy for flower and seed production, and then it dies.

- **Perennials** survive for many years and the process of resuming growth year after year is called *perennation*. Perennials are of two kinds. *Woody perennials* are trees and shrubs, in which the aerial parts remain alive in winter (though in deciduous plants they lose their leaves). Growth is resumed each spring from buds on the previous year's aerial shoots, which get taller each year. In these plants, energy reserves are stored above ground as well as below. In *herbaceous perennials* the aerial parts die back at the end of each year. Growth is resumed in spring from buds near or below ground level. In these plants energy reserves are stored below ground. Note that many plants have stems with 'woody' texture, but which die back at the end of the year, and so are strictly speaking herbaceous.

	Gametophyte	Sporophyte	External water requirement for fertilisation
Moss	Independent, long-lived, cannot retain water content in dry conditions	Never independent, short-lived, produces one kind of spore	Yes
Fern	Independent, short-lived, cannot survive dry conditions	Independent, long-lived, most produce one kind of spore	Yes
Selaginella	Develops apart from sporophyte but depends of energy received from it	Independent, long-lived	Yes
Conifer	Microscopic, develops inside the sporophyte	Independent, long-lived	No
Flowering plant	Microscopic, develops inside the sporophyte	Independent, long-lived	No

Table 13.2 Summarising some trends in plant life cycles

Summary of key facts and ideas in this chapter

○ In the plant life cycle there is an alternation of a diploid *sporophyte* (spore-producing) generation and a haploid *gametophyte* (gamete-producing) generation. In plants, spores are produced by *meiosis* and gametes are produced by *mitosis*.

○ In mosses the gametophyte is fully independent and survives indefinitely. Though the sporophyte can photosynthesise it depends on the gametophyte for water and minerals and is short-lived. The gametophyte can only reproduce in damp conditions, as external water is required for fertilisation.

○ In ferns the opposite is true; the sporophyte lives independently and indefinitely. Though the gametophyte is independent it lives for a limited period only. External water is required for fertilisation.

○ In conifers the megaspore is retained within the megasporangium, which is now called an *ovule*. The female *prothallus* develops within the ovule.

○ In conifers the *microspores* (pollen grains) are dispersed by wind. If one lands on an ovule it may germinate to produce a *pollen tube*. This grows down to the female gametophyte, carrying the male gametes with it. The evolution of the pollen tube removes the need for external water in fertilisation. After fertilisation the ovule develops into a *seed*.

○ In flowering plants the male gametophyte is reduced to two nuclei within the pollen grain, and the female gametophyte is reduced to the embryo sac within the ovule. The ovules are protected by highly modified leaves called *carpels*, and the ovules develop within the protection of an *ovary*. After fertilisation, this develops into a *fruit*.

○ In most flowering plants, pollen is transferred by wind or by animals such as insects (adaptations associated with each are described in the text).

○ *Self-pollination* leads to inbreeding and loss of genetic diversity, and most flowering plants have ways of avoiding this by *cross-pollination*.

○ After fertilisation the ovary wall develops into the fruit wall or *pericarp*, which may play a key role in dispersal of the seed or fruit. Dispersal may be by wind, attachment to animals, by being eaten by animals, or by water or by violent opening of the fruit.

○ Germination requires water, oxygen and a suitable temperature (and in some very small seeds, light).

○ Germination begins with the uptake of water and the digestion of energy reserves by enzymes, followed by the emergence of the radicle and then the plumule.

○ With the emergence of the plumule, photosynthesis begins and the plant begins to gain organic matter. Prior to this the seedling depends on energy reserves derived from the parent.

Test your basics

Copy and complete the following sentences. In some cases the first and last letters of a missing word are provided.

1. All plants have a life cycle consisting of two alternating phases: a h__*__ __*__ generation that produces gametes, and a d__*__ __*__ generation that produces __*__.

2. In mosses the generation that persists from year to year is the __*__. The male organs are the __*__ and the female organs are the __*__. Fertilisation depends on external __*__ in which the male gametes can __*__. After fertilisation the __*__ divides __*__ally and develops into the __*__. This plant can photosynthesise, but remains attached to and dependent on the __*__ for water and minerals. After it has produced __*__ by __*__ the plant dies. The spores are dispersed by wind. If a spore lands in a suitably __*__ habitat, it germinates to form a thread-like __*__. This develops buds, each of which can develop into a new moss plant.

3. In ferns the situation is opposite to that in mosses; the __*__ is the persistent generation, and is fully independent. Though the __*__ is independent, it only lives long enough to reproduce, and then dies. The __*__ of a fern produce sporangia on the undersides of their leaves, and inside each __*__ spores are produced __*__ by __*__ic divisions. The __*__ are dispersed by wind, and if one lands in a suitably damp habitat it germinates and grows into a tiny, heart-shaped __*__. This produces sex organs on the underside; male organs or __*__, and female organs or __*__. Male gametes swim to the female organs, and after fertilisation the zygote divides __*__ and grows into a __*__ plant. Though this depends on the __*__ at first for nutrition, it soon develops roots and becomes __*__.

4. Conifers are more independent of an external water supply than ferns, for two reasons. Whereas a fern plant produces single-celled __*__, conifers produce multicellular *seeds*. A seed contains far more energy reserves than a spore, so it can thus germinate deeper in the soil (i.e. without light), where the __*__ supply is more reliable. Also, the male gametes do not need __*__ water to reach the __*__ gamete; they are carried to the female gamete by a __*__ __*__.

5. A flower such as a buttercup consists of four layers (whorls) of modified leaves. The outermost layer consists of the __*__, which protect the flower in bud. Inside these are the brightly coloured __*__, which make the flower conspicuous to __*__. Inside these are the male organs, the __* Each has an __*__, inside which __*__ grains are produced by __*__ divisions. The innermost whorl is the female part of the flower, the __*__. In the buttercup this consists of many separate __*__, but in many flowers these are joined together or there may be only one of them. The tip of each carpel is the __*__ which secretes sugar necessary for __*__ grains to germinate. Below is a short stalk, the __*__, which connects with the ovary below. This contains an __*__ which has the potential to develop into a seed.

6. Before fertilisation, one or more __*__ grains must be transferred from anther to __*__, either on the same plant (a process called __*__-__*__), or a different plant (__*__-__*__). In the buttercup this is achieved by __*__, which are 'rewarded' by sugary __*__ secreted at the bases of the petals.

7. In some flowers __*__ is transferred by wind. These flowers are usually small and inconspicuous. The pollen grains are very __*__, non-sticky and produced in __*__ quantities.

8. After pollination, ___*___ secreted by the ___*___ stimulates each pollen grain to develop a pollen ___*___. Inside the pollen grain the nucleus divides ___*___, and two male gametes are produced. This grows down the ___*___, secreting digestive ___*___ as it does so.

9. Eventually the tip of the pollen tube reaches an ___*___. The tip ruptures, releasing the male gametes. One of them fuses with the ___*___ gamete in the ovule, producing a _____. The other joins with two other nuclei to form an ___*___ nucleus.

10. The zygote divides repeatedly and eventually develops into an ___*___ plant consisting of young root or ___*___, a young shoot or ___*___, and (in the buttercup) two tiny leaves or ___*___. The endosperm nucleus divides repeatedly to form an energy store or ___*___. In some seeds, such as peas and beans, the energy reserves are transferred from the endosperm to the ___*___ and the endosperm has disappeared by the time the seed is mature. The coats around the ovule develop into the seed coat or ___*___.

11. After fertilisation the ovary develops into a ___*___. The ovary wall develops into the ___*___ and each ovule develops into a ___*___. In ___*___ fruits the ___*___ is fleshy and eaten by animals. In *dry* fruits the pericarp is tough and inedible. Dry fruits containing more than one seed usually open to release them and are said to be ___*___.

12. Fruits and seeds that are dispersed by wind all have a high ___*___ to ___*___ ratio. This is usually achieved either by having wing-like or feathery outgrowths. Some seeds have a high ___*___/ ___*___ ratio because they are very ___*___, for example orchids.

13. Fruits and seeds may be dispersed by being eaten by animals, in which case the seed is protected from the digestive ___*___ of the animal. Such fruits are ___*___ and usually 'advertised' by their bright ___*___. Fruits may also become ___*___ to animals by hooks as the animal walks past.

14. A seed consists of an ___*___ plant and a store of ___*___, surrounded by a seed coat or ___*___. The embryo consists of a young shoot or ___*___, a young root or ___*___, and one or two young leaves or ___*___. At dispersal, the ___*___ content of most seeds is much lower than in active plant tissue, and metabolism is un___*___. All seeds store ___*___, and in addition, either ___*___ (e.g. cereals) or ___*___.

15. Germination begins with the absorption of ___*___, and a consequent increase in live ___*___. Next, ___*___ digest the energy reserves. Starch is converted to ___*___, and protein into ___*___ ___*___. In fat-storing seeds, the fat is converted first into ___*___ acids and ___*___, and then into carbohydrate.

16. The products of digestion are transported to the growing points in the r___*___ and ___*___, where they are used either as raw materials for ___*___, or as fuel for ___*___.

17. The _____ emerges from the seed first, followed by (in the pea) the ___*___. The radicle grows downwards, showing p___*___ ___*___, and the ___*___ grows upward, showing n___*___ ___*___.

18. Until the ___*___ reaches the light the ___*___ mass of the seed has been ___*___, but when photosynthesis begins, it ___*___.

14 Genetic Variation

In publishing his book *On the Origin of Species* in 1859, Charles Darwin was not the first to propose that life had evolved from simpler forms. He was, however, the first to suggest a plausible mechanism as to how it has occurred. He called his idea **natural selection** and based it on a series of observations (O) and deductions (D):

- All species have the potential to produce far more offspring than are needed to replace the parents (O).

- Over many generations the numbers of individuals in a species do not change much (O).

- Therefore most individuals must die without reproducing (D).

- Amongst the members of a species, no two individuals are exactly alike (O).

- Therefore some individuals are better adapted than others and are thus more likely to survive (D).

- To at least some extent, variation is inherited (O).

- On average, the better-adapted individuals will therefore leave more offspring than the less well-adapted ones (D).

- Over many generations, species will tend to change and become better-adapted (D).

Darwin's ideas had one major weakness — he did not understand the mechanism of inheritance (and neither did his contemporaries). Many people believed that the genetic material — the information handed on from parents to offspring — is a kind of fluid. Like fluids, the contributions of the two parents were thought to mix, so that the offspring would have characteristics half way between those of the parents. The trouble with this idea of *blending inheritance* was that variation should steadily decrease over the generations until all members of a species were the same. So how could variation — the raw material for selection — be maintained?

Darwin did not know it, but the beginnings of a solution to his problem emerged a few years after his book was published. It came from two quite different directions: by breeding experiments, and by the study of chromosomes and cell division.

Twenty years after Darwin's death it became apparent that the results of breeding experiments could be explained by the behaviour of chromosomes. The *Chromosome Theory of Heredity*, as it became known, drew largely on the behaviour of chromosomes in meiosis.

Sources of genetic variation

Genetic variation can arise in two quite different ways:

- Reshuffling existing genes into new combinations in *meiosis* and *fertilisation*.
- Creation of new genes as a result of *mutation*.

The next section in this chapter deals with meiosis as a source of variation.

Meiosis

Meiosis is more complicated than mitosis, but it is essential to an understanding of the way heredity works. The fine details are not so important — the important thing is its results. These differ from those of mitosis in two essential ways:

- The number of chromosomes is reduced from diploid to haploid.
- The daughter nuclei are genetically different from each other.

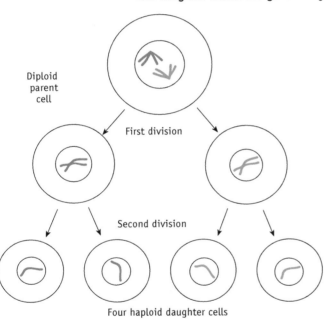

Diploid parent cell

First division

Second division

Four haploid daughter cells

Fig. 14.1 How meiosis reduces chromosome number

In animals, meiosis occurs in ovaries and testes and yields *gametes* (eggs and sperm). In flowering plants it occurs in the stamens and ovules and produces *spores*.

It is best explained in two stages. Fig. 14.1 shows how the chromosome number is reduced from diploid to haploid. Only one pair of chromosomes is shown; the other pairs behave similarly.

In meiosis there are two divisions, called meiosis I and meiosis II. In meiosis I the two members of each homologous pair move to opposite poles of the spindle. This 'parting of the ways' is called **segregation**. Each daughter nucleus therefore only receives *one* of each pair. Another point to note is that in meiosis I the segregating chromosomes are *double*-stranded.

Now for the second consequence of meiosis — *variation*. To understand how this happens, we need to consider more than one pair of chromosomes (Fig. 14.2). Each chromosome pair segregates *independently* of every other pair. This means that there are two different ways in which two pairs of chromosomes can segregate. An animal cell with two pairs of chromosomes can thus produce four kinds of gamete, and a plant cell could produce four kinds of spore. Every additional chromosome pair doubles the number of possible kinds of product. A human cell (23 pairs) could produce 2^{23} = over 8 million kinds of gamete.

Crossing over

Independent assortment reshuffles genes on *different* chromosome pairs. Genes on the same chromosome pair are recombined by *crossing over*, which Fig. 14.2 does not show. During meiosis I, chromosomes come together in homologous pairs, forming *bivalents*. Fig. 14.3 shows a simple case involving a single crossover. Two breaks occur at points opposite each other on *non-sister* chromatids (chromatids of different chromosomes). The broken ends of non-sister chromatids are then joined together. As a result the two strands of each chromosome are no longer identical — each now consists of both paternal and maternal sections.

What this means is that the chromosome number 3 you inherited from your mother actually consists of bits of chromosome number 3 from each of *her* parents. Your mother's chromosome number 3 (and every one of the other 22 pairs) had exchanged bits in the formation of her eggs. Similarly, each of your

ISBN: 9780170214094

paternal chromosomes is a kind of mosaic of your father's parents' chromosomes, and so on, for as many generations as you want to go.

Without crossing over, all the genes in a chromosome would be inherited together as a kind of permanent, indivisible 'lump'. Likewise, genes in homologous chromosomes would always go their separate ways as they segregate in meiosis. Crossing over enables genes that were originally in different homologues to be brought together into a more permanent relationship in the same chromosome strand. This may be advantageous because some genes work well together.

Mitosis and meiosis compared

Mitosis	Meiosis
One division	Two divisions
Daughter cells are genetically identical	Daughter cells are genetically different
Chromosome number does not change	Chromosome number is halved
Homologous chromosomes do not pair up	Homologous chromosomes pair up

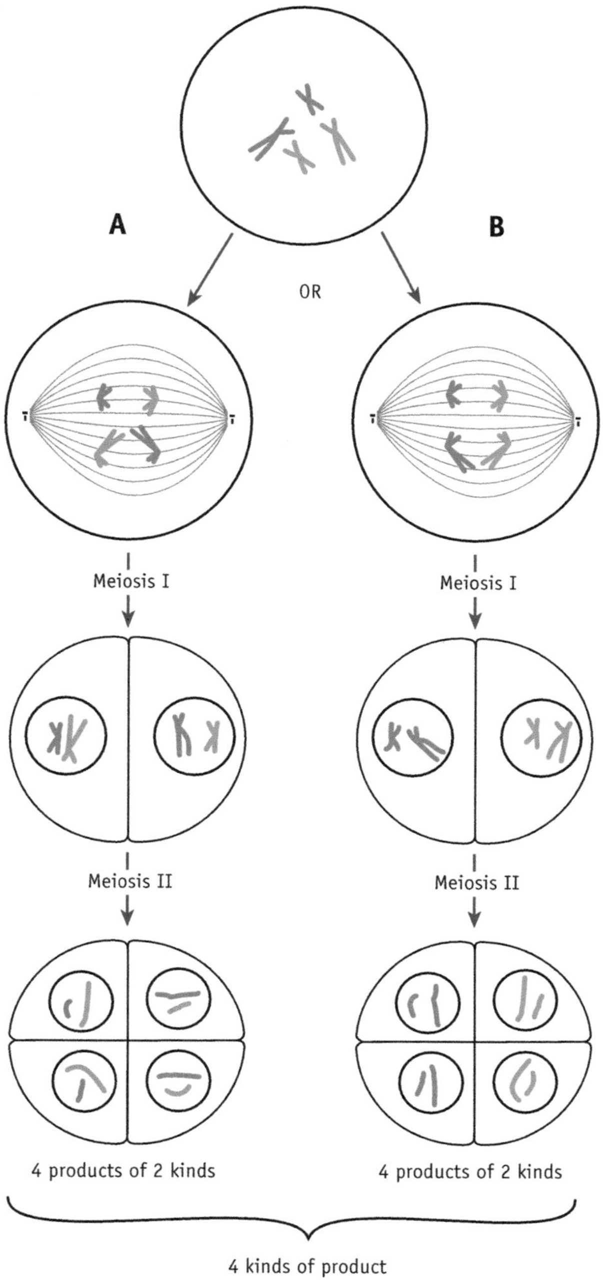

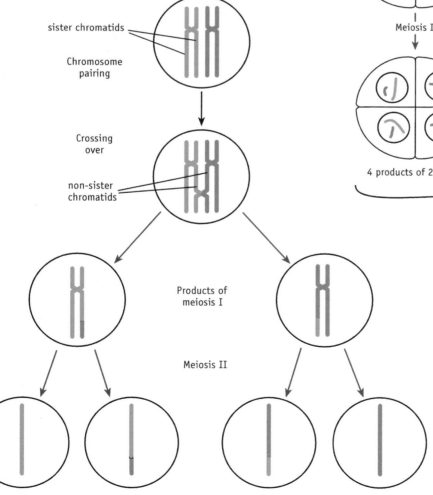

Four different products of meiosis II

Fig. 14.2 Variation by independent assortment. In a cell with two pairs of chromosomes there are two ways in which the chromosomes can segregate, with equal probabilities. In a population of cells, about half will divide as in A, and the other half as in B, giving four kinds of product.

Fig. 14.3 Crossing over

The work of Mendel

At the very time Darwin's book was making sensational headlines, a little-known monk was beginning a series of experiments that would eventually lead to the answer Darwin sought. The man responsible was Gregor Mendel, who lived in what is now the Czech Republic. Mendel was performing breeding experiments on varieties of garden pea (*Pisum sativum*), and his results were published in an obscure horticultural journal in 1865.

Mendel used varieties of pea that differed in clearly distinct ways, for example, seeds *round* or *wrinkled*, stems *tall* (about 2 metres) or *short* (about 0.5 metres), flowers *red-violet* or *white*. These are examples of **discontinuous variation**, in which individuals can be placed in distinct groups with no intermediate forms. Human blood groups are another example.

There are two advantages in using this kind of character in breeding experiments:

- Because individuals can be sorted into groups, the numbers in each group can be *counted*. Mendel was not the first to do breeding experiments, but he was the first to work out the *ratios* between the different kinds of offspring. He was the first mathematical biologist, and other biologists couldn't understand him!

- Environmental factors play little or no part (your blood group cannot be changed by environmental influences), so Mendel could be sure that heredity was the only factor involved.

Although it was not realised in Mendel's day, *continuously varying* characteristics (e.g. human height) are controlled by several or many genes, making interpretation of results much more difficult than with discontinuously varying characters.

Mendel's simplest experiments were **monohybrid** crosses. Here he considered only one pair of contrasting traits (characteristics) at a time, such as tall stem *x* short stem, and red-violet flower *x* white flower. The garden pea is normally self-fertilising, so Mendel had to transfer pollen artificially between varieties. There were two ways he could do this, since any given variety could be used as a male or as a female. A flower that was being used as a female had first to have its anthers (male part) removed to ensure that it did not pollinate itself. After artificial cross-pollination Mendel had to cover the flowers with little bags to protect them from 'foreign' pollen.

The varieties Mendel used were all *'true-breeding'*. This means that if allowed to self-fertilise over many generations, a true-breeding tall variety only gives rise to tall descendants.

Fig. 14.4 shows the results of one of Mendel's monohybrid experiments, in which he crossed a true-breeding red-violet-flowered variety with a true-breeding white-flowered variety.

He did **reciprocal crosses** (both ways around):

- using red-violet flowers as male, and
- using white flowers as male.

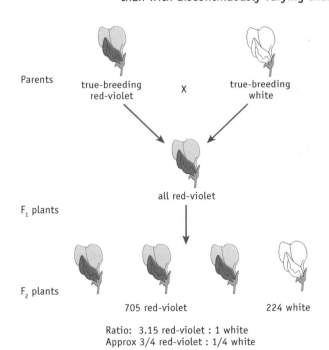

Parents
true-breeding red-violet   X   true-breeding white

F₁ plants

all red-violet

F₂ plants

705 red-violet   224 white

Ratio: 3.15 red-violet : 1 white
Approx 3/4 red-violet : 1/4 white

Fig. 14.4 Results of one of Mendel's monohybrid experiments

Regardless of which way he did the cross, he found similar results. The offspring, which Mendel called the *first filial generation* or **F₁**, had red-violet flowers. When he inbred the F₁ by allowing them to self-fertilise, approximately 3/4 of the second filial generation or **F₂** had red-violet flowers. When he inbred some of the F₂, he found that about 1/3 of the red F₂ plants bred true, the other 2/3 giving rise to a mixture of offspring in a ratio of 3/4 red : 1 white. The white-flowered F₂ on the other hand, all bred true.

Excellence in Biology Level 2   ISBN: 9780170214094

From these results we can draw the following conclusions, using some terms that were introduced after Mendel:

- Since in one cross the F_1 resembled the female parent and in the reciprocal cross they resembled the male parent, both parents must make an equal genetic contribution to the offspring.

- Although the F_1 plants resembled their red parents, they must have differed in their genetic makeup since they did not breed true. The F_1 plants had the same **phenotype** or physical characteristics, but had a different **genotype** or genetic makeup.

- The information for white flowers must have been present in the F_1 plants (since some of the F_2 had white flowers), but it was not expressed. Mendel said that the red trait (symbol R) is **dominant** whilst the white flowered trait (symbol r) is **recessive**.

- The F_1 plants must have inherited two factors (later called genes), one from each parent. They must therefore have had the genotype Rr. The white-flowered plants must have had the genotype rr and the red-violet-flowered parent must have had the genotype RR. After Mendel, the alternative forms of a gene (R and r in this case) were called **alleles**.

- To produce a white-flowered offspring (rr), some of the gametes produced by the F_1 plants must have carried only an r allele. Similarly, some of the F_1 gametes must have carried the R allele only. Before the F_1 plants made their gametes, the R and r alleles must have separated, or **segregated**. A gamete can therefore carry only *one* of each gene pair.

- Since both contrasting phenotypes were present in the F_2, the R and r alleles had not mixed like fluids, but must have remained distinct. This showed that the genetic material behaves like particles rather than blending like fluids. If only Darwin had known this, his theory of natural selection would have gained acceptance more quickly.

Mendel's work was ignored, even by some of the most distinguished scientists of the day — he was simply too far ahead of his time. His work was rediscovered independently by three scientists in 1900, and the science of genetics was born. Two other terms were introduced:

- **Homozygous**, meaning having the same allele twice (another name for true-breeding), e.g. RR, rr.

- **Heterozygous**, meaning having different alleles (and thus not true-breeding), e.g. Rr.

Fig. 14.5 shows the genetic explanation for Mendel's results, using the standard genetic shorthand. The Punnett square shows the four different ways in which gametes can combine at fertilisation (Punnett was an early geneticist). Two of these (R sperm with r egg and r sperm with R egg) give the same result (Rr).

Notice that the ratio of 3/4 red : 1/4 violet was not exact. This is to be expected, since it is always a matter of chance which kind of sperm joins with any given kind of egg.

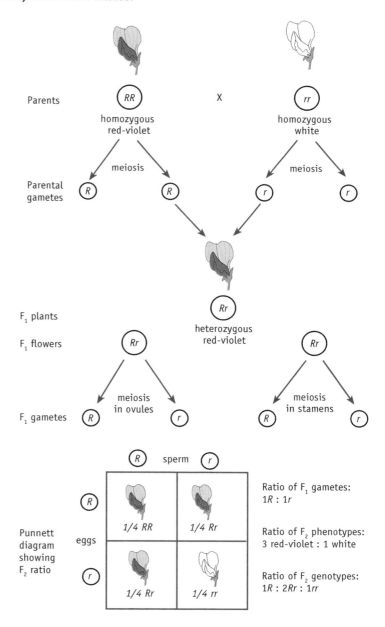

Ratio of F_1 gametes:
$1R : 1r$

Ratio of F_2 phenotypes:
3 red-violet : 1 white

Ratio of F_2 genotypes:
$1R : 2Rr : 1rr$

Fig. 14.5 Explanation for Mendel's monohybrid experiment

Testcrosses

The 3:1 F_2 ratio Mendel obtained was consistent with the idea that the F_1 plants produced two kinds of sperm and two kinds of egg, each in a 1:1 ratio. The trouble is that gametes do not express the alleles they are carrying, so we do not actually see a 1:1 ratio in the gametes. The alleles are not expressed until after fertilisation, and the *r* allele is only expressed in homozygotes.

Mendel solved the problem by doing a **testcross** or *backcross*. This involves crossing the F_1 plant with the parent showing the recessive trait. In this case, white flowered plants can only produce gametes carrying the *r* allele. Any variation in the offspring must therefore be due to variation in the gametes produced by the F_1 plant (Fig. 14.6). Hence the ratio of phenotypes in the testcross offspring must be *the same as the ratio of the different kinds of gamete produced by the F_1 plant.*

Mendel's testcross ratio of 1:1 showed that the F_1 plants were indeed producing two kinds of gametes in equal numbers.

Altogether, Mendel used seven pairs of contrasting characters, and obtained the results shown in Fig. 14.7.

In every case he obtained an F_2 ratio close to the theoretical 3:1.

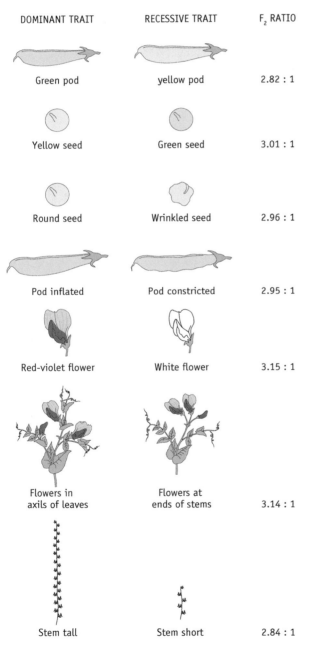

DOMINANT TRAIT	RECESSIVE TRAIT	F_2 RATIO
Green pod	yellow pod	2.82 : 1
Yellow seed	Green seed	3.01 : 1
Round seed	Wrinkled seed	2.96 : 1
Pod inflated	Pod constricted	2.95 : 1
Red-violet flower	White flower	3.15 : 1
Flowers in axils of leaves	Flowers at ends of stems	3.14 : 1
Stem tall	Stem short	2.84 : 1

Fig. 14.7 The seven pairs of contrasting traits used by Mendel

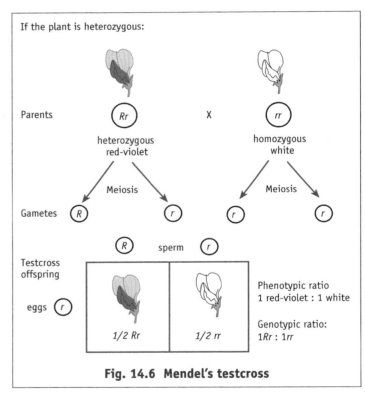

Fig. 14.6 Mendel's testcross

Improper use of a Punnett square

The purpose of a Punnett square is to show the different ways in which gametes can combine at fertilisation. When Mendel's F_1 plants fertilised themselves in his monohybrid experiment, there were four different kinds of fertilisation event, shown in Fig. 14.8. When a homozygote (e.g. *RR*) fertilises itself, there is only one way in which gametes can combine — in this case an *R* sperm and an *R* egg. It is quite improper to use a four-cell Punnett in this case. Likewise, in a testcross *Rr* x *rr*, there are only two ways in which gametes can combine, so the Punnett should have two compartments.

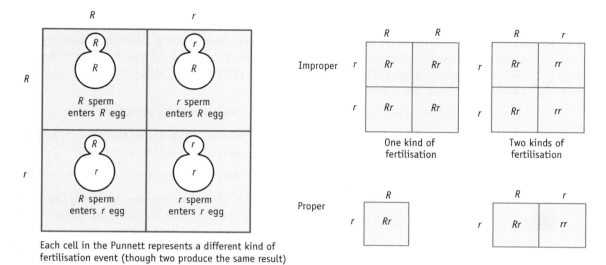

Each cell in the Punnett represents a different kind of fertilisation event (though two produce the same result)

Fig. 14.8 Proper and improper use of a Punnett square

Genes and chromosomes

Although Mendel had never seen a gene, he had been able to make two important deductions about them:

- Each individual pea plant has *two* genes for each characteristic he studied.

- Each gamete only carries *one* of each pair, so that before gametes could be formed, the two members of each pair must *segregate*.

Chromosomes were not known when Mendel did his work, but in the following decades a great deal was learned about them. In particular, it was realised that:

- In each species there are two of each kind of chromosome.

- During meiosis the two members of each pair segregate, so that gametes only carry one of each pair.

These and other observations led scientists to the conclusion that the genes are carried on the chromosomes. Each gene occupies a particular position or **locus**, on a particular chromosome. Mendel's monohybrid cross can thus be diagrammed showing the location on the chromosomes of the genes concerned, as in Fig. 14.9.

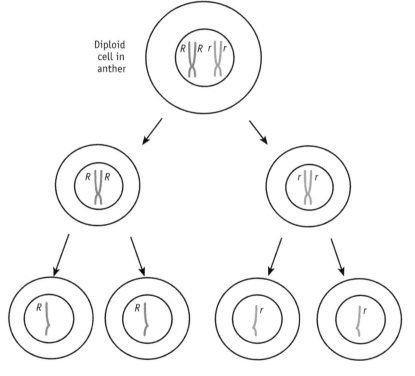

Fig. 14.9 How the genes in Mendel's monohybrid experiment are carried on the chromosomes

Pedigrees

Breeding experiments cannot be done with humans, since we prefer to select our own mating partners. Heredity in humans had therefore to be studied using pedigrees or family trees. Studies like this are the opposite of breeding experiments since they are *retrospective* — looking backwards in time.

In pedigree diagrams, circles represent females and squares represent males. Individuals affected by the condition in question are represented by black shapes, whilst unaffected people are shown in white.

When solving pedigree problems, remember these key things:

- To find out which is the recessive trait, look for a case where a child differs from *both* its parents. A recessive allele is only expressed in homozygotes, so a child with the recessive trait must be homozygous. Both parents must carry this allele, yet it is not expressed, so the parents must show the dominant trait.

- A recessive trait often appears to skip a generation; dominant traits do not.

- Recessive traits often appear after mating between relatives. A recessive allele that is rare in the general population is much more likely to be present in two related people.

- Although recessive traits are often less common than dominant traits, this is not always so.

Fig. 14.10 shows a pedigree involving *albinism*. In this condition the body cannot make melanin, the pigment that gives protection against ultra-violet light. Albinos have pink skin, white hair and the iris of the eye is very pale.

Sam is albino, yet neither of his parents is, so albinism must be recessive. Paul and Sam must be *aa*, and both Sam's parents must be *Aa*. Since Paul is *aa*, all his children must be heterozygous. Lloyd, Barbara and Ann must be either *AA* or *Aa*. The Punnett square for Karen and Robin's family shows that 2/3 of the fertilisations produce heterozygotes. We can therefore say that Lloyd is twice as likely to be heterozygous as homozygous.

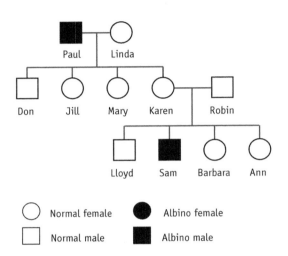

Fig. 14.10 **Pedigree involving a recessive allele**

Fig. 14.11 shows a pedigree for polydactyly, in which affected individuals have extra fingers and toes.

In this pedigree there is no family in which a child differs from both parents, so we cannot be *certain* which trait is dominant. However, if polydactyly were recessive, it would mean that both Joan and Mavis would have to be heterozygous. Since polydactyly is uncommon in the general population and Joan and Mavis are both 'outsiders', this is unlikely. The pedigree is therefore consistent with polydactyly being a dominant trait.

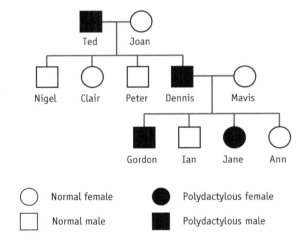

Fig. 14.11 **Pedigree involving a dominant allele**

Incomplete dominance

In some cases neither allele is completely dominant over the other, with the result that heterozygotes are intermediate between the two homozygotes. An example is the 'frizzle' condition in fowls, in which the feathers are brittle and curly and easily fall out; the birds may be nearly bald. When a normal fowl is crossed with a frizzle fowl, the F_1 are 'mild frizzle', and when these are inbred the F_2 consists of normal, mild frizzle, and frizzle in a 1:2:1 ratio (Fig. 14.12).

Multiple alleles

Some genes can exist as more than two alleles, for example the alleles determining the ABO blood groups. When blood of incompatible groups is mixed, the red cells clump together, with serious results.

There are three alleles, I^o, I^A and I^B. I^o is recessive to both I^A and I^B. I^A and I^B are both expressed in heterozygotes, and are said to be *co-dominant*. Each person has two genes for these blood groups, one on each chromosome number 9.

Fig. 14.13 shows the possible blood groups among the children if a Group A heterozygote marries a Group B heterozygote.

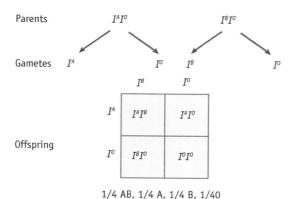

1/4 AB, 1/4 A, 1/4 B, 1/40

Fig. 14.13 Inheritance in human blood groups

Sex-linked inheritance

Some inherited conditions are more common in males than in females, for example, red-green colour-blindness. The genes responsible for these conditions are located on the X chromosome, of which males only have one, and females two (Fig. 14.14).

The allele for red-green colour-blindness is recessive. Males therefore only need one copy for it to be expressed. Females need two, so they only have the condition if *both* parents carry the allele. A man gives his X chromosome to his daughters and his Y chromosome to his sons. Males therefore inherit the allele from their mothers, who are usually normal, but heterozygous carriers. Fig. 14.15 shows the inheritance of red-green colour-blindness.

Fig. 14.14 Sex determination in humans

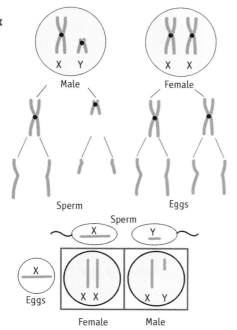

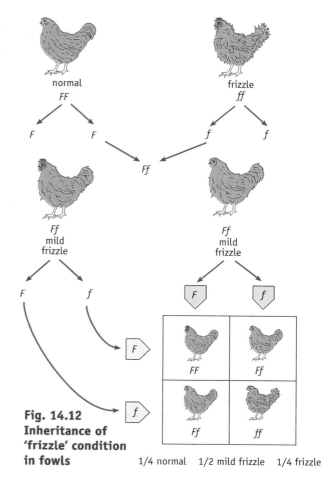

Fig. 14.12 Inheritance of 'frizzle' condition in fowls

1/4 normal   1/2 mild frizzle   1/4 frizzle

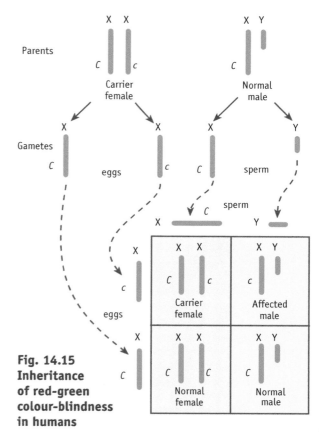

Fig. 14.15 Inheritance of red-green colour-blindness in humans

ISBN: 9780170214094

A small number of X-linked traits are dominant, for example a rare kind of rickets that is not due to dietary deficiency of vitamin D. X-linked dominant conditions are more common in females. (Can you suggest why?)

Chromosomes not concerned with the determination of sex are called **autosomes**, and characteristics due to genes on these chromosomes are *autosomal*.

Probability in genetics

Most of the animals and plants used in genetics have reasonably large families, so it is reasonable to talk of ratios. Humans have small families; it would be meaningless to talk about a 3:1 ratio in a family of two! It is more sensible to speak of the *probability* that a given child will have a particular characteristic. For example, approximately half of newborn children are girls. The probability that any given child will be a girl is 0.5.

When calculating probabilities, there are two rules:

The product rule. This is used when we want to know the probability that *both* of two independent events will occur. For example, we might want to know the probability that the first two children in a family will both be boys. These two events are independent of the other (the sex of the first child has no effect on the sex of the second). To find the probability that the first and the second child will both be boys, we multiply the two probabilities: 0.5 x 0.5 = 0.25.

The sum rule. This is used when we want to know the probability that *either* one event *or* the other will occur. In this case we add the two probabilities. What is the probability that in a two-child family there will be one of each sex? Fig. 14.16 shows that, as far as sexes of the children are concerned, there are four kinds of two-child family. Two of these four kinds have one of each sex. To find what proportion of two-child families contain one of each sex, we need to know the probability of a family consisting of girl followed by boy *or* boy followed by girl. The probability of each of these is 0.5 x 0.5 = 0.25. Hence the probability that a family will consist of a boy and a girl = 0.25 + 0.25 = 0.5.

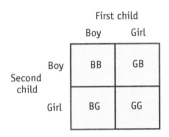

Fig. 14.16 The various possible two-child families

Dihybrid crosses — independent assortment

Mendel also performed *dihybrid* crosses, in which he considered two pairs of contrasting characteristics at the same time. In one such experiment he crossed two parents differing with regard to seed *shape* and seed *colour*.

One parent was true-breeding for round, yellow seeds and the other was true-breeding for wrinkled, green seeds. All the F_1 seeds were round and yellow, showing that these were the dominant traits. He therefore expected that 3/4 of the F_2 would be round and 1/4 wrinkled, and also that 3/4 of the F_2 would be yellow and 1/4 green. What he did not know was what proportion of the round F_2 seeds would also be yellow, and what proportion of the wrinkled seeds would also be green. In the event, his results were as shown in Fig. 14.17.

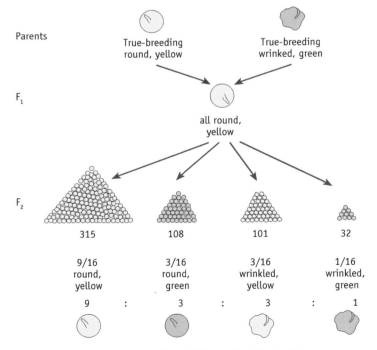

Fig. 14.17 Results of Mendel's dihybrid cross

COLOUR

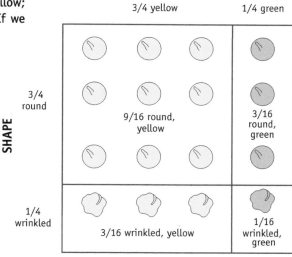

Fig. 14.18 A ratio of 9:3:3:1 is really two independent 3:1 ratios

There were four phenotypes, in the proportions 9/16 round, yellow; 3/16 round, green; 3/16 wrinkled, yellow; 1/16 wrinkled, green. If we consider one pair of traits at a time, then for shape we have:

Round: 9/16 + 3/16 = 12/16 = 3/4
Wrinkled: 3/16 + 1/16 = 4/16 = 1/4

and for colour:

Yellow: 9/16 + 3/16 = 12/16 = 3/4
Green: 3/16 + 1/16 = 4/16 = 1/4

So as expected, there was a 3:1 ratio for shape and a 3:1 ratio for colour. Less obvious is the fact that these two ratios are *independent of each other*. To see why, we need to remember one of the rules of probability, that if two events are independent, the probability that both will occur is the product of the two individual probabilities. The probability that any F_2 seed picked at random will be round is 3/4. Similarly, the probability that it will be yellow is 3/4. So, the probability that it will be both round *and* yellow = 3/4 x 3/4 = 9/16.

The proportions for all four phenotypes are:

round *and* yellow = 3/4 x 3/4 = 9/16
round *and* green = 3/4 x 1/4 = 3/16
wrinkled *and* yellow = 1/4 x 3/4 = 3/16
wrinkled *and* green = 1/4 x 1/4 = 1/16

If, like me, you find maths difficult, Fig. 14.18 shows the same idea pictorially.

Fig. 14.19 shows the whole thing, with genotypes. When the F_1 plants undergo meiosis, each allele pair segregates independently of the other pair. Thus, if a gamete receives R, it is equally likely to receive Y or y. Thus there will be four kinds of gamete produced, in approximately equal numbers. There will therefore be 16 different ways in which four kinds of sperm can fertilise four kinds of egg, so we need a Punnett square with 16 compartments. The result is not 16 different genotypes but nine, since some genotypes can be produced in more than one way.

Mendel obtained direct evidence that the alleles for shape and colour were segregating independently by performing a *testcross*. In this he backcrossed the F_1 plants with the double recessive parental variety (Fig. 14.20). Since the *rryy* parent only contributes recessive

Fig. 14.19 Explanation for Mendel's dihybrid cross

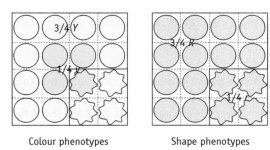

Colour phenotypes   Shape phenotypes

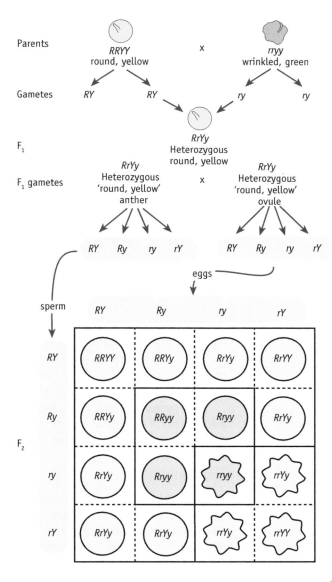

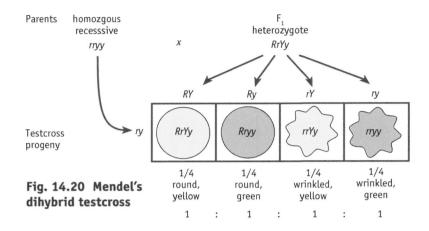

Parents homozygous recessive
rryy

x

F₁ heterozygote
RrYy

| RY | Ry | rY | ry |

Testcross progeny
ry

| *RrYy* | *Rryy* | *rrYy* | *rryy* |

| 1/4 round, yellow | 1/4 round, green | 1/4 wrinkled, yellow | 1/4 wrinkled, green |
| 1 | : 1 | : 1 | : 1 |

Fig. 14.20 Mendel's dihybrid testcross

alleles to the offspring, the variation in the offspring must entirely be due to variation in the gametes supplied by the F₁ heterozygote. The phenotypic ratio of 1:1:1:1 must therefore be the same as the ratio of the four kinds of gamete produced by the F₁ plants.

Genes and chromosomes again

Fig. 14.21 shows the chromosomal explanation of Mendel's dihybrid results. For each chromosome pair, two kinds of gamete can be produced. For two chromosome pairs 2 x 2 = 4 kinds of gamete can be produced. The diagram shows two of the chromosome pairs in the process of segregating during meiosis I.

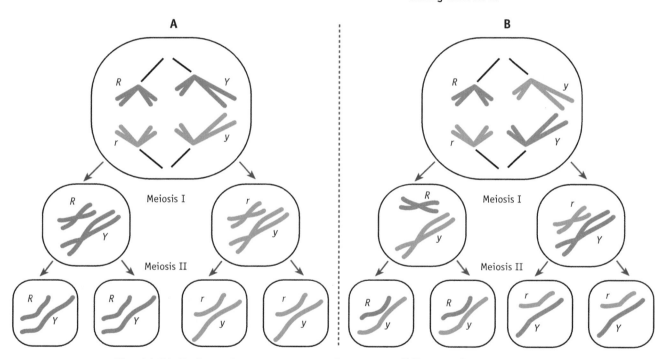

Fig. 14.21 Independent assortment of genes on different chromosome pairs

There are two different ways in which two chromosome pairs can segregate in meiosis I (A and B in Fig. 14.21). Approximately half the cells behave as in (A) and half as in (B).

The garden pea has seven pairs of chromosomes. If we take all of them into account, then 2^7 = 128 kinds of gamete could be produced by independent assortment. But this is ignoring crossing over, which gives untold trillions of possibilities (the number of noughts would extend more than the width of this page!).

Each gene occupies a characteristic position or *locus* on a particular chromosome. The entire genetic information of an organism is its *genome*. Scientists estimate that the human genome consists of about 23 000 genes, and the positions of many of them are now known.

Gene mutation

Between them, meiosis and fertilisation reshuffle genes into new combinations. But this only applies to existing genes — no new genetic material is produced. For significant evolutionary change to occur, new alleles must be produced by **mutation**. When DNA is replicated, it does so with extraordinary accuracy, but 'mistakes' do sometimes occur. Mutations are such 'mistakes' in replication.

ISBN: 9780170214094

- They are *rare*, typically occurring in 1 in 30 000 to 1 in 100 000 divisions.
- They are *harmful* more often than they are beneficial. Given *changed* environmental circumstances however, a mutant gene may become beneficial. For example, many insects have become resistant to some of the insecticides we use to kill them. The mutant genes responsible were of no advantage until humans developed insecticides.
- They are *random*, meaning that they occur without regard to *need*. If a mutation is beneficial, then it is just good luck.
- Most are *recessive*.
- Mutations can occur in any cells, but only those in the germ line — i.e. in the gametes or in cells that give rise to gametes — can be inherited. Mutations in body cells are called **somatic** mutations and are of no evolutionary significance.

Mutations occur spontaneously in nature (without external cause), but are also induced by a variety of agents called **mutagens**. There are three main categories of mutagen:

- Radiation, such as gamma-rays, X-rays, and ultraviolet radiation.
- Chemicals, such as nitrous acid, benzene.
- Viruses. Some viruses can cause cancer, which is a result of mutation of genes controlling cell division. For example the hepatitis B virus can cause liver cancer.

In organisms that do not recombine genetic material by sex, mutation is the only source of variation.

Summary of key facts and ideas in this chapter

- ○ Genetic variation is a key requirement for evolution.
- ○ Genetic variation can arise either by reshuffling existing alleles in meiosis and fertilisation, or by producing new alleles by mutation.
- ○ Meiosis results in a reduction in chromosome number, and the formation of genetically different products.
- ○ In animals, meiosis produces gametes; in plants it produces spores.
- ○ In meiosis genetic variation results from *independent assortment* and *crossing over*.
- ○ In his experiments using the garden pea, Mendel chose discontinuously varying characters.
- ○ A monohybrid cross involves one pair of contrasting characters; a dihybrid cross involves two pairs.
- ○ By performing *reciprocal crosses*, Mendel showed that both parents make an equal genetic contribution to the next generation.
- ○ An organism's *phenotype* is its physical characteristics.
- ○ An organism's *genotype* is its genetic makeup.
- ○ Inherited characteristics are determined by units of genetic material called *genes*.
- ○ An animal or a plant has *two* genes for each characteristic; a gamete has *one*.
- ○ Different forms of a gene for a given characteristic are called *alleles*.
- ○ An organism with two of the same allele is *homozygous*.
- ○ An organism with two different alleles for a given characteristic is *heterozygous*.
- ○ A characteristic or trait that is expressed in homozygotes and heterozygotes is said to be *dominant*.
- ○ A characteristic that is expressed only in homozygotes is said to be *recessive*.
- ○ In a monohybrid cross the offspring (F_1) show the dominant trait.
- ○ If the F_1 are mated among themselves (inbred), approximately ¾ of the offspring (F_2 generation) show the dominant trait and ¼ show the recessive trait — a 3:1 ratio.
- ○ In a Punnett square the number of cells should be the same as the number of different ways in which the gametes can combine.
- ○ To find out if an organism showing a dominant trait is homozygous or heterozygous, a *testcross* is performed, by crossing it with an organism showing the recessive trait. If the organism under test is heterozygous for one trait, a 1:1 ratio of phenotypes is obtained.

○ When traits show incomplete dominance, the F_2 are in a 1:2:1 ratio.

○ Some genes exist as *multiple alleles*.

○ Genes are carried on chromosomes, each gene occupying a particular position or *locus*.

○ In mammals, sex is determined by one pair of sex chromosomes. In males these are different and are called X and Y chromosomes. Females have two X chromosomes. It is therefore the male parent that determines the sex of an offspring.

○ Sex-linked traits are due to genes carried on the X chromosome.

○ In a dihybrid cross involving genes on different chromosome pairs, the F_2 consist of four phenotypes in a 9:3:3:1 ratio. This is two independent 3:1 ratios.

○ In a dihybrid testcross in which the organism under test is heterozygous for two traits, a 1:1:1:1 ratio of phenotypes is obtained.

Test your basics

Copy and complete the following sentences. In some cases the first or last letters of a missing word are provided.

1. Genetic variation can result from two processes – reshuffling existing genetic material in m____*____ and ____*____, and by creating new genetic material as a result of ____*____.

2. Meiosis has two essential results: it reduces the c____*____ number from ____*____ to ____*____, and it produces genetically ____*____ nuclei.

3. In animals meiosis occurs in the o____*____ and ____*____ and produces ____*____, and in flowering plants it occurs in the a____*____ and ____*____ and produces ____*____.

4. The reduction of chromosome number occurs in the ____*____ division of ____*____. The second division of meiosis is superficially like mitosis, but is actually different because the nuclei produced are genetically different as a result of c____*____ o____*____

5. In meiosis there are two mechanisms for generating variation. Genes on different chromosome pairs are reshuffled into new combinations by ____*____ ____*____. Genes on the same chromosome pair are recombined by ____*____ ____*____.

6. Mendel's first experiments on the garden pea were ____*____ crosses, involving a single pair of contrasting characters.

7. When two different pure-breeding varieties are crossed, the offspring are called the ____ generation. If these are inbred (self-fertilised in the case of the garden pea) or mated among themselves, the offspring are the ____ generation.

8. The physical characteristics of an organism are its ____*____, whereas its genetic makeup is its ____*____.

9. Whereas flowering plants and animals carry ____*____ genes for a given character, the gametes carry ____*____.

10. Genes can exist in two or more alternative forms or ____*____. If an organism carries two identical ____*____ it is said to be ____*____, and if it carries two different ____*____ it is ____*____.

11. An allele that is expressed in heterozygotes is said to be ____*____. An allele that is only expressed in homozygotes is ____*____.

12. An approximate 3:1 phenotypic ratio is obtained when two organisms, ____*____ for the same gene pair showing complete dominance, are mated.

13. To determine an organism's genotype, a ____*____ or backcross is performed. This is a cross between the organism of unknown ____*____ and one that is ____*____ ____*____.

14. In mammals, sex is determined by a pair of sex ____*____. In males these are different and are called ____*____ and ____*____. Females have two ____*____ chromosomes.

15. X-linked recessive traits are more common in ____*____s. X-linked dominant traits are more common in ____*____s.

16. A cross between two organisms true-breeding for two pairs of contrasting traits is a ____*____ cross. The offspring of such a cross are ____*____ for two gene pairs, and if they are testcrossed and dominance is complete, the phenotypic ratio is approximately ____*____.

17. If the F_1 of a dihybrid cross are inbred and if dominance is complete, the F_2 phenotypic ratio in such a cross is approximately ____*____.

18. The ultimate source of new ____*____ is mutation. These can be produced by r____*____, by c____*____, or by ____*____. Though most mutations are harmful, there is a small chance (especially under new ____*____ conditions) that a mutation may be beneficial.

15 Genetic Change

Almost all natural populations consist of individuals of different genotypes. As explained in Chapter 14, genetic variation results from two processes: the creation of new alleles by mutation, and the reshuffling of existing alleles in sexual reproduction.

Evolution is a genetic change in a population over successive generations. To understand what this means, we need to introduce the ideas of *gene pool* and *allele frequency*.

Whereas the genome is the sum total of all the genes carried by an individual, a **gene pool** is all the genes carried by a population of interbreeding individuals. Every time an organism dies, its genes are removed from the population. Every time a new individual is produced by fertilisation, genes are added to the gene pool.

Although each individual carries at least one allele of every gene locus, many gene loci exist as two or more alleles. Such genes are said to be *polymorphic* ('having many forms'). Since one individual can only carry two alleles for each locus, no individual can possesses more than a proportion of the genes in the gene pool.

In a population, each allele exists at a particular proportion or *frequency*. If a gene exists as two alleles *A* and *a*, which are equally common in the population, the frequency of each is 0.5. If 9/10 of the alleles of that locus are *A*, its frequency is 0.9.

Calculating allele frequency

In most cases, the frequency of an allele cannot be calculated from the frequency of the corresponding phenotype, though it can be *estimated* (see below). This is because recessive alleles are 'hidden' by dominance. Where there is incomplete dominance, however, we can distinguish between homozygotes and heterozygotes, and we can work out the allele frequencies from the phenotypic frequencies.

Imagine a flowerbed containing 100 snapdragons. 64 are red, 32 pink, and 4 white. Flower colour in snapdragons is determined by a pair of alleles showing incomplete dominance. The frequencies (proportions) of the genotypes are thus as follows:

> *RR* = 0.64
> *Rr* = 0.32
> *rr* = 0.04

Each plant carries two genes for flower colour. The proportions of the alleles are worked out as shown in Fig. 15.1.

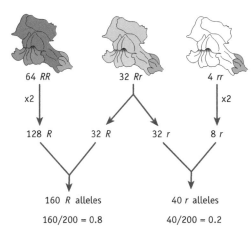

Fig. 15.1 Calculating allele frequencies from phenotypic frequencies (with incomplete dominance)

ISBN: 9780170214094

We can only calculate allele frequencies from phenotypic frequencies when we know the frequencies of the heterozygotes — that is, when there is incomplete dominance. In most cases this is not possible, and the allele frequencies have to be *estimated* using a method (Hardy-Weinberg equation) that is beyond the scope of this book.

When allele frequencies change over successive generations, evolution is occurring. There are several ways in which a gene pool can change — *selection, drift, founder effects, 'bottlenecks', migration,* and *mutation.*

Selection

Selection occurs when some genotypes are more successful than others. By this we mean that some genotypes have more surviving offspring than others, so they make a greater contribution to the gene pool. The successful genotypes not only have to be able to survive to reproductive age — they must also be able to reproduce. For many animal species, this means having the ability to attract a mate and rear offspring.

The phenotype is the 'target' of selection

It is important to remember that only alleles with a phenotypic effect (i.e. are expressed) can be selected for or against — an allele that is not expressed might as well not be there. In heterozygotes, recessive alleles are protected from selection by dominant alleles. For example in New Zealand, about one baby in 3000 is born with cystic fibrosis, yet about one person in 50 carries the allele. The overwhelming majority of carriers mate with non-carriers, so their children are unaffected. Only in rare cases do *both* parents carry the allele, and even then only about one in four of the children is affected.

Conditions like cystic fibrosis are much rarer than the allele frequencies might suggest because, generally speaking, people do not mate with close relatives. Mating between relatives is called **inbreeding**. It often results in children with harmful, recessive traits. Alleles that are rare in the general population are much more likely to be present in two people who are related than two who are unrelated. As a result, recessive alleles that would normally be protected from selection by dominant alleles are 'unmasked' by production of homozygotes. Selection is then likely to be intense.

The most intense form of inbreeding is self-fertilisation. For any given gene locus, self-fertilisation of a heterozygote results in half the offspring being homozygous and half heterozygous (Fig. 15.2).

After many generations of inbreeding, a population becomes homozygous for most (and eventually all) of its genes. Any harmful recessive genes will thus be eliminated by selection. This is why it is often believed that inbreeding causes mutations. It doesn't — it simply causes recessive alleles that are normally 'hidden' in the general population to be expressed in homozygotes.

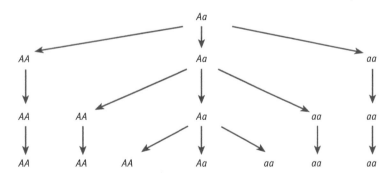

Fig. 15.2 How repeated self-fertilisation increases the frequency of homozygotes

Though selection can only act on phenotypic differences, these differences must have a *genetic* basis. This was shown by Johannsen in a classic experiment just after Mendel's work was rediscovered (it was Johannsen who first coined the term 'gene'). He took a sample of seeds from a pure-breeding variety of kidney beans and divided them into four groups according to mass. The following year he sowed each group separately and collected the seeds and weighed them. He found that the average mass of seeds produced by plants grown from heavy seeds did not differ significantly from those grown from lighter seeds (Fig. 15.3). He concluded that the variation in seed size was not inherited, but was due to variations in the environment.

ISBN: 9780170214094

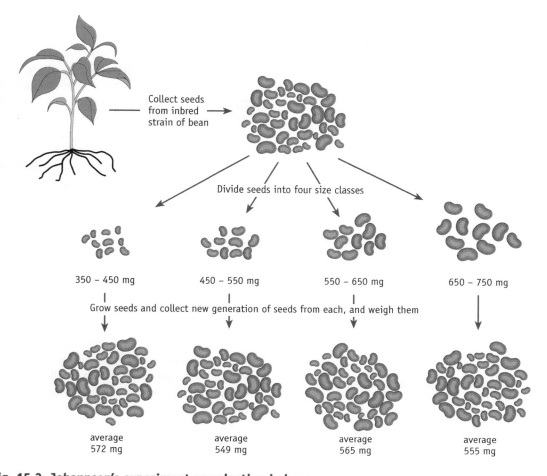

Fig. 15.3 Johannsen's experiment on selection in beans

What this means is that all our efforts to 'improve' ourselves as individuals has no inherited effect. We can build up strength by training, but this has no effect on our genes.

Types of selection

There are three kinds of selection: *stabilising, directional,* and *disruptive* (Fig. 15.4).

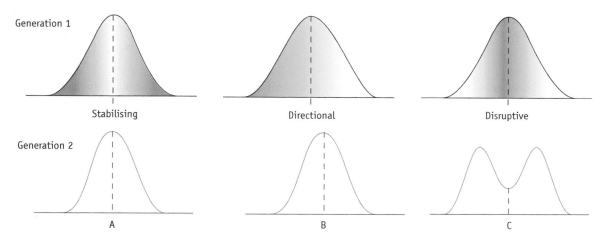

Fig. 15.4 Different kinds of selection. Each shape represents a frequency distribution for a phenotypic character. Red-shaped areas represent the phenotypes selected against, the probability of reproductive failure being indicated by the intensity of shading.

ISBN: 9780170214094

Stabilising selection

The parents of all living organisms have passed the test of natural selection — otherwise they would not have become parents. Provided that the range of environments experienced by the present generation is similar to that experienced by their parents, the 'best' genotypes will be the same as those of the previous generation. This results in stabilising selection, which acts *against* change.

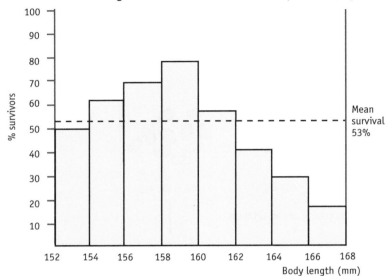

Fig. 15.5 shows a classic example of a study by Hermon Bumpus in New England, USA. He found 136 sparrows that had been incapacitated by a severe winter storm. He took them back to his laboratory, where 72 survived and 64 died. He measured all of them to see if there was any relation between size and survival. He found that those at the extremes had a higher death rate than those near the mean length. Being excessively large or excessively small was evidently a disadvantage in winter storms.

Fig. 15.5 Histogram showing stabilising selection in sparrows incapacitated after a storm

It is important to realise that storms are not the only hazard with which sparrows have to contend. In a period of severe cold, in which heat loss would be a major factor, one might expect the larger birds, with a smaller surface/volume ratio, to come through more successfully. The 'best' genotype is actually the result of a compromise between many different conflicting environmental challenges.

This is illustrated by selection for birth weight in humans. Fig. 15.6 shows the results of a study on the relationship between infant mortality and birth weight at a London hospital. It was carried out in the mid 20th century, before modern medicine enabled seriously premature babies to survive. Babies nearest the average birth size had much the best survival rate than those at the extremes. Babies born too early tended to die because their lungs, gut and immune systems were immature. Those born too late would have had mechanical problems as the head passed through the birth canal.

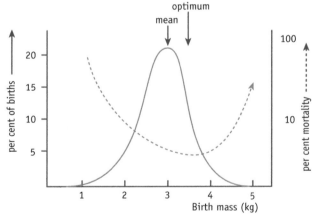

Fig. 15.6 Stabilising selection in babies in a London hospital 1935-45. Note the log scale for mortality.

Directional selection

In directional selection the result is a change towards one end of a range of phenotypic variation. The expansion of the brain during human evolution is an example. Change on this timescale (three million years in this case) is too slow for us to witness. However, since the Industrial and Technological revolutions, humans have been changing the environments of most other organisms at an increasing pace. The result has been a series of evolutionary bursts, for example:

- Resistance to insecticides in insects.

- Resistance to antibiotics in bacteria.

- Resistance to warfarin in rats.

- Resistance to heavy metals in grasses.

- Industrial melanism in moths.

All these examples are responses to chemicals produced by humans. The important thing to note is that all the species involved have short life-spans, and thus can respond quickly to environmental change.

Industrial melanism

One of the best-known and most fully investigated examples of natural selection is that of industrial melanism in the European peppered moth, *Biston betularia*. The normal form has speckled wings which renders it almost invisible to bird predators against a background of lichen-covered bark of trees. Studies of old moth collections showed that in the nineteenth century a black, melanic form appeared in urban areas and eventually replaced the normal speckled form. In rural areas the speckled form remained the predominant type.

This change in phenotype coincided with drastic environmental changes brought about by industrial smoke. Lichens are extremely sensitive to sulfur dioxide, and they virtually disappeared from urban areas. Moreover, the tree trunks became covered with soot, against which the melanic form blended almost invisibly. In urban areas, the speckled phenotype was a distinct hazard, whilst the melanic form was strongly advantageous.

Industrial melanism is an example of **polymorphism**. This is the occurrence of two or more distinct forms of a species, the rarest form being more common than could be accounted for by mutation alone.

Though it might seem self-evident, the selection explanation was submitted to experimental testing in the 1950s by H.B. Kettlewell. He marked and released both kinds of moth in urban Birmingham and in rural Dorset, and then recaptured as many as possible by trapping. Some of his results are shown in the table below.

		Speckled	Melanic	Total
Dorset	Released	496	473	969
	Recaptured	62	30	92
	% recaptured	12.5	6.3	
Birmingham	Released	64	154	218
	Recaptured	16	82	98
	% recaptured	25.0	52.3	

That predation by birds was the cause was suggested by extensive observation through binoculars. In Dorset, birds were seen to capture nearly six times as many of the dark form as the specked form. In Birmingham about three times as many of the speckled form were captured. Moreover, clean air legislation passed in the 1950s was followed by a return of the speckled form as lichens recolonised urban areas.

A later evaluation of Kettlewell's work has cast doubt on some of his conclusions. Kettlewell released his moths by placing them on tree trunks, but in nature very few have been found in such a position. Their natural resting places are unknown. Also, the moths were released during the day, but they normally settle at night.

Selection for a recessive allele

Melanism in moths is a *dominant* trait. Although the first moths to carry the allele would have been heterozygotes, they would have been black. Selection would thus have begun to operate as soon as the moths emerged from their pupae.

What happens when an advantageous mutation is recessive? Imagine a population in which a locus is represented by one allele only, *A*, so its frequency is 1.0. Suppose a new advantageous allele, *a*, is produced by mutation. To start with, its frequency will be extremely low and it will only be present in heterozygotes, *Aa*. As more mutant alleles are produced by mutation, the frequency of *a* rises until heterozygotes are sufficiently common to meet other heterozygotes and mate with them. Some of the offspring will be homozygotes, *aa* (Fig. 15.7). These are at an advantage and so their survival rate is higher. From this point on the frequency of the allele increases as a result of selection. Eventually the frequency of *a* approaches 1. It never quite reaches 1, however, because

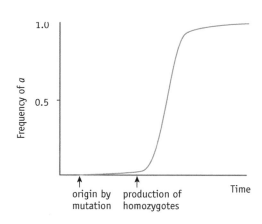

Fig. 15.7 The effect of selection on the frequency of an advantageous, recessive allele

of occasional back-mutation from *a* to *A*. There is an equilibrium between the rate at which *A* is lost from the gene pool by selection and the rate at which it is added by mutation.

Disruptive selection

This is by far the least common of the three types. An example is selection for size in male Coho salmon, which live in the Pacific Northwest of the United States. Most males, called 'hooknoses', return to their stream three years after hatching. Here they fight to fertilise the females, and the largest ones usually win. Some males, called 'jacks', mature after two years and are only half the size of the hooknoses. They are able to reach the females by sneaking rather than fighting, using their small size to hide. The most successful males are the largest hooknoses and the smallest jacks. The least successful males are the largest jacks and the smallest hooknoses. Whether a male matures early as a jack or later as a hooknose is partly genetically determined.

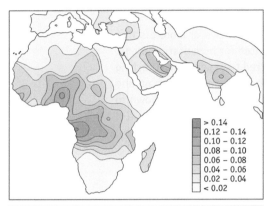

Fig. 15.8 Frequency of the *Hbˢ* allele (top) and the distribution of malaria in Africa and Western Asia

Selection for heterozygotes

Whether a gene is beneficial, harmful or neutral depends on the external environment, and also on the other genes in the genotype. An example of the latter is the inherited blood disease **sickle cell anaemia**. It is caused by a mutation in one of the two genes for haemoglobin. It results in just one 'wrong' amino acid in one of the two kinds of polypeptide in the haemoglobin molecule. This apparently trivial difference causes the haemoglobin to crystallise at the low oxygen concentrations in the capillaries. As a result the red cells become sickle-shaped and block the capillaries. The affected red cells are attacked by white cells, causing serious anaemia.

The normal allele (Hb^A) is incompletely dominant to the sickle cell allele (Hb^S). The condition occurs only in homozygotes. Heterozygotes are relatively unaffected, and have the additional benefit of being highly resistant to the most severe form of malaria.

The result is an example of *polymorphism*. Homozygous normal individuals are more likely to die of malaria, whilst those homozygous for the sickle cell allele always die in childhood. Although Hb^S alleles are eliminated from the gene pool, so too are Hb^A alleles.

We therefore cannot say that the sickle cell allele is 'bad'. In malarial areas it is undoubtedly beneficial, which explains its high frequencies in those parts of the world (Fig. 15.8). In African Americans whose ancestors came from malarial areas, the advantage of the sickle cell allele is no longer present, but the disease persists at a low frequency.

Chance effects in the gene pool

In large populations, chance plays little part in changing allele frequencies, but in small populations chance effects can be considerable. Chance can play its part in three ways: the *founder effect*, *bottlenecks*, and *genetic drift*. All are essentially cases of 'sampling error'. It is what pollsters mean when the say that the margin of error in a public opinion poll is, say, 3%. When a public opinion poll is conducted, there is always a 'margin of error'. The larger the number of people sampled, the smaller the margin of error.

Founder effects

When a small number of people leaves the main population and starts a new colony, it takes with it a random sample of the 'parent' gene pool. The frequencies of the alleles in the founder gene pool are almost certain to differ from the parent one. By chance, some alleles will be lost and others will increase their frequency (Fig. 15.9).

ISBN: 9780170214094

Suppose there are ten founders and one person (who carries 10% of the gene pool) is heterozygous for an allele that is very rare in the parent population. Its frequency in the founder population will be 0.05.

Something like this has happened on numerous occasions in human populations. An example is the incidence of porphyria in South Africa. Porphyria is a severe skin condition resulting from extreme sensitivity to UV light. It is caused by an autosomal dominant allele that is rare in most populations. Amongst South Africans of Dutch descent about one in 300 people is affected — a far higher incidence than in Holland.

Another example is the very high frequency of blood Group O (over 95%) among South and Central American Indians. It seems likely that they are all descended from a small band that walked across from Asia during the last ice age (when sea levels were low enough to link the two continents).

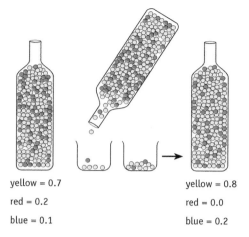

yellow = 0.7
red = 0.2
blue = 0.1

yellow = 0.8
red = 0.0
blue = 0.2

Fig. 15.9 A model of the Founder Effect

Genetic bottlenecks

Similar to the founder effect is the 'bottleneck effect', in which a population crashes to near-extinction, but recovers. The few survivors carry only a proportion of the original gene pool, and they provide the alleles for later generations. Though the numbers may recover to their original value, the population remains genetically uniform. Only over long periods of time can mutation restore genetic variability.

A well-known example is the cheetah. Individuals are so alike that skin can be grafted from one individual to another without rejection. At some time in the past the numbers must have declined to a very low value.

The evidence for a genetic bottleneck is indirect in the case of the cheetah. A number of other species have been pushed to near-extinction by humans, but conservation efforts have led to recovery — of numbers. In the nineteenth century hunting reduced the population of northern elephant seals to about 20 individuals. Since then the numbers have recovered to about 50 000, but the genetic diversity has not.

Genetic drift

Each generation carries a 'sample' of the gene pool of the previous generation. In large populations (and therefore large 'samples', the gene pool of each generation differs little from the previous one. In very small populations, the effect of chance is much greater. As a result of repeated 'sampling error', allele frequencies can fluctuate a great deal. These random changes in allele frequency are called **genetic drift** (Fig. 15.10).

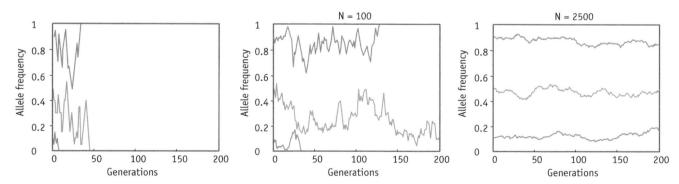

Fig. 15.10 Computer simulation of genetic drift over 200 generations, for a gene with three alleles (after Cavalli-Sforza)

Small populations and inbreeding

Drift, bottleneck and founder effects can actually increase the effects of selection. Whenever numbers reach very low levels, inbreeding occurs, with the production of homozygotes. Recessive alleles that were previously protected in heterozygotes become exposed to selection. If an allele is even mildly harmful, it is more likely to be eliminated if combined with the effect of drift.

An allele that is mildly deleterious in one environment may be beneficial in another. A large population carries a considerable number of alleles that may be potentially useful in the event of significant environmental change. Severe population reduction can negate this potential benefit, leaving a population genetically too unvarying to adapt to change.

Immigration and emigration

If a population is small, any immigration (which adds to the gene pool) or emigration (which removes alleles) will affect allele frequencies. In a large population, immigration will only affect the gene pool if it is from an area in which allele frequencies are different, perhaps due to different selection pressures. In large populations emigration can only affect allele frequencies if emigration is non-random i.e. if some genotypes are more likely to emigrate than others.

Summary of key facts and ideas in this chapter

○ A *gene pool* is all the genes carried by a population of interbreeding individuals.

○ The *frequency* of an allele in a population is its proportion out of the total alleles of that gene locus.

○ *Evolution* is a change in the frequency of alleles in a population over successive generations.

○ Allele frequencies can be calculated directly when dominance is incomplete.

○ Selection occurs when some genotypes produce more surviving offspring than others, and thus make a greater contribution to the gene pool than others.

○ Only alleles that are expressed in the phenotype can be selected.

○ Inbreeding increases the frequency of homozygotes, and consequently the expression of normally rare recessive alleles.

○ The most common form of selection is *stabilising selection*, which acts *against* change.

○ *Directional selection* occurs when one end of a range of variation is favoured, and results from a change in environmental conditions. The best examples result from rapid, human-induced environmental changes.

○ *Disruptive selection* occurs when forms that are intermediate between two extremes are at a disadvantage, and is the rarest type of selection.

○ Whether an allele is advantageous or disadvantageous may depend on which other genes are present. An allele may be advantageous in heterozygotes and disadvantageous in homozygotes.

○ *Founder effects* occur when a *small* number of individuals leaves a 'parent' population and establishes a new population elsewhere.

○ *Bottleneck* effects occur when a population becomes *almost* extinct and then recovers. The tiny number of survivors carry only a small proportion of the original gene pool and are forced to inbreed, which may cause further loss of alleles from the population.

○ Genetic drift occurs in very small populations and is a change in allele frequency due to chance.

Test your basics

Copy and complete the following sentences.

1. The sum total of all the genetic material in a population is its ___*___ ___*___; the sum total of all the genes in an individual is its ___*___.

2. The frequency of an allele in a population is its ___*___ in the gene pool.

3. When some genotypes are reproductively more ___*___ than others, selection occurs. An allele can only be selected if it is expressed in the ___*___.

4. ___*___ selection occurs when the most successful phenotypes are near the 'average'.

5. The evolution of resistance to pesticides is an example of ___*___ selection.

6. Selection against the mean of a character distribution and in favour of the two extremes is ___*___ selection.

7. Sickle cell anaemia is an example of ___*___. It occurs when selection favours a heterozygote over both homozygotes.

8. When a population is reduced to very small numbers, allele frequencies are more likely to be affected by chance events (as when an individual is in the wrong place at the wrong time). Such random changes in allele frequencies are called genetic ___*___.

9. The founder effect occurs when a ___*___ number of individuals becomes detached from a much ___*___ population and establishes a new gene pool in a new territory.

10. A genetic bottleneck results when a population experiences near ___*___, the survivors carrying only a small proportion of the ___*___ in the original population. Though the population may recover with regard to ___*___, it can take thousands of years before it recovers its genetic ___*___.

16 Cells Under the Light Microscope

The cellular basis of life

The word 'cell' was first used by Robert Hooke in 1665 to describe the little 'compartments' he saw when he examined a thin slice through cork. During the next two centuries microscopists gradually realised that all living things are composed of cells (or of only one cell). By the mid-nineteenth century this idea that cells are the units of life became known as the **Cell Theory**. It contained three essential principles:

1. All living things consist of cells. Some organisms are said to be *unicellular* because they consist of only one cell. Gametes (eggs and sperm) are cells.

2. The activities of living things are simply the outward signs of processes occurring in their cells. For example, the tears that are secreted over the eyes are made in the cells of the tear glands, and the pumping of the heart is due to the contraction and relaxation of its muscle cells.

3. Just as all organisms have parents, so do all cells arise from pre-existing cells.

An exception to the first principle is when many nuclei share the same cytoplasm. This situation can arise in two ways:

- In a *syncytium*, many originally separate cells join to form larger units. An example is the fibres of skeletal muscle.

- In a *coenocyte*, nuclei divide but the cytoplasm does not, for example the common bread fungi *Mucor* and *Rhizopus*.

Parts of a school microscope

A school microscope (Fig. 16.1) has two kinds of lens:

- One or more *objectives*, mounted on a rotatable *nosepiece*. By turning this, different objectives can be used. Because the high power objective is the longest, it is much closer to the specimen. The objective produces an *inverted* image (one that is upside down and back to front). This is why, to move the image upwards and to the right, you have to move the slide downwards to the left.

ISBN: 9780170214094

- The *eyepiece*, at the top of the microscope. This magnifies the image produced by the objective. If the objective magnifies 20 times and the eyepiece magnifies ten times, the total magnification is 200. The final image is produced on the retina of the eye.

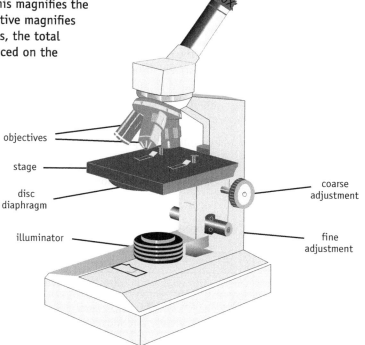

The specimen is normally mounted on a glass *slide* which is placed on the *stage* and is held by two clips. Most modern school microscopes have a built-in illumination under the stage (older types use a mirror that reflects light from a separate bench lamp). In the best microscopes there is also a *condenser* lens below the stage. This focuses light on the specimen, making it as bright as possible.

There are two focusing knobs:

- A large, *coarse adjustment. This must not be used with high power* because of the risk of scratching the high power objective on the coverslip, if it is lowered too far.

- A smaller, *fine adjustment*, which is used with high power objectives.

Fig. 16.1 A school microscope

Beneath the stage is a device for adjusting the light intensity. In modern school microscopes this is a rotating **disc diaphragm** with holes of various sizes.

Mounting a specimen

Living specimens are mounted in water, which helps to keep them alive. A thin glass *coverslip* is carefully lowered on to a drop of water containing the specimen (Fig. 16.2).

The coverslip has several important functions:

- It holds the specimen steady.

- It keeps the specimen flat so that you do not have to vary the focusing as you move the slide around.

- It stops the high power objective getting wet, and prevents it misting up due to condensation.

- A drop of water has a curved surface which would act as a lens, distorting the image.

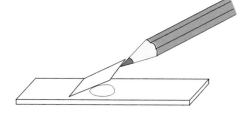

Fig. 16.2 Mounting a specimen

It is important to use the right amount of water. Too much and the specimen cannot be held steady; too little and a delicate animal may be crushed by the surface tension pulling the coverslip down onto the slide.

Staining specimens

Many living cells are transparent and it often makes it easier to see details if they are first *stained*. A disadvantage of this is that most stains kill cells, so if you want to observe living cells they should not be stained.

Setting up the microscope

1. Switch on the lamp.

2. Making sure the microscope is on low power, place the slide over the hole in the stage.

3. With the low power objective as close as possible to the stage, move the slide until the specimen comes into view (it is easier to find a specimen on low power as the field of view is wider).

4. Use the *coarse adjustment* knob to bring the specimen into focus, and then use the fine adjustment if necessary. Because the field of view is smaller at higher magnifications, you will need to move the specimen into the centre of the field of view before switching to a higher power.

5. If you wish to use high power, use middle power first, and readjust the fine focusing and re-centre the specimen in the field of view.

6. With your eye level with the slide, carefully rotate the nosepiece to bring the high power objective into position. *Make sure that the high power objective does not touch the coverslip.*

7. Focus using the *fine adjustment.*

8. If necessary, adjust the light intensity using the disc diaphragm.

Symptom	Fault
Dirt clearly visible and does not move when slide is moved	Dirty eyepiece
Field of view is hazy	Dirty objective
Dirt clearly visible and moves with slide	Dirty slide or coverslip
Field of view is darker on one side	Nosepiece not clicked into position
Insufficient contrast, resulting in 'glare'	Hole in disc diaphragm too wide

Table 16.1 Common problems encountered when using a school microscope

Units for microscopic objects

Most of the things you see under the light microscope are so small that it is best to measure their sizes in **micrometres (µm)**. 'µ' is the Greek letter *mu* (pronounced 'mew'). An even smaller unit, the **nanometre (nm)**, is needed when dealing with the very small things seen under the electron microscope.

$$1 \ \mu m = 0.001 \ mm \ (10^{-6} \ m)$$
$$1 \ nm = 0.001 \ \mu m \ (10^{-9} \ m)$$

Many microscopic objects are so small that it is difficult to appreciate just how tiny they are. Table 16.2 gives an idea of relative sizes of some biological things you may come across.

Object	Approx. shortest dimension
Liver cell	20 µm
Photosynthetic leaf cell	40 µm
Chloroplast	5 µm
Mitochondrion	1 µm
Ribosome	20 nm
Thickness of plasma membrane	7 nm
Haemoglobin molecule	7 nm
Glucose molecule	1 nm
Water molecule	0.2 nm

Table 16.2 The approximate sizes of some biological objects

To put these sizes into perspective, imagine a glucose molecule to be the size of a pea. The plasma membrane surrounding a cell would be about 5 cm thick, an average bacterium would be about 7 metres across, and a liver cell would be about 150 metres across!

Extension: Estimating the size of a specimen

To estimate the diameter of a cell you are seeing under high power, you can use the field of view as a kind of ruler, as follows. Suppose the high power field of view is 300 µm wide, and that the cell fits about ten times across it. The cell would be about 300/10 = 30 µm wide.

But first, of course, you have to find the diameter of the high power field of view, as follows. First measure the diameter of the low power field of view by placing a transparent plastic ruler across the stage, and focusing on it. Suppose this is 4 mm, or 4000 µm, across. To find the high power field of view you need to know how many times more powerful high power is than low power. Suppose the low power objective magnifies four times and the high power objective magnifies 40 times. The high power magnification is thus ten times as powerful as low power. The high power field of view is therefore a tenth of the low power field of view – in this imaginary case, 0.1 x 4000 µm = 400 µm.

An animal cell under the light microscope

Any multicellular organism consists of many different kinds of cell, each *specialised* for carrying out a particular function. There is therefore no such thing as a 'typical' animal cell — each kind of cell has one or more features that make it different from others. Fig. 16.3 shows cells from the lining of the cheek, which are specialised for *protection*. Notice that the cells are flattened, so their appearance depends on the angle of viewing.

With a school microscope you can see the following parts:

- The **cytoplasm.** This is a transparent, jelly-like material, which forms most of the cell. It is here that most of the cell's activities occur. The outermost part of the cytoplasm is the **plasma membrane** (the modern name that has replaced the older term 'cell membrane'). Its function is to regulate the movement of materials into and out of the cell (Chapter 21). It is too thin to see even with the most powerful light microscope — what you actually see is not the plasma membrane itself but the *boundary* between the cytoplasm and the outside of the cell.

- The **nucleus.** This is the 'control centre' of the cell. In the living cell it is transparent, but it can be seen more easily when stained brown with iodine. It contains **chromosomes**, which are not distinguishable in non-dividing cells, even after staining. The chromosomes contain the genetic material in the form of a complex substance called **DNA**.

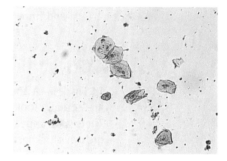

Fig. 16.3 Human cheek cells

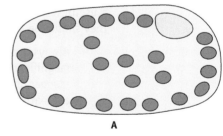

A

Plant cells under the light microscope

Fig. 16.4 shows an entire plant cell (A), and (B) shows a similar cell cut down the middle.

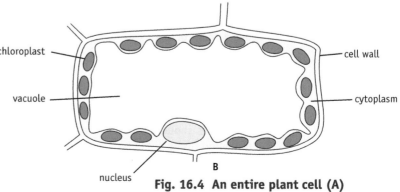

chloroplast

vacuole

nucleus

cell wall

cytoplasm

B

Fig. 16.4 An entire plant cell (A) and a section through a plant cell (B)

Besides the nucleus and cytoplasm, a number of other parts can be seen:

- The **cell wall**, consisting of a framework of **cellulose** with some protein. It acts like the outer layer of a football — it *resists stretching,* enabling the cell to withstand high internal pressure. In some plant cells the cytoplasm and nucleus die, leaving the cell wall as a hollow shell.

- Most living plant cells have small structures called **plastids** in the cytoplasm. These can be of several kinds. **Chloroplasts** contain the green light-trapping pigment **chlorophyll** and carry out **photosynthesis**. In this process light energy is used to convert CO_2 and water to sugar and oxygen. Plant cells that do not carry out photosynthesis often have starch-storing **amyloplasts** ('starch grains'), or orange-red coloured **chromoplasts** as in many flowers and fruits. The various kinds of plastid all develop from *proplastids* — small self-reproducing structures present in all dividing plant cells.

- A large space called the **vacuole**, which makes up most of the cell. It contains **cell sap**, a solution of salts, sugars and other substances. It is separated from the cytoplasm by a membrane called the **tonoplast**. As with the plasma membrane, it is not actually visible under the light microscope, but its presence can be inferred from the fact that there is a boundary between the vacuole and the cytoplasm. Because the cytoplasm is displaced to the edge of the cell, the chloroplasts are close to the source of CO_2 (which diffuses very slowly in solution). Also, the salts dissolved in the sap produce a strong tendency for water to enter the cell by osmosis (Chapter 21), which results in the cell becoming tightly inflated and firm, helping to support soft parts of the plant.

Cells in three dimensions

In Fig. 16.4A the chloroplasts *seem* to fill the entire cell, but they don't really. This is because you are actually looking through two thin 'windows' of cytoplasm separated by a thick layer of cell sap. There also appear to be more chloroplasts at the sides of the cell because you are looking through a greater thickness of cytoplasm (i.e. edge on).

Drawing specimens

Drawing a specimen is a useful exercise because it makes you look more closely at it and may provide a useful record for later reference. Biological drawings do not need to be artistic masterpieces, but they should be *clear and accurate*. Here are some useful tips:

- Always include a title that gives as much information as possible, for instance the magnification of the drawing. This is not the magnification of the microscope *but the number of times bigger the drawing is than the actual specimen* (see above for how to estimate the size of the specimen). If the specimen is a thin slice, then may be useful to state whether it is a transverse (cross-) section (T.S.), or a longitudinal (lengthwise) section (L.S.).

- To show fine structures in proportion the drawing should be *large,* with fine pencil lines, so use a *sharp* pencil.

- Lines should be single and clear — the fuzzy lines of a sketch obscure detail.

- Shading is best avoided unless you have real artistic talent.

- Do not draw anything you cannot see.

- Guidelines should not have arrowheads and should not cross each other. Use only one guideline for each kind of structure. Guidelines that are in ruled pencil are best because they stand out just enough from the rest of the drawing to be distinct. Ink guidelines stand out too much and tend to hide the lines of the drawing.

- Labels should be *outside* the drawing.

ISBN: 9780170214094

Differences between animal and plant cells under the light microscope

Some of the most fundamental differences between animals and plants are in their cell structure. Even under a school microscope you can see most of these differences clearly. The most important are set out in Table 16.3.

	Plant cells	Animal cells
Nucleus	present	present
Cell wall	present	absent
Large vacuole	usually present	absent
Plastids	present	absent

Table 16.3 Comparison between animal and plant cells

Summary of key facts and ideas in this chapter

○ Objects viewed under the light microscope are measured in *micrometres* (μm). 1 μm = 10^{-6} of a metre, or 0.001 of a millimetre.

○ Structures that can be seen under a school microscope include plant cell wall, nucleus, and chloroplasts.

○ As the magnification increases, the diameter of the field of view decreases in proportion. If the magnification is increased from 40x to 400x, the diameter of the field of view decreases to 1/10.

○ The diameter of the low power field of view can be measured directly by viewing a ruler with a millimeter scale.

○ The size of a specimen can be estimated if the diameter of the field of view is known.

  Many other details cannot easily be summarised without simply repeating them – see text.

Test your basics

Copy and complete the following sentences. In some cases the first letter of a missing word is provided.

1. Objects seen under the light microscope are usually measured in ___*___, abbreviated to μm. 1 μm = ___*___ of a metre, or ___*___ of a millimetre.

2. When dealing with objects visible only under the electron microscope, ___*___ (nm) are used. 1 nm = ___*___ of a metre or ___*___ of a millimeter.

3. As the magnification of a microscope increases, the diameter of the field of view ___*___.

4. The magnification of a light microscope is calculated by ___*___ the magnification of the ___*___ and o___*___.

5. The part of a cell that contains most of the genetic material is the ___*___, which contains threads of DNA called ___*___.

6. The outermost layer of the cytoplasm is the ___*___ membrane. Though this cannot be seen under a light microscope, it appears as a boundary.

7. In plant cells the cytoplasm is surrounded by a ___*___ ___*___ made mainly of ___*___.

8. Photosynthetic plant cells have ___*___ in their cytoplasm; they are the site of ___*___. They are near the edge of the cell because of a large fluid-filled ___*___, which pushes the cytoplasm into a thin layer lining the cell wall.

17 Cells Under the Electron Microscope

Limits to the light microscope — resolving power

The most powerful light microscopes magnify about 1400 times. Though higher magnifications are possible, no more *detail* can be seen — the picture just becomes more and more blurred. You get similar 'empty' magnification when you look at a newspaper photograph with a lens. The photograph is made up of thousands of dots; the more dots per square centimetre, the more detailed the picture. Rather than revealing more dots, a lens just enlarges the dots you can already see.

The amount of detail that can be revealed by a microscope is called the **resolving power**. This is expressed as *the closest two points can be to each other without them appearing as one*. The resolving power of a microscope depends on the objective, because the eyepiece cannot add detail that the objective has failed to pick up. The resolving power of the human eye is about 0.1 mm or 100 μm. Thus two points 90 μm apart are seen as one, while two points 110 μm apart appear separate. A good school microscope has a resolving power of about 0.5 μm (about half the width of an average bacterium), giving a useful magnification of about 400. The best light microscopes can resolve points about 0.2 μm (200 nm) apart.

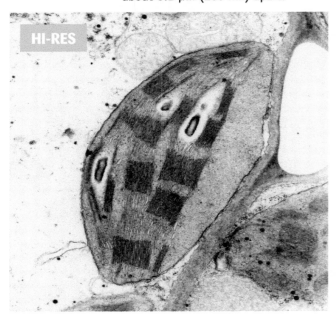
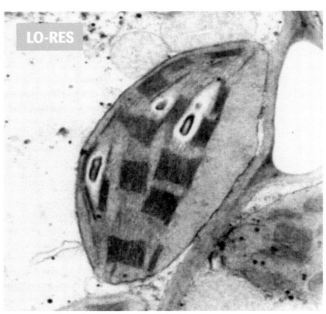

Fig 17.1 Chloroplasts as seen under similar magnification but different resolution

ISBN: 9780170214094

The reason why things smaller than about 200 nm cannot be seen with a light microscope is related to the fact that light travels in *waves*. The distance between successive wave peaks is called the *wavelength*. Objects closer than about half the wavelength of the light cannot be distinguished. Of visible light, blue has the shortest wavelength — about 400 nm — so using blue light enables objects about 200 nm to be distinguished. The thickness of a DNA molecule is about 2 nm — about 100 times too small to be seen using the light microscope.

Cells under the electron microscope

In the second half of this century a new kind of microscope was invented, called the **electron microscope**. It uses the fact that a beam of electrons has wavelike properties. A high voltage electron beam has a much shorter wavelength than light, and electron microscopes can resolve points as close together as 0.5 nm, with useful magnifications of up to 500 000 times. The electron microscope differs in several important ways from the light microscope (Fig 17.2):

- Instead of using light rays focused by glass lenses, it uses a beam of electrons, focused by powerful electromagnets.

- Electrons do not pass easily through air, so the interior of the microscope has to be a vacuum.

- Because organisms cannot survive in a vacuum, only dead specimens can be studied.

- The specimens must be much thinner than those used with light microscopes.

- Instead of staining the specimen with a chemical that absorbs light, it is 'stained' with a substance that absorbs electrons, such as a heavy metal.

- Since the eye cannot see electrons, the image is produced on a fluorescent screen instead of the retina of the eye.

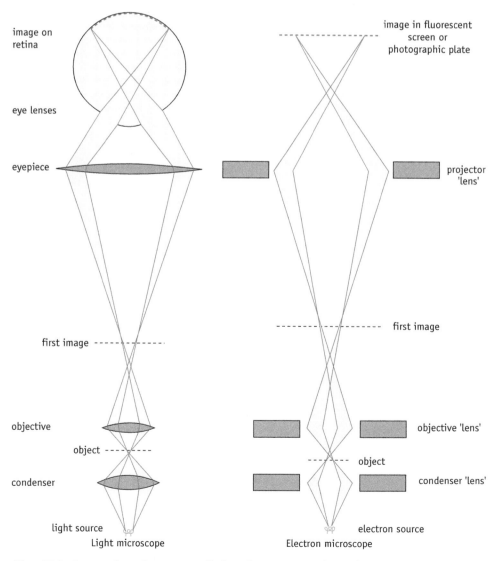

Fig. 17.2 Comparison between a light microscope and an electron microscope

ISBN: 9780170214094

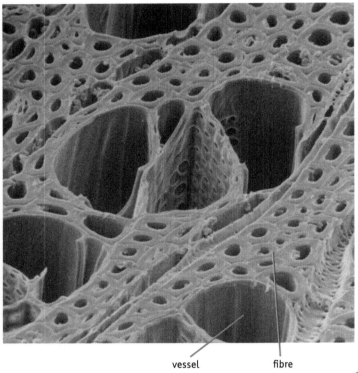

Fig. 17.3 Scanning electron micrograph of cut surface of wood of black beech (Courtesy of Dr Brian Butterfield, University of Canterbury)

The most powerful electron microscopes are called *transmission electron microscopes* because they use beams of electrons that pass through the specimen. Another kind of electron microscope uses electrons that are scattered by the specimen. Though less powerful, these *scanning electron microscopes* give a three-dimensional image of a surface.

The invention of the electron microscope enabled biologists to probe more deeply than ever before into the structure of cells. They were shown to have a complex organisation, with distinct membrane-bound parts called **organelles**, specialised for different functions. Some organelles can be seen under the light microscope (e.g. the nucleus and chloroplasts), but most can only be seen under the electron microscope. An electron microscope view of an animal cell is shown in Fig 17.4, and a plant cell is shown in Fig 17.5.

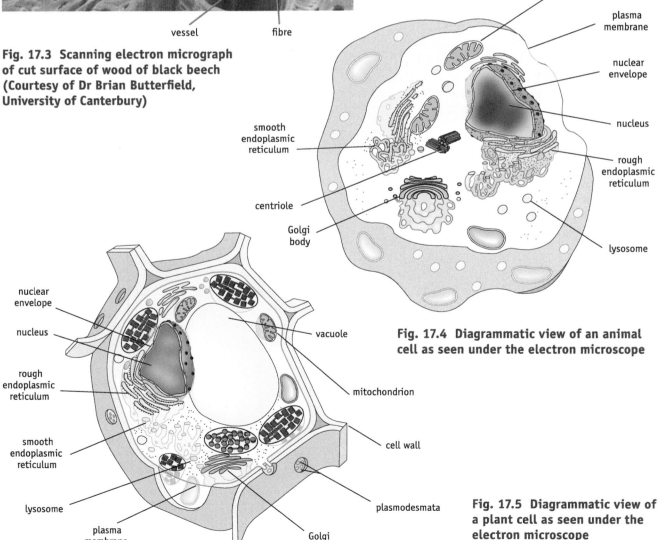

Fig. 17.4 Diagrammatic view of an animal cell as seen under the electron microscope

Fig. 17.5 Diagrammatic view of a plant cell as seen under the electron microscope

ISBN: 9780170214094

Eukaryotic and prokaryotic cells

When bacteria began to be studied under the electron microscope, it became clear that their cell structure is fundamentally different from that of animals, plants and fungi. As a result, organisms were divided into two groups:

- **Prokaryotes**, in which the DNA is not enclosed in a nucleus; they include bacteria.

- **Eukaryotes**, in which the chromosomes are open-ended and enclosed in a distinct nucleus, enclosed in a double layered, nuclear *envelope*. There is also a complex internal membrane system (endoplasmic reticulum), together with organelles such as Golgi bodies and mitochondria. Eukaryotes include animals, plants and fungi.

The fundamental differences between a prokaryotic and a eukaryotic cell are shown in Fig. 17.6 and summarised in Table 17.1. The rest of this chapter applies to eukaryotes.

	Prokaryote	Eukaryote
Plasma membrane	Present	Present
Size	Generally less than 2 μm	Generally larger than 10 μm
Nuclear envelope	Absent	Present
Chromosomes	Open-ended or linear	Closed loop
Endoplasmic reticulum	Absent	Present
Mitochondria	Absent	Present
Golgi bodies	Absent	Present
Ribosomes	Smaller	Larger

Table 17.1 Similarities and differences between prokaryotic and eukaryotic cells

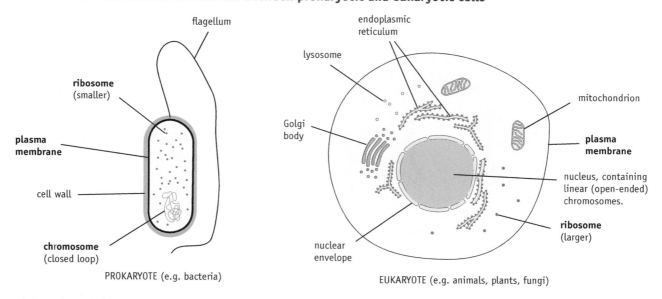

Fig. 17.6 Similarities and differences between prokaryotic and eukaryotic cells. Structures labeled in bold are universally present (or nearly so).

The plasma membrane

The plasma membrane forms the boundary between the cell and its surroundings, and used to be called the 'cell membrane'. It has three important functions:

- It acts as a kind of 'molecular sieve', allowing certain molecules to enter and leave the cell (e.g. oxygen and CO_2), while preventing the movement of others (see Chapter 21). It is therefore *partially permeable*.

- It is concerned with *active transport* of substances into and out of the cell (see Chapter 21).

- It enables the cells of multicellular organisms to 'recognise' each other, and in more complex animals, to distinguish the body's own cells ('self') from foreign cells ('non-self').

- It may contain protein molecules that act as 'receptors' for hormones.

The plasma membrane is about 7 nm thick, and is thus far too thin to be resolved by the light microscope. It consists of four kinds of constituent (Fig 17.7):

- **Phospholipid** molecules, forming a two-layered sheet or **bilayer**. The polar (water-attracting) 'heads' of the phospholipids always face out of the membrane into the water, while the 'tails' lie buried in the interior. Although they keep this orientation, they move freely past each other — two molecules that are next to each other one moment but a second later may have moved 1 μm (the width of a bacterium) apart.

- **Globular proteins**, which float like icebergs in the bilayer. Some of the proteins span the whole thickness of the membrane and are involved in transporting substances into and out of the cell (Chapter 21). Others are confined to one half of the membrane.

- **Carbohydrates**. These are anchored to proteins on the outer surface of the membrane, forming *glycoproteins*. They serve as 'markers', enabling cells to 'recognise' one another. They may also be anchored to lipids, forming *glycolipids*.

- **Cholesterol** molecules are wedged between the tails of the phospholipids. Despite its bad publicity, it is an essential constituent of all cell membranes except in prokaryotes. It helps to prevent the bilayer becoming solid at low temperatures, while helping to make it less fluid at higher temperatures.

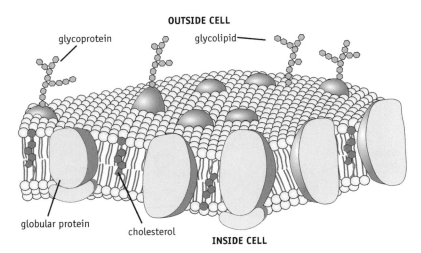

Fig. 17.7 Structure of the plasma membrane

The plasma membrane is but the outermost of a complex system of membranes, all of which have a similar structure. Added together, the area of these internal membranes is far more extensive than that of the plasma membrane — in a liver cell, for example, it is about 50 times greater.

Just as the plasma membrane separates the inside of the cell from the outside, the membranes inside a cell make it possible to have distinct compartments, in which different chemical processes occur.

The nucleus

The nucleus is the 'information library and control centre' of the cell. It contains the chromosomes, which consist of the genetic material (DNA) and also protein. Except during cell division, the chromosomes cannot be seen even with the electron microscope because they are too thin and tangled.

The layer separating the nucleus from the cytoplasm was originally called the 'nuclear membrane', but the electron microscope has shown that it is actually two-layered, and is now called the **nuclear envelope** (Fig 17.8). It is penetrated by many pores through which there is a constant two-way traffic of molecules between nucleus and cytoplasm. For example, RNA copies of the genes pass from nucleus to cytoplasm, while raw materials pass in the reverse direction.

The endoplasmic reticulum

The invention of the electron microscope showed that the cytoplasm has a highly organised internal structure. One of the most complex structures in the cytoplasm of all cells except bacteria is the **endoplasmic reticulum** (*reticulum* means 'network'). The endoplasmic reticulum (ER) is invisible under the light microscope, but under the electron microscope it appears as a complex system of membrane-bound spaces or **cisternae** (Fig. 17.8).

In electron micrographs the cisternae appear to be separate from each other, but they are believed to be interconnected, forming of a single continuous compartment, the *ER lumen*. The liquid surrounding the ER is the **cytosol**. As shown in Fig 17.9, it communicates with the nucleus via the pores in the nuclear envelope. The ER lumen is continuous with the cavity of the nuclear envelope.

The endoplasmic reticulum can be divided into two regions:

- In the *rough* ER the cisternae are flattened. It is so-called because its membrane is studded with many particles called **ribosomes**. Their function is to make proteins by 'translating' RNA copies of the genes. Ribosomes are about 20 nm across and consist of proteins and RNA. They are absent from the cisternae, which have no direct connection with the nucleus. The ribosomes are assembled in the nucleus and reach the cytosol via the pores in the nuclear envelope.

- The *smooth* ER is continuous with the rough ER and consists of tubular spaces without ribosomes (Fig. 17.8). It is concerned with making steroids and other lipids, and is particularly abundant in cells making steroid hormones such as the sex hormones testosterone, oestrogen and progesterone.

Table 17.2 summarises the differences between rough and smooth ER.

Rough ER	Smooth ER
Concerned with protein synthesis	Concerned with lipid synthesis
Ribosomes present	Ribosomes absent
Cisternae are flattened	Cisternae are tubular

Table 17.2 Differences between rough and smooth endoplasmic reticulum

Not all ribosomes are associated with the rough endoplasmic reticulum; some are free in the cytosol. These *free ribosomes* make enzymes and other proteins that function in the cytosol, such as the enzymes involved in glycolysis.

The total area of the endoplasmic reticulum may be very large — in liver cells, it is about 25 times larger than that of the plasma membrane.

Fig. 17.9 The nuclear envelope with part of the endoplasmic reticulum, showing how they are continuous with each other

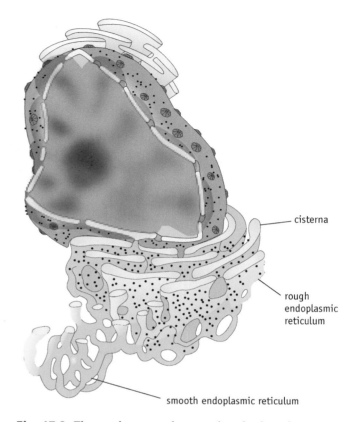

cisterna

rough endoplasmic reticulum

smooth endoplasmic reticulum

Fig. 17.8 The nuclear envelope and endoplasmic reticulum

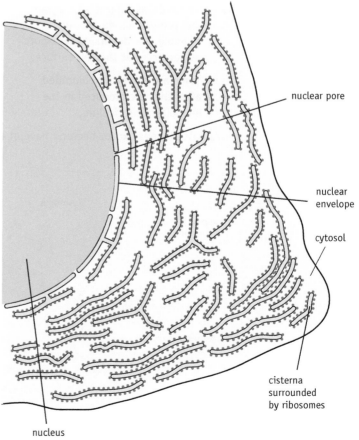

nuclear pore

nuclear envelope

cytosol

cisterna surrounded by ribosomes

nucleus

Golgi bodies

Just before the end of the last century Camillo Golgi discovered the organelle that bears his name. Much later, the electron microscope revealed a Golgi body to consist of a stack of flattened, membrane-bound sacs or cisternae, with small vesicles round the edges (Fig 17.10).

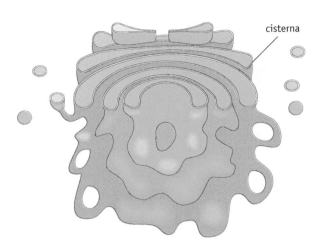

The two sides of the Golgi body are different from each other (Fig 17.11). The vesicle nearest one side of the stack is continually receiving proteins by fusing with vesicles budded off the ER. On the other side, new vesicles are being budded off for dispatch to other parts of the cell. Each cisterna is continually fusing with vesicles on one side and budding them off from the other. Proteins are thus passed from one cisterna to another, eventually reaching the 'exporting' side.

Fig. 17.10 Structure of a Golgi body

On their journey across the Golgi stack, proteins are subjected to two kinds of process:

- They are 'sorted' and enclosed in vesicles destined for different destinations. Digestive enzymes and protein hormones, for example, are destined for 'export' by secretion at the cell surface.

- Some are chemically modified after they are produced in the endoplasmic reticulum.

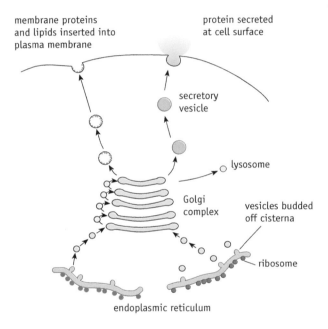

Fig. 17.11 The Golgi complex as a sorting centre

Almost all cells except bacteria have at least one Golgi body, but gland cells may have many.

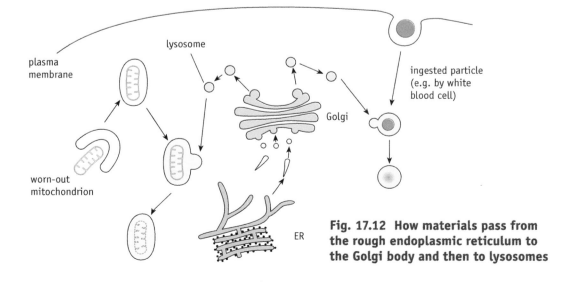

Fig. 17.12 How materials pass from the rough endoplasmic reticulum to the Golgi body and then to lysosomes

ISBN: 9780170214094

Lysosomes

Lysosomes are tiny vacuoles containing digestive enzymes, and are budded off from the Golgi bodies. They have a number of functions, for example:

- In most cells they are used to break down worn-out organelles such as mitochondria, and in some white blood cells to digest bacteria (Fig 17.12).

- They are also used in the digestion of food by single-celled organisms such as *Paramecium*.

- The shrinkage of the tail of a tadpole as the animal changes into a frog is due to the digestion of cells by enzymes released by lysosomes. A similar process occurs in a pupa as it changes into a butterfly or moth. Here many of the muscles of the caterpillar are broken down to form a soupy liquid that provides the raw materials to build the adult organs.

Mitochondria

Mitochondria (singular, *mitochondrion*) are present in almost all eukaryotic cells. They are sausage-shaped organelles about 1 µm thick and are just visible under the best light microscopes. They are known as the 'powerhouses of the cell', because they provide chemical energy in the process of respiration (Chapter 19). Each cell has many mitochondria, and the most active cells can have more than a thousand.

A mitochondrion consists of two membranes enclosing an inner space (Fig 17.13). The inner membrane contains proteins that play a key part in respiration and its area is extended by folds called **cristae** (singular, *crista*). The total area of cristae in a liver cell is about 15 times that of the plasma membrane, and in an entire human liver they have a total area of about 3000 m² — more than twice the area of an Olympic-sized swimming pool!

Mitochondria contain their own DNA and ribosomes and reproduce by dividing into two.

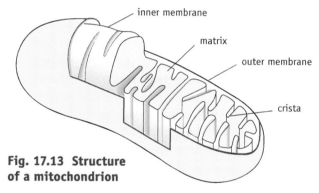

Fig. 17.13 Structure of a mitochondrion

Chloroplasts

These are the little 'factories' in which plants make sugar from CO_2 and water in **photosynthesis**. Like mitochondria, chloroplasts reproduce by division, and contain their own DNA and ribosomes.

A chloroplast contains a complex system of internal membranes surrounded by two outer membranes (Fig 17.14). The internal membranes are packed with chlorophyll and other pigments, and consist of piles of flattened sacs or **thylakoids**. Each pile is called a **granum** (plural, *grana*). It is in the thylakoids that light is converted into chemical energy. This is then used by enzymes to convert CO_2 to carbohydrate in the liquid part of the chloroplast, or **stroma**.

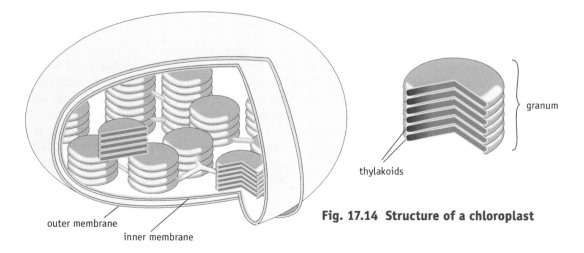

Fig. 17.14 Structure of a chloroplast

Cilia and flagella

These are hair-like outgrowths of the cytoplasm found in certain cells of animals, and in the male gametes of mosses and ferns, as well as in many single-celled organisms. They are completely absent from the kingdom Fungi. With a width of about 200 nm, they can just be seen under the light microscope. In larger animals cilia are used to move fluid over surfaces such as the lining of breathing tubes of mammals, and the gills of mussels, pipi and other bivalve molluscs.

Flagella (singular, *flagellum*) are also common in single-celled organisms. They are longer than cilia and occur in smaller numbers per cell, and also beat in a different way (Fig 17.15). Flagella also occur in larger animals as sperm tails.

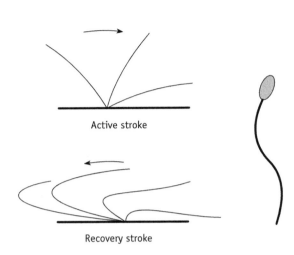

Active stroke

Recovery stroke

Fig. 17.15 The beat of a cilium (left) and a flagellum (right)

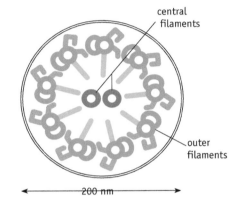

central filaments

outer filaments

200 nm

Fig. 17.16 Cross-section of a cilium

One of the most interesting things about cilia and flagella is that their internal structure is almost identical in all eukaryotic cells that have them (the flagella of bacteria are fundamentally different). In cross-section they have a ring of nine pairs of fine tubules surrounding two central ones (Fig 17.16). The mechanism of bending is complicated, but involves the two members of each outer pair sliding past each other in a coordinated sequence.

In organisms as diverse as seaweeds, ferns, earthworms, starfish and pandas, cilia and flagella have this same '9:2' arrangement of tubules.

Centrioles

Centrioles are minute, cylindrical bodies that occur as a pair just outside the nucleus. They occur in animal cells and in plants with motile gametes (e.g. mosses and ferns). They are absent from the cells of most seed-bearing plants (in which even the sperm lack flagella). The function of centrioles is to organise the formation of the *spindle* during cell division (Chapter 22).

The plant cell wall

Plant cells have a cell wall immediately outside the plasma membrane. In face view under the electron microscope the wall looks like coarse fabric, consisting of a meshwork of very fine threads of **cellulose**, called **microfibrils** (Fig 17.17). These are very strong and enable the wall to resist tension. The spaces between the microfibrils are wide enough for most molecules to pass through.

The wall that is laid down while the cell is still enlarging is the **primary wall**. After the cell has stopped increasing in size, further wall material is laid down, forming the **secondary wall**.

In many plant cells other substances may be deposited between the cellulose microfibrils. One of these is **lignin**, which binds the cellulose

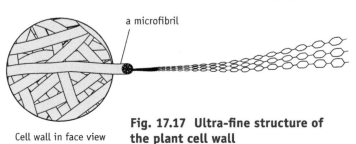

a microfibril

Cell wall in face view

Fig. 17.17 Ultra-fine structure of the plant cell wall

microfibrils together and makes cell walls stiff and resistant to buckling. As a result the cell does not cave in easily when its contents are under tension. This an important property in the water-conducting xylem cells of land plants. Another such substance is **cutin**, a greasy substance that is deposited in the outer walls of the epidermis ('skin') of leaves and stems.

Although separated by cell walls, plant cells are not completely isolated from each other. All living plant cells are interconnected by very thin strands of cytoplasm called **plasmodesmata**, which pass through the walls from one cell to the next (Fig 17.18).

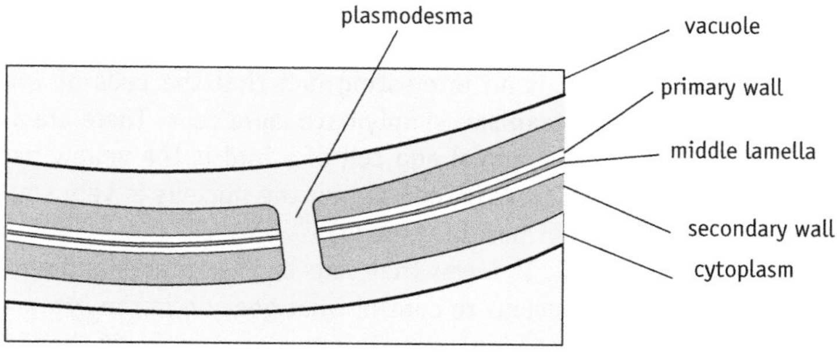

Fig. 17.18 How plasmodesmata connect adjacent plant cells

Cell specialisation — tissues

There are over 200 different kinds of cell in the human body, and a considerable (though smaller) number in flowering plants. Each kind of cell concentrates on doing a particular kind of job, and is said to be *specialised* for that function.
Figs 17.19 and 17.20 show some examples of cell types in flowering plants and in humans.

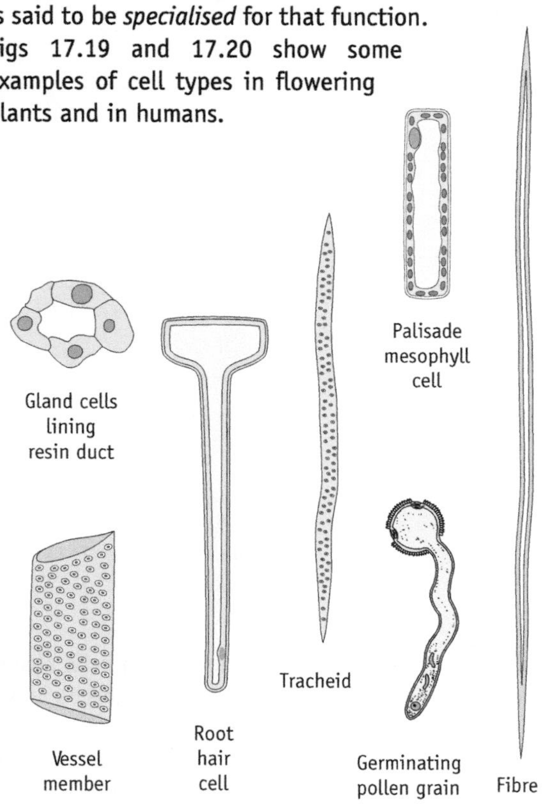

Fig. 17.19 Some examples of plant cells

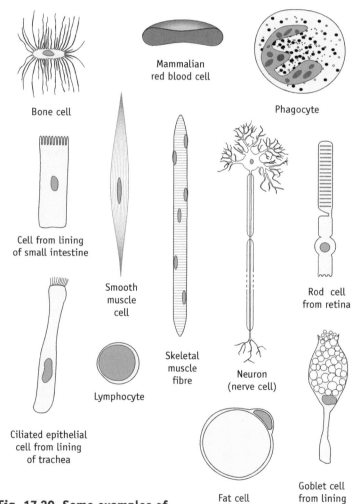

Fig. 17.20 Some examples of animal cells (not to the same scale)

Many cell types occur in groups called *tissues*. A tissue is a group of cells that cooperate together to perform a function that in many cases the individual cells cannot do. For example, the cells lining the breathing passages are organised in that they all have cilia on the same side, beating in the same direction. The cells of the epidermis of a leaf are not a random collection, but fit tightly together and secrete a cuticle on the same side. Only in such an organised group can the function be carried out.

Why are cells so small?

It is an interesting fact that the cells of whales are about the same size as those of mice. Larger organisms simply have more cells. There are a few very large cells, such as the yolky eggs of animals. The actual egg cell of a bird is the yellow part, over 99.95% of which is inactive yolk — the bit of cytoplasm containing the nucleus is very small (the albumen and shell surrounding the egg cell are not part of the egg).

It seems that cells can't get large and remain active, because it gets increasingly difficult for the nucleus to control what goes on in the cytoplasm. Chemicals have to pass from nucleus to cytoplasm, and vice versa. This two-way traffic of chemicals occurs through pores in the nuclear envelope.

The rate at which chemicals can pass from nucleus to cytoplasm depends on the *area* of the nuclear envelope. On the other hand, the amount of cytoplasm to be controlled depends on the *volume* of cytoplasm.

As a cell increases in size, the volume of the cytoplasm increases more quickly than the area of the nuclear envelope. This is because area has *two* dimensions, while volume has *three*. So, if a cell doubles in length and keeps the same shape, then the other two dimensions will also double. The volume therefore increases $2 \times 2 \times 2 = 2^3 = 8$ times. The area of the nuclear envelope only increases $2 \times 2 = 2^2 = 4$ times (since area only has two dimensions). So, if a cell doubles in its length, height and thickness, there is eight times as much cytoplasm to control, but the controlling chemicals only have four times as much nuclear envelope to get through. Fig 17.21 shows these effects for an imaginary cubical cell (since cubes are easy to work with). Thus as cells get larger the nucleus becomes increasingly unable to control what goes on in the cytoplasm.

Fig. 17.21 How surface and volume change at different rates

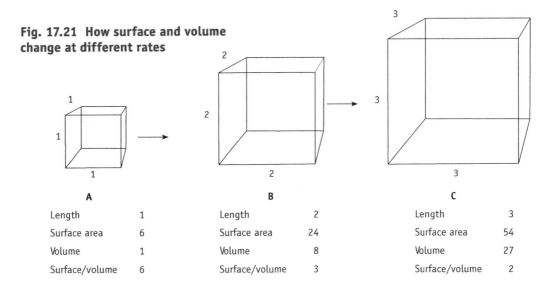

	A		**B**		**C**
Length	1	Length	2	Length	3
Surface area	6	Surface area	24	Surface area	54
Volume	1	Volume	8	Volume	27
Surface/volume	6	Surface/volume	3	Surface/volume	2

Another thing that increases with cell size is the plasma membrane, through which cells absorb oxygen and nutrients. In a unicellular organism the rate at which oxygen diffuses into the body will be limited by the area of the plasma membrane. The demand for oxygen, however, will depend on the volume of the body. Hence, if the animal were to double in length while maintaining the same shape, it would need eight times as much oxygen but it would only have four times as much plasma membrane through which to obtain it.

In multicellular organisms the situation is not so simple, because cells are very close together. If the gap between them is too narrow for liquid to circulate freely, oxygen may have to enter one cell by diffusing from its neighbour.

Summary of key facts and ideas in this chapter

○ Objects viewed under the electron microscope are usually measured in *nanometers* (nm). 1 nm = 10^{-9} of a metre, or 0.001 of a µm.

○ The most important feature of a microscope is its *resolving power*, or ability to distinguish two points as separate. The best light microscope cannot distinguish points closer together than 200 µm.

○ The electron microscope can resolve points as close together as a quarter the width of a DNA molecule, giving useful magnifications of up to 500 000 times.

○ Based on cell structure, organisms fall into two groups: *prokaryotes*, which include bacteria, and *eukaryotes*, which include animals, plants and fungi.

○ All cells are surrounded by a *plasma membrane*, which consists of two layers of phospholipid in which are embedded proteins and other molecules.

○ The plasma membrane acts as a selective barrier, allowing certain substances to pass through while preventing others.

○ The nucleus of a eukaryotic cell is bounded by a two-layered *nuclear envelope* which communicates with the cytoplasm via large numbers of pores.

○ Eukaryotic cells contain various structures called *organelles*, specialised for particular functions. Those that can only be seen under the electron microscope include the plasma membrane, nuclear envelope, endoplasmic reticulum, ribosomes and lysosomes. In addition, fine details of mitochondria, chloroplasts and Golgi bodies can be seen.

○ The *endoplasmic reticulum* (ER) is a complex network of membrane-bound spaces or *cisternae*, and is concerned with the synthesis and processing of proteins (rough ER) and the synthesis of lipids (smooth ER).

○ The endoplasmic reticulum is surrounded by the *cytosol*, a liquid that communicates with the nucleus via pores in the nuclear envelope.

○ The rough ER is so-called because of the large number of *ribosomes* in the cytosol; these tiny granules are the site of protein synthesis.

○ Golgi bodies are concerned with the modification of newly-synthesised proteins and with their subsequent destinations.

○ Mitochondria are the site of *respiration*, in which most of the cell's ATP is produced.

○ Chloroplasts are only present in plant cells and are the site of photosynthesis. A chloroplast contains *grana*, each granum consisting of a pile of flattened sacs called *thylakoids*. The membranes of the thylakoids contain chlorophyll and other photosynthetic pigments. The grana are surrounded by a liquid *stroma*.

○ *Lysosomes* are tiny vesicles containing digestive enzymes, whose functions include the destruction of worn-out mitochondria.

○ The *cilia* and *flagella* of eukaryotic cells have a characteristic internal structure, consisting of two central hollow tubules surrounded by nine pairs of tubules.

○ *Centrioles* are present in animal cells and in cells of those plants with motile male gametes (mosses and ferns), and are concerned with the organisation of the spindle in cell division.

○ The cell wall of plant cells is permeable to most substances but is strongly resistant to tension. As a result plant cells can develop considerable internal pressure.

○ Adjacent plant cells communicate with each other by slender threads of cytoplasm called *plasmodesmata*.

○ In plants and animals, most cells are organised into groups called *tissues*. Each tissue is specialised for carrying out a particular function.

○ Most cells are small. The reason is probably connected with the fact that as an object increases in size, its various dimensions increase at different rates. For example if a cell doubles in its length (keeping the same shape), the area of plasma membrane and nuclear envelope increases four times but the volume of the cell increases eight times (see text for more detail).

Test your basics

Copy and complete the following sentences. In some cases the first letter of a missing word is provided.

1. In the performance of a microscope, the most important thing is not its magnification but its ___*___ ___*___, or its ability to distinguish ___*___. It is normally expressed as the closest together two points can be yet still remain distinct. For the light microscope this is about 200 nm. The best electron microscopes can distinguish points 0.5 nm apart.

2. A eukaryotic cell contains a number of membrane-bound structures called ___*___, specialised for particular functions.

3. Organisms in which the chromosomes are enclosed in a distinct nucleus are called ___*___. Organisms in which the chromosomes are not so enclosed are called ___*___. Examples of these organisms are ___*___.

4. The membrane that regulates the movement of substances into and out of the cell is the ___*___. It cannot be resolved by the light microscope because it is only about ___*___ nm thick. Like other membranes in cells, it consists mainly of a ___*___ bilayer, in which are embedded ___*___ proteins.

5. The nucleus is separated from the cytoplasm by a double-layered nuclear ___*___. It contains hundreds of ___*___ through which substances can pass.

6. The ___*___ ___*___ (ER) consists of a system of interconnected membrane-bound spaces in the cytoplasm. The spaces are called ___*___ and the liquid in which they are suspended is the ___*___.

7. The cisternae are continuous with the cavity inside the nuclear envelope, and the cytosol is continuous with the nucleus via the pores in the nuclear envelope.

8. The ___*___ ER contains large numbers of granules called ___*___, and is particularly well developed in cells active in making ___*___.

9. The ___*___ ER lacks ribosomes and is particularly well developed in cells actively making ___*___.

10. The ___*___ body is a series of flattened membrane-bound sacs and is concerned with the modification of ___*___ after their production, and also their 'packaging' into vesicles according to their destinations.

11. Lysosomes are tiny vesicles containing ___*___ enzymes, and are used to break down worn-out organelles. They are also responsible for the digestion of bacteria engulfed by white blood cells.

12. Mitochondria are sausage-shaped organelles concerned with the production of the ___*___ used to 'drive' energy-requiring processes in cells. Their inner membrane is extended into folds called ___*___, which greatly increase its ___*___ ___*___.

13. Chloroplasts are the ___*___ concerned with ___*___. A chloroplast contains many ___*___, each consisting of a pile of flattened sacs called ___*___. Like mitochondria, they contain their own DNA in the form of a closed loop.

14. Cilia and flagella are hair-like extensions of the ___*___ ___*___, containing a bundle of ___*___ paired hollow tubes surrounding ___*___ central tubes. They are used to ___*___ male gametes of animals and some plants.

15. In the cells of animals and some plants, the spindle that is formed in cell division develops under the influence of two cylindrical structures called ___*___, which lie just outside the nucleus.

16. The cell wall in plants consists largely of a meshwork of ___*___, consisting of the carbohydrate ___*___. In living plant cells the cytoplasm of adjacent cells is connected by threads of ___*___ called ___*___.

17. In most animals and plants, cells are organised in groups called ___*___.

18. The reason why cells are so small is probably connected with the fact that small structures have a large ___*___ compared with their ___*___.

ISBN: 9780170214094

18 The Chemicals of Life

There are thought to be at least ten million species of organism, each unique. Yet despite this diversity, they all consist of the same few chemical types. The main ones are:

- Water
- Proteins
- Carbohydrates
- Nucleic acids
- Fats and phospholipids
- Inorganic ions
- Steroids

and a few others we don't need to mention.

Water

Water (H_2O) is not only the smallest molecule in living matter, it is the major constituent, forming about 80% of most living cells. It is so important to life that most cells cannot tolerate much change in their water content. Here are some of its most important properties:

- It acts as a *solvent* in which the chemical reactions of metabolism occur and in which substances are transported round the body and wastes excreted.

- It plays an essential part in maintaining the structure of cell membranes and globular proteins.

- Its incompressibility enables it to provide *hydrostatic* support in living plant cells and in many soft-bodied animals such as sea anemones, earthworms and snails.

- It has a high heat capacity, meaning that it can absorb a lot of heat without much rise in temperature. As a result lakes and oceans provide a much more thermally stable environment than land habitats.

- It needs a lot of heat to evaporate water (notice how cool your finger feels if you wet it and wave it in the air). This enables many terrestrial (land-living) organisms to keep cool by the evaporation of water.

- Ice is less dense than water, so it floats. Because of this, lakes and ponds freeze from the top downwards. The layer of ice helps to insulate the water below, protecting it and the organisms in it from freezing.

- Wherever water meets air, the water behaves as if it has an elastic 'skin'. This is called *surface tension,* and enables aquatic insects such as pond skaters to use their waxy feet to walk on the surface of the water. Other insects such as mosquito larvae are able to breathe at the surface by hanging from it. Surface tension is the reason why a drop of water cannot easily enter greasy pores such as the stomata of plants.

ISBN: 9780170214094

Water is a *very* peculiar substance. It is liquid at room temperature, when other substances with similar-sized or larger molecules are gases (e.g. methane, ammonia, hydrogen sulfide). This is because water molecules are strongly attracted to each other. Although each water molecule is electrically neutral, the oxygen is so 'greedy' for electrons that it takes more than its 'fair' share away from the two hydrogen atoms. The result is that the oxygen has a slight negative charge (⁻) and each hydrogen atom has a slight positive charge (⁺). A molecule that is neutral as a whole but has differently charged ends is said to be **polar** (Fig. 18.1).

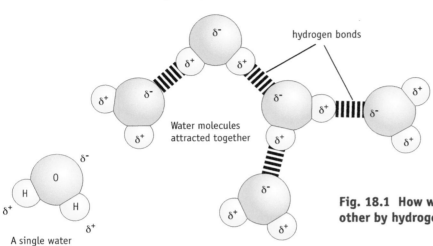

hydrogen bonds

Water molecules attracted together

A single water molecule

As a result the hydrogen atom of each water molecule is attracted to an oxygen of a neighbouring water molecule. This is an example of **hydrogen bonding**, which also plays an essential part in maintaining the shape of protein molecules and in the structure of cell membranes (see later in this chapter).

Fig. 18.1 How water molecules are attracted to each other by hydrogen bonding

Organic molecules

Organic molecules contain carbon and nearly always hydrogen and (in organisms) oxygen. Many organic molecules also contain nitrogen, sulfur and phosphorus as well. Of all the elements essential for life, carbon is perhaps the most special. It is the only element whose atoms easily form long chains and rings. This makes it possible for large, complex molecules to be built up.

Most organic compounds belong to a few families, the most important being carbohydrates, lipids (e.g. fats), proteins, and nucleic acids.

Carbohydrates

Carbohydrates contain the elements carbon, hydrogen and oxygen, the hydrogen and oxygen being in the ratio of 2 : 1. Because this is the same ratio as in water, biochemists used to think that carbohydrates were formed by combination between carbon and water. *Hydrate* means 'containing water', and the general formula for carbohydrate approximates to CH_2O.

Carbohydrates have three main functions in organisms:

- They can act as fuels to provide energy, e.g. glucose.

- They may provide *support,* for example **cellulose**.

- As constituents of the plasma membrane they are important in the way cells 'recognise' one another (see Chapter 17).

Carbohydrates can be divided into three groups: simple sugars, complex sugars and polysaccharides.

Simple sugars (monosaccharides)

Examples of simple sugars are *glucose* and *fructose*. They are called **hexose** sugars because each molecule has six carbon atoms ('hexa' means six). Although they both have the formula $C_6H_{12}O_6$, their atoms are arranged differently.

ISBN: 9780170214094

Glucose is the main fuel used by cells, and is also the carbohydrate transported in the blood. *Ribose* ($C_5H_{10}O_5$) is a **pentose** because it has five carbon atoms, and is a constituent of ATP (see Chapter 19). Deoxyribose is another pentose, and is a constituent of DNA (Chapter 23). These sugars have ring-shaped molecules, as shown in Fig. 18.2.

All simple sugars give a red precipitate when boiled with Benedict's solution (blue). In doing so they are said to be *reducing sugars*.

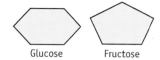

Fig. 18.2 Molecular shapes of glucose and fructose

Complex sugars

Most complex sugars consist of two simple sugars joined together and are called **disaccharides**. Like simple sugars, complex sugars are soluble and taste sweet. A common example is *sucrose* or cane sugar, which is the main form in which energy is transported around a plant. A sucrose molecule is formed by joining glucose and fructose and removing a molecule of water.

$$C_6H_{12}O_6 \ + \ C_6H_{12}O_6 \quad \longrightarrow \quad C_{12}H_{22}O_{11} \ + \ H_2O$$
$$\text{glucose} \quad \text{fructose} \qquad\qquad \text{sucrose} \quad \text{water}$$

The joining together of two molecules with the production of water is called a *condensation* reaction. The reverse — splitting a molecule into smaller molecules by the addition of water — is called *hydrolysis*. All the reactions of digestion are hydrolytic reactions. Other disaccharides are *lactose* (milk sugar) and *maltose* (malt sugar).

All common disaccharides except sucrose are reducing sugars.

Polysaccharides

These are very complex carbohydrates, formed by linking thousands of simple sugars together (*poly* means 'many'). Polysaccharides are insoluble in cold water, have no sweet taste, and do not reduce Benedict's solution. The most important are *starch*, *glycogen* and *cellulose*.

Fig. 18.3 Part of an amylose molecule

Starch is the main energy storage carbohydrate of plants and is stored in many tubers and seeds. 'Starch' is actually a mixture of two polysaccharides:

- **Amylose** consists of unbranched helical chains of glucose units (Fig. 18.3). The helix is just wide enough for an iodine molecule to fit inside. This forms a deep blue starch-iodine complex, which is used to test for starch.

- **Amylopectin** consists of branched chains. **Glycogen** is similar to amylopectin and is stored in the liver and skeletal muscles (Fig. 18.4).

Cellulose forms the main constituent of plant cell walls. It consists of thousands of unbranched chains of glucose units, linked in a different way from those in starch. Adjacent cellulose molecules are cross-linked by hydrogen bonds to form rope-like bundles or *microfibrils*, which can be seen under the electron microscope (Fig. 18.5). Very few animals can produce a cellulose-digesting enzyme; most plant eaters rely on microorganisms in their guts to do this.

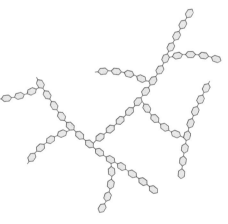

Fig. 18.4 Part of a glycogen molecule

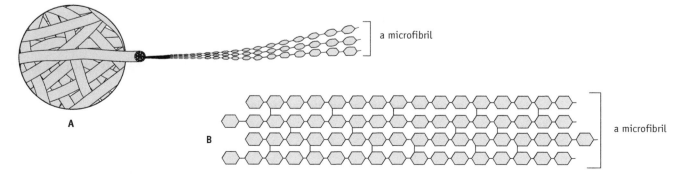

Fig. 18.5 (A) Part of a cellulose wall shown in face view, with one microfibril enlarged. (B) A short section of four cellulose molecules, cross-linked by hydrogen bonds.

ISBN: 9780170214094

Chitin is similar to cellulose except that it also contains nitrogen. It forms the main constituent of the cuticle of insects and other arthropods, and also the cell walls of fungi.

Like disaccharides, polysaccharides are built up by condensation reactions. They can also be hydrolysed into simple sugars by boiling with water and a little acid as a catalyst.

Extension: More about starch and cellulose

Although glucose exists mainly as ring-shaped molecules, it actually exists in more than one molecular form, as shown in Fig. 18.6. In solution, a tiny proportion is an open-chain, but this can give rise to two alternative ring forms, called   and   glucose. Formation of the ring results in a new –OH group on carbon number 1. In -glucose this is formed *below* the ring, and in -glucose it is formed above.

Whereas starch and glycogen consist of -glucose units, cellulose consists of -glucose units. This seemingly trivial difference accounts for the fact that enzymes that can break down starch cannot digest cellulose.

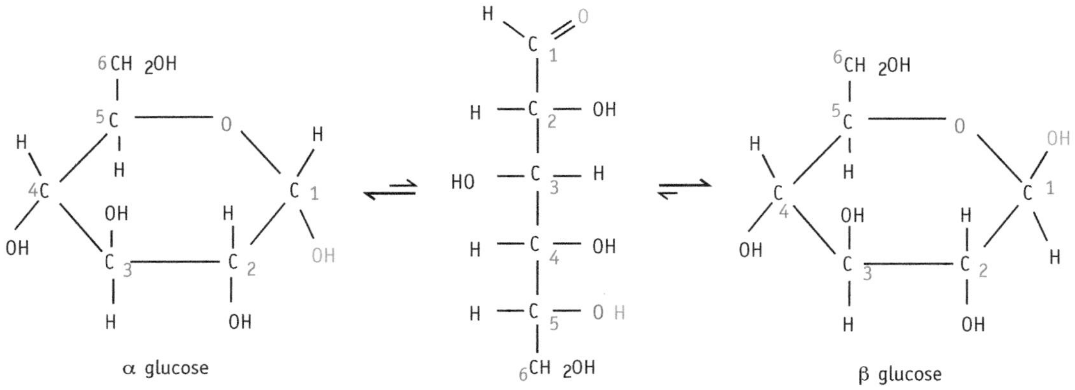

α glucose   Chain form   β glucose

Fig. 18.6 Three different forms of glucose

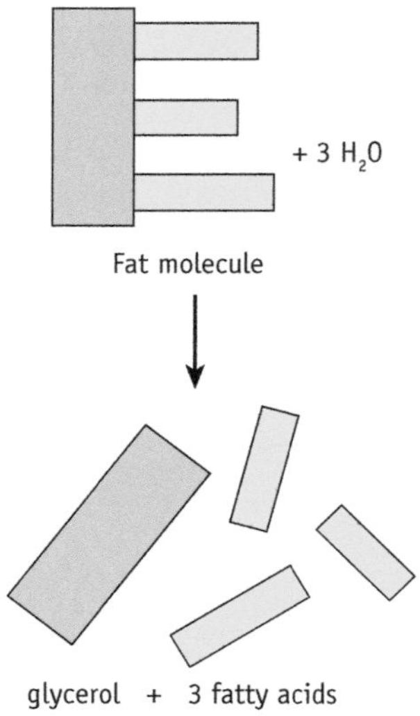

Lipids

Fats, oils, phospholipids and steroids are collectively called *lipids*. Their common feature is that they will dissolve in organic solvents such as alcohol. Though much less diverse than proteins they play a variety of roles, some of which are essential to life, for example:

- Phospholipids form the structural framework of *cell membranes*.

- Fats are an important *energy store* in animals and many seeds.

- Some *hormones*, for example the sex hormones, are steroids.

- In mammals, fat is stored under the skin in a layer of fat-storage cells which act as a *heat insulator*. It also helps protect against bruising.

- In some single-celled organisms fat droplets give *buoyancy*.

Fat molecule

+ 3 H_2O

glycerol + 3 fatty acids

Fig. 18.7 A fat can be broken down into fatty acids and glycerol

Fats

Fats contain carbon, hydrogen and oxygen, but there is proportionately much less oxygen than in carbohydrates. It therefore needs more oxygen to burn fat than carbohydrate, and more energy is released in the process.

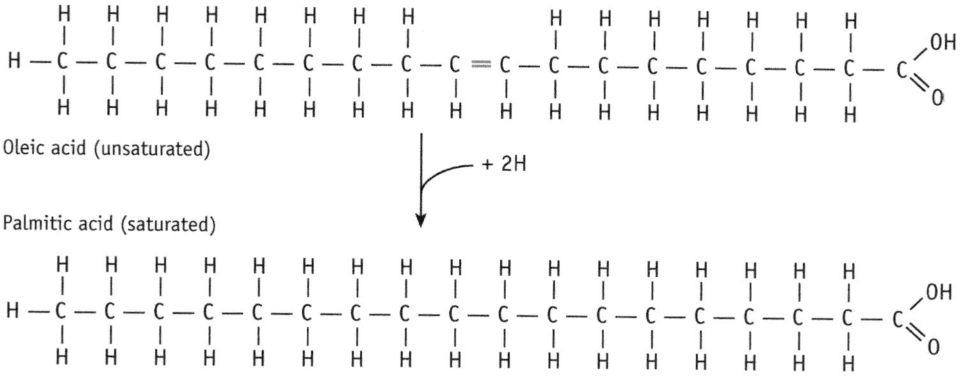

Oleic acid (unsaturated)

+ 2H

Palmitic acid (saturated)

Fig. 18.8 Saturated and unsaturated fatty acids

Fig. 18.7 shows that a fat molecule can be split by hydrolysis into a molecule of glycerol and three fatty acid molecules.

A fatty acid molecule consists of a long hydrocarbon chain with an acid (COOH) group at one end. The many different kinds of fatty acid differ in the length of the hydrocarbon chain and also in whether or not they are *saturated* with hydrogen. A saturated fatty acid contains as much hydrogen as it can (because the hydrocarbon chain has no double bonds). An unsaturated

$$CH_2 - OH \quad\quad HO-\overset{\overset{O}{\|}}{C} - R_1$$

$$CH - OH \quad\quad HO-\overset{\overset{O}{\|}}{C} - R_2$$

$$CH_2 - OH \quad\quad HO-\overset{\overset{O}{\|}}{C} - R_3$$

$3H_2O$

$$CH_2 - O \quad -\overset{\overset{O}{\|}}{C} - R_1$$

$$CH - O \quad -\overset{\overset{O}{\|}}{C} - R_2$$

$$CH_2 - O \quad -\overset{\overset{O}{\|}}{C} - R_3$$

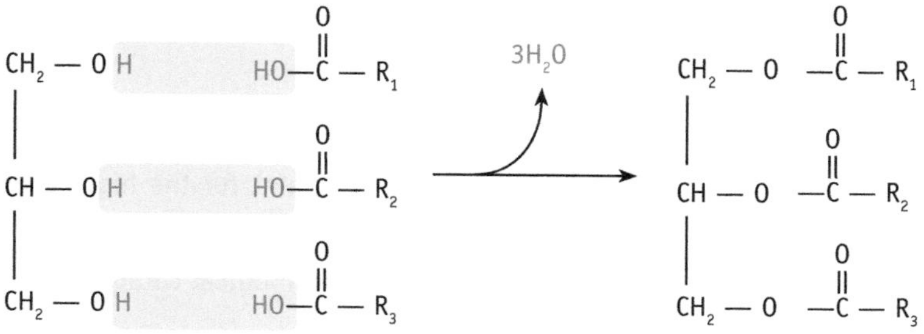

Fig. 18.9 Formation of a fat

fatty acid contains one or more double bonds and can combine with more hydrogen (Fig. 18.8). Plant fats contain mainly unsaturated fatty acids. They are usually liquid at room temperature and are called *oils* (e.g. palm oil, sunflower oil). Animal fats contain mainly saturated fatty acids and are usually solid at room temperature (e.g. lard, butter).

A fat is formed by condensation, the reverse of hydrolysis (Fig. 18.9). Each –OH group of the glycerol combines with a –COOH group of a fatty acid. In the scheme shown, the hydrocarbon chains of the fatty acids are represented by R_1, R_2 and R_3.

Phospholipids

Phospholipids differ from fats in that one of the three fatty acids is replaced by a group containing phosphate. Because this is electrically charged it is attracted to water, but the rest of the molecule is not (Fig. 18.10).

As a result, phospholipid molecules tend to form themselves into a double sheet or **bilayer**, the water-attracting ends facing out into the water and the other ends facing inwards towards each other (Fig. 18.11). All cell membranes consist partly of a lipid bilayer.

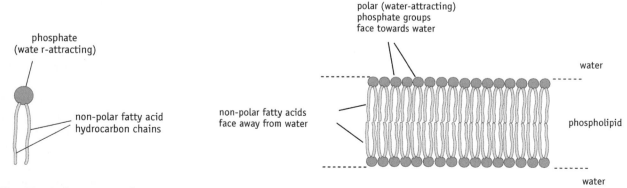

phosphate
(wate r-attracting)

non-polar fatty acid hydrocarbon chains

polar (water-attracting) phosphate groups face towards water

water

non-polar fatty acids face away from water

phospholipid

water

Fig. 18.10 Structure of a phospholipid molecule

Fig. 18.11 How phospholipid molecules associate to form a sheet-like bilayer

ISBN: 9780170214094

Steroids

Though grouped with fats because they are insoluble in water but soluble in alcohol, steroids are chemically quite unrelated to fats and cannot be hydrolysed into smaller molecules. Cholesterol is a constituent of cell membranes, and some hormones are steroids, e.g. the *sex hormones*, oestrogen, progesterone and testosterone. Vitamin D is also a steroid.

Fig. 18.12 shows a familiar steroid molecule — cholesterol. Like all steroids it consists of four joined rings, the 'corner' of each ring is occupied by a carbon atom.

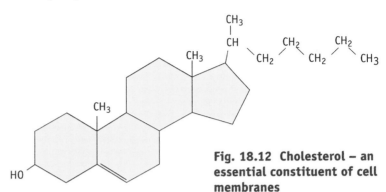

Fig. 18.12 Cholesterol – an essential constituent of cell membranes

Proteins

Next to water, proteins account for the highest proportion of living cells. They perform a very wide range of functions, for example:

- Nearly all *enzymes* (biological catalysts) are proteins. Without enzymes, chemical reactions in cells would not occur.

- Proteins form an important part of all *cell membranes*.

- Some have a *structural* function, for example *collagen* in tendons and ligaments, and *keratin* in hair, nails and feathers.

- Some *hormones* are proteins, for example insulin.

- The *antibodies* that help defend more complex animals against disease are all proteins.

- Some proteins are important in *transport;* for example haemoglobin transports oxygen in many animals.

- They are important in *movement*; the pulling apart of chromosomes, the beating of cilia and the contraction of muscle are all due to the action of proteins.

Proteins are among the most complex substances known. An average sized protein molecule has a diameter of about 5–7 nm, or about the thickness of a cell membrane, but some are much bigger. As well as carbon, hydrogen and oxygen, all proteins contain nitrogen, and most (if not all) contain sulfur. Proteins are *polymers*, consisting of long chains of **amino acids**. They are extremely complex because each amino acid subunit can be any one of 20 different kinds. Fig. 18.13 shows the general formula for an amino acid.

The $-NH_2$ group is called an **amino** group and is *basic*. The $-COOH$ is a **carboxyl** group and is *acidic*. The part indicated by 'R' differs from one kind of amino acid to another. In two of the amino acids the R-group contains sulfur.

Two amino acids can join together, the amino group of one amino acid reacting with the carboxyl group of the other to form water (Fig. 18.14).

This is a *condensation* reaction and the result is a *dipeptide*. The

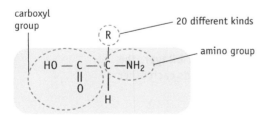

Fig. 18.13 General formula for an amino acid

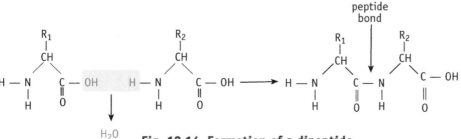

Fig. 18.14 Formation of a dipeptide

ISBN: 9780170214094

link between two amino acids is called a **peptide bond**. If more amino acids are added, *polypeptides* are produced, one water molecule being formed for every peptide bond formed. A protein molecule is simply one or more long polypeptides. Most polypeptides consist of a few hundred amino acids, but some contain over a thousand

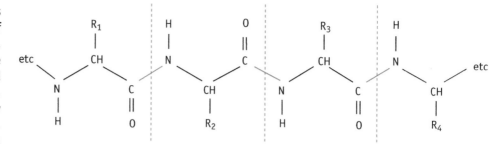

Fig. 18.15 Part of a polypeptide chain

(the largest known is the muscle protein *titin*, consisting of 26 926 amino acids). Fig. 18.15 shows part of a polypeptide chain, the dotted lines passing through the peptide bonds.

The 'backbone' of the peptide chain is the same for all proteins. How can there be so many different proteins? The answer lies in the R-groups. Since each can be of 20 different kinds, the number of different amino acid sequences is unimaginably large.

Take insulin, for example. This is a small protein consisting of 51 amino acids. The number of ways these same 51 amino acids could be rearranged, without using any others, is about 10^{48}. If each of the 51 amino acids could be *any* of the 20 types, the number of possible sequences would be 20^{51}, or 10^{66}. Compare this with the estimated number of electrons in the universe — about 10^{80}.

Although there are millions of different proteins in nature, these are obviously only a minute proportion of the possible proteins that could, in principle, exist.

One of the most important things about a protein molecule is its *shape*. This depends on its amino acid sequence, *which is unique for each protein*. To make a protein, then, a cell must have information for joining the amino acids *in the correct order*. This information is contained in DNA (see Chapter 23).

Some proteins consist of more than one polypeptide chain, loosely held together. Haemoglobin for instance consists of four polypeptide chains of two kinds — two   chains and two   chains.

Proteins can be divided into two main groups:

- Fibrous proteins. These have long, rope-like molecules, and most have a *mechanical* function, providing strength. Examples are **collagen**, the main constituent of tendon and ligament, and **keratin**, which occurs in nails, hair and the outer layer of the skin.

- Globular proteins. In a globular protein the peptide chain is coiled and folded into a ball-like molecule, e.g. enzymes. Many globular proteins consist of more than one polypeptide chain, loosely held together, e.g. haemoglobin (Fig. 18.16).

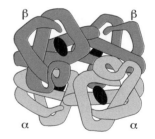

Most globular proteins take part in chemical reactions of some sort. Globular proteins are large molecules and usually only a few R-groups are actually chemically active. These 'special' R-groups are usually clustered close together in a small part of the molecule. In enzymes this region is called the *active site*.

Fig. 18.16 Haemoglobin, a protein consisting of four polypeptide chains of two kinds

Although the R-groups at the active site are close together, they are usually a long way apart along the peptide chain. It is the special folding pattern that brings them close together (Fig. 18.17).

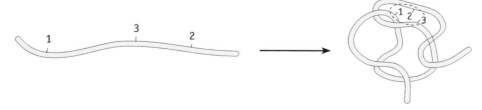

Fig. 18.17 R groups at the active site of an enzyme (shown by dotted line) are brought close together by folding of the polypeptide chain

In every globular protein the polypeptide chain is folded in a particular way — any major change in the folding pattern and the molecule will no longer work. The coils and folds are held in place by hydrogen bonds and other weak forces. Since they are weak forces they are easily broken. At temperatures of about 50–60°C or above, the molecules collide and vibrate violently enough to break the weak bonds that hold the molecule in shape. This inactivates the protein and usually makes it insoluble, as for instance when an egg is cooked. This change is called **denaturation**, and is permanent (try un-cooking egg white!).

Extension: How protein molecules keep in shape

Every globular protein has a molecular shape that differs from that of every other kind of protein. The protein can only continue to work if it stays in this shape, so what keeps it in the right shape? The answer lies in the R-groups of the amino acids. Of the 20 kinds of amino acid, some have R-groups that are attracted to water, either because they are electrically charged or because they are neutral as a whole but polar. Other R-groups are non-polar and are not attracted to water. Now the polypeptide chain is 'floppy' enough to enable it to take up that shape which allows most of the polar and charged R-groups to face outwards into the water, leaving the non-polar groups buried inside (Fig. 18.18).

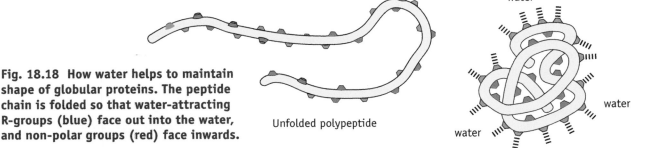

Fig. 18.18 How water helps to maintain shape of globular proteins. The peptide chain is folded so that water-attracting R-groups (blue) face out into the water, and non-polar groups (red) face inwards.

Unfolded polypeptide

Polar and non-polar R-groups also keep membrane proteins in position (Fig. 18.19). The polar R-groups face away from the membrane into the water, whereas the non-polar R-groups face into the non-water-attracting fatty acid chains.

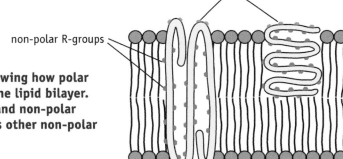

polar R-groups

non-polar R-groups

Fig. 18.19 Structure of a cell membrane showing how polar and non-polar R-groups anchor proteins in the lipid bilayer. Polar groups (blue) face towards the water and non-polar groups (red) face into the bilayer or towards other non-polar groups of the polypeptide chain.

Nucleic acids

Nucleic acids are complex substances concerned with the manufacture of proteins, and are dealt with in Chapters 23 and 24.

Inorganic ions (mineral salts)

About 27 elements are known to be essential to at least some organisms, and most of these are essential to all life. A few non-metals (carbon, hydrogen, oxygen, nitrogen, phosphorus and sulfur) account for most of the organic matter of cells. Most of the other essential ions are metals and are present in very small amounts — some in parts per *billion*. Some, for example copper, are poisonous at higher concentrations. Inorganic ions play diverse roles, for instance:

- Some are constituents of larger molecules; for example iron forms part of the oxygen-carrying blood pigment haemoglobin.

- Some have a structural function, for example calcium phosphate in bones and teeth.

- Metals such as magnesium may be essential for the action of certain enzymes.

- Some are important in maintaining the water balance between the inside and outside of cells, for example sodium chloride in the blood.

Enzymes

In any living cell, hundreds or even thousands of different chemical reactions are going on at a given moment. Collectively, these chemical processes are called **metabolism**. Though we cannot see metabolism, we can observe its outward effects, such as the contraction of muscle and the growth and division of cells. Most of the reactions of metabolism cannot occur without the presence of globular proteins called **enzymes**. They are so-called because the first enzymes to be extracted from living cells were obtained from yeast (*enzyme* means 'in yeast').

Enzymes are **catalysts** — that is, they speed up chemical reactions without being used up. Since enzyme molecules can be used millions of times over, only minute amounts are needed.

Enzymes differ from other catalysts in several important ways:

- They work thousands of times faster than 'ordinary' catalysts. For example, to break down starch into sugar using acid as a catalyst, it has to be boiled for many hours. The enzyme in saliva does a similar job in a few minutes at room temperature.

- They are *specific*, meaning that they only act on a particular kind of substance (the **substrate**), or a group of chemically similar substrates. For instance, amylase can only act on starch, breaking it down into maltose.

- They are extremely sensitive to conditions such as temperature and pH, and are easily inactivated by certain chemicals.

How enzymes work

Although enzymes are not used up in the processes they catalyse, *they do take part in the reactions*. They do this by combining with the substrate to form an **enzyme-substrate complex**. The substrate combines with a part of the enzyme molecule called the **active site**. This is sometimes likened to a key fitting into a lock. Obviously, only substrates with molecules of the right shape would be able to fit into the active site.

While the substrate is combined with the enzyme, the substrate is changed into the product (or products). The product molecule then separates from the enzyme, which is then free to combine with another substrate molecule, and so on. At ordinary temperatures, most enzymes can convert several hundred thousand substrate molecules into products every second.

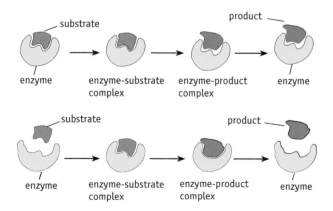

Fig. 18.20 Above: The (out-dated) lock and key mechanism of enzyme action. Below: The modern induced-fit mechanism of enzyme action.

Though the lock and key hypothesis explains a number of experimental observations, the mechanism is actually more complex. The active site is not as rigid as a lock, but changes its shape slightly when combined with the substrate. The effect of this flexible grasp or *induced fit* is to distort the substrate molecule, making it more reactive (Fig. 18.20).

The enzyme molecule is able to change shape slightly as it clasps the substrate because the hydrogen bonds holding it in shape are *weak*. Unfortunately, it is the weakness of these forces that causes enzymes to be so easily inactivated by heat (see below).

Factors that affect the activity of enzymes

The most important factors that affect enzyme activity are:

- temperature
- pH
- substrate concentration
- co-factors
- inhibitors.

Temperature

Chemical reactions go faster at higher temperatures because molecules move faster and collide more often and more violently. At moderate temperatures enzyme reactions are no exception, but at higher temperatures the situation becomes more complicated. Fig. 18.21 shows how temperature affects the activity of salivary amylase, which hydrolyses starch to maltose.

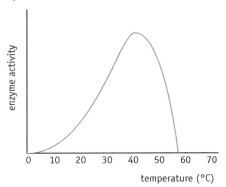

Fig. 18.21 **The effect of temperature on the activity of salivary amylase**

Up to about 35–40° C an increase in temperature causes the enzyme to work faster because the starch molecules collide harder and more frequently with the enzyme molecules. Notice that the enzyme doesn't just work faster; on the left side of the graph the curve gets steeper. This is because a given rise in temperature increases the activity of the enzyme by about the same *percentage* (it roughly doubles for every 10° C rise in temperature).

At higher temperatures molecular collisions and vibrations become violent enough to break the hydrogen bonds holding the enzyme in shape, and it becomes inactivated or **denatured**. This effect is permanent, in contrast to the effect of low temperature, which is temporary.

The temperature at which the enzyme works fastest is the **optimum** temperature. Organisms that live in cold environments have enzymes with optimum temperatures as low as 5–10° C. Those living in hot climates have enzymes with much higher optima — some hot springs bacteria flourish at 90° C. Endothermic ('warm-blooded') animals regulate their body temperatures at between 37° and 40°, and their enzymes work best at these temperatures. A rise of only a few degrees above the optimum may cause heat stroke or death.

These chemical effects explain why insects and other 'cold-blooded' animals are active on warm sunny days and sluggish in cooler weather. Mammals and birds on the other hand have body temperatures that are maintained at or very near the optimum, allowing them to remain active in widely varying environmental temperatures.

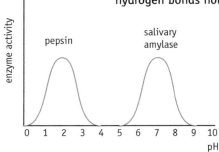

Fig. 18.22 **The effect of pH on enzyme activity**

pH

This a measure of the hydrogen ion concentration of a solution, and therefore of its acidity or alkalinity. Water has a pH of 7 and is neutral. Solutions with pH values above 7 are alkaline and solutions of pH less than 7 are acidic. The pH scale is logarithmic — pH 1 is ten times as acidic as pH 2, a hundred times as acidic as pH 3 and a thousand times as acidic as pH 4.

Most enzymes work inside the cells that make them. Since the pH inside cells is normally between 7.2 and 7.4, it is not surprising that most enzymes work best under these conditions. Enzymes secreted into the gut work outside the cells that make them, and can operate under different pH conditions. Some work best under acid conditions; for example the pepsin of gastric juice made by the stomach has an optimum pH of about 1.5 (Fig. 18.22).

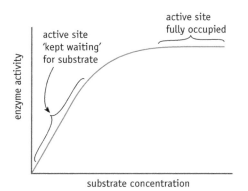

Fig. 18.23 **The effect of substrate concentration on enzyme activity**

ISBN: 9780170214094

Substrate concentration

The higher the substrate concentration the more likely its molecules are to collide with the enzyme, so the faster the reaction goes. But the active site can only deal with one substrate molecule at a time. At high substrate concentrations, substrate molecules are colliding with the active site of the enzyme when it is already 'occupied'. Under these conditions the enzyme cannot work any faster and is said to be 'saturated' with substrate (Fig. 18.23).

Extension: Enzyme 'activity' and the distinction between production and productivity

It is often stated that enzyme activity is affected by enzyme concentration (with a graph like Fig. 18.24 to illustrate). This is not so. The activity of an enzyme is expressed as the amount of product formed (or the amount of substrate used up) per unit time *per unit mass of enzyme*. While doubling the number of enzyme molecules does double the rate of the reaction, it has no effect on the output of each enzyme molecule. Temperature, pH and substrate concentration on the other hand do affect the rate of working of each individual enzyme molecule.

Think of enzyme molecules as typists. Doubling the number of typists doubles the production, but has no effect on the *productivity*, which is the output of each individual typist.

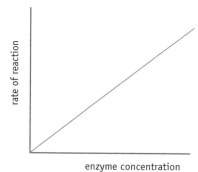

Fig. 18.24 The non-effect of the concentration of an enzyme on its activity

Co-factors

Many enzymes cannot work on their own, but need an additional, non-protein substance or *co-factor*. Co-factors may be quite complex organic molecules, in which case they are called **co-enzymes**. Most, if not all, vitamins are enzyme co-factors (or are used to make them). Some of the B vitamins, for example, are used to make co-enzymes that are important in the reactions of respiration. Some co-factors are simple inorganic ions. This is probably why certain elements are needed by both plants and animals in minute amounts. These are called *trace elements,* examples cobalt and manganese.

Inhibitors

Many enzymes are prevented from working by inhibitors, which are therefore *poisons*. Cyanide, for example, inhibits the working of one of the enzymes involved in respiration. Heavy metals such as lead and mercury combine with and inactivate enzymes in the cells of the nervous system, leading to serious effects.

Extension: What enzymes can and cannot do

Though enzymes are rather special catalysts, they are not magicians. Enzymic reactions are subject to the same two chemical laws as other chemical reactions. A chemical reaction will only occur if two conditions are satisfied:

- The molecules must collide with a certain minimum energy, called the *activation energy*.

- The products must have less energy than the raw materials.

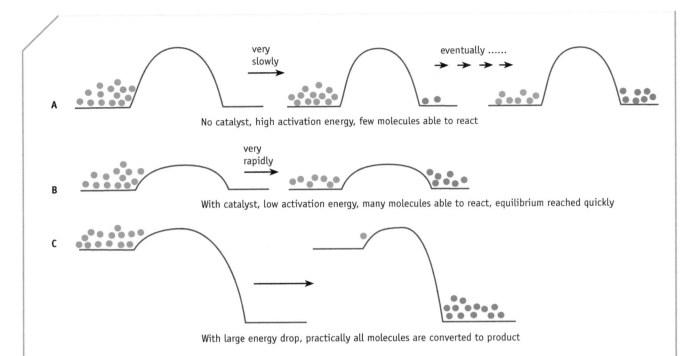

very
slowly →

eventually ......
→ → → →

A

No catalyst, high activation energy, few molecules able to react

very
rapidly →

B

With catalyst, low activation energy, many molecules able to react, equilibrium reached quickly

C

→

With large energy drop, practically all molecules are converted to product

Fig. 18.25 Model showing the effect of a catalyst on a reaction. In A and B there is no net energy change, so an equilibrium is reached with equal numbers of reactant and product molecules. In A there is no catalyst present so activation energy is high and reaction is slow. In B a catalyst is present so activation energy is low and reaction is rapid. In C, products have less energy than the raw materials, so most molecules are converted to product.

Fig. 18.25A shows a model that illustrates these ideas. On either side of the hill are many randomly moving balls, which are assumed to be perfectly elastic so they don't lose their 'bounce'. The balls are moving randomly, some faster than others. To begin with, all the balls are on left side, and represent the raw materials. From time to time, one of the faster balls is moving fast enough (i.e. is 'hot' enough, with the necessary activation energy) to get over the hill, and becomes a 'product' molecule. Eventually, there are roughly equal numbers of balls on either side of the hill — an equilibrium is established, with as many balls crossing from left to right as in the reverse direction. This is because the energy level of reactants and products is the same.

Fig. 18.25B shows the effect of adding an enzyme. The 'hill' (activation energy) is *lower*, so balls can cross over more frequently. But, there is no effect on the relative numbers balls on either side of the barrier once equilibrium has been established, because the energy level of reactants and products are the same. This illustrates the fact that enzymes do not add energy, and have no effect on the position of equilibrium — they simply speed up the attainment of equilibrium.

Fig. 18.25C shows the same situation except that there is a big energy drop from left to right. Once a ball has bounced from left to right it is very unlikely to get back again because the activation energy is so much greater. Hence at equilibrium there are far more product molecules than raw material molecules.

Summary of key facts and ideas in this chapter

○ The most abundant compound in cells is *water*.

○ The key property of water is the *polarity* of its molecules, causing them to be attracted to other polar molecules.

○ Most of the other chemical compounds in cells are *organic*, in which carbon atoms may form chains or rings.

○ The chief classes of organic compounds are *carbohydrates*, *proteins* and *lipids*.

○ Carbohydrates have three main functions. They
  - are a source of *energy* (e.g. glucose, starch, glycogen),
  - provide mechanical *support* (e.g. cellulose),
  - are important constituents of the plasma membrane.

○ Carbohydrates fall into three groups: simple sugars such as glucose (*monosaccharides*), complex sugars (e.g. *disaccharides* such as sucrose), and *polysaccharides*. Polysaccharides (e.g. starch and cellulose) are *polymers*, consisting of many monosaccharides linked together.

○ Lipids are essential constituents of *cell membranes* and, as fats, are important *energy storage* compounds.

○ Proteins are polymers of *amino acids* and have many functions, especially as enzymes and as constituents of cell membranes.

○ There are 20 kinds of amino acid in proteins, and in any particular protein they are joined in a particular sequence, called the *primary structure*. A chain of amino acids forms a *polypeptide chain*.

○ In globular proteins the polypeptide chain is is further folded into an irregular *ball shape*. In any particular protein the pattern of folding is specific, and is held in place by hydrogen bonds and other weak forces.

○ Some globular proteins (such as haemoglobin) consist of more than one polypeptide chain, held loosely together.

○ The specific properties of a globular protein such as an enzyme depend on its shape. Since the forces maintaining this are weak, the shape is easily changed by heat. This is called *denaturation* and results in permanent loss of activity of the protein.

○ Enzymes are almost all globular proteins. As catalysts, they speed chemical reactions in cells without being used up in the process. They are thus effective in minute concentrations.

○ The catalytic activity of an enzyme is due to a small part of its surface called the *active site*. This is the part of the enzyme that temporarily combines with the substrate.

○ It used to be thought that the substrate fits into the active site like a key in a lock. It is now known that the shape of the active site changes as it combines with the substrate, 'clasping' it in an *induced fit*.

○ Enzyme activity is affected by temperature, pH, substrate concentration, co-factors and inhibitors.

Test your basics

Copy and complete the following sentences. In some cases the first or last letters of missing words are provided.

1. Although a water molecule has no net charge, the oxygen atom has a slight ___*___ charge and the two hydrogen atoms have a slight ___*___ charge. A neutral molecule with an uneven charge distribution is said to be ___*___.

2. Carbohydrates contain ___*___, ___*___ and ___*___, the last two in the same ratio as in ___*___.

3. Glucose is an example of a ___*___ sugar or ___*___. Sucrose is a ___*___ sugar or ___*___. It can be ___*___ into glucose and ___*___.

4. The most complex carbohydrates are ___*___; they consist of many ___*___ joined together. Examples are ___*___ which is an energy store in plants, ___*___ which is an energy store in animals, and ___*___ which forms the chief constituent of cell walls.

5. Starch is actually two carbohydrates. ___*___ consists of unbranched chains and gives an intense ___*___ colour with ___*___. The other constituent is ___*___, which consists of branched chains (like glycogen).

6. A fat molecule consists of ___*___ joined to three ___*___ acids. When oxidised, fats yield over twice as much energy as an equal mass of carbohydrate.

7. ___*___ form the main constituent of cell membranes. They are like fats except that one of the fatty acids is replaced by a group containing ___*___, which is ___*___ to water. As a result they spontaneously tend to form double sheets or ___*___.

8. Proteins are a major constituent of ___*___ ___*___, and almost all enzymes are proteins.

9. A protein consists of one or more ___*___, each of which consists of up to several hundred or more ___*___ ___*___ joined together in a particular ___*___.

10. All amino acids have an acidic —COOH group and a basic —NH_2 group. In addition, there is an 'R' group that is specific for each amino acid. There are ___*___ kinds of amino acid in proteins, each with a different R group. It is the sequence of R groups that determines the properties of the protein.

11. In ___*___ proteins the polypeptide chains form rope-like bundles, for example ___*___ in tendon, ligament and bone, and ___*___ in the outermost layer of the ___*___, and in hair, nails, feathers.

12. In ___*___ proteins the polypeptide chain is irregularly folded into a ball-like structure. The polypeptide chain is held in its irregular but specific shape by ___*___ forces such as hydrogen bonds.

13. The properties of a globular protein depend on the 3-dimensional arrangement of its ___*___ groups. The specific ___*___ of the ___*___ chain can easily be broken by moderate heat. This changes the ___*___ of the molecule, inactivating the ___*___. This process is called ___*___ and when caused by heat, is permanent.

14. Nearly all enzymes are ___*___ proteins, and work much ___*___ than other catalysts. Unlike other catalysts, they are highly ___*___, meaning that they only act on one particular kind of substance, called the ___*___. They also differ from other catalysts in that they are easily ___*___ by heat and extremes of pH.

15. When an enzyme catalyses a reaction it first joins with the ___*___ at a small part of its surface called the ___*___ ___*___. As it does so the enzyme undergoes a temporary change of shape, 'clasping' the substrate to form an ___*___-___*___ ___*___. In combination with the enzyme, the substrate is then changed into the ___*___. This then breaks free, leaving the enzyme free to combine with another substrate molecule.

16. At moderate temperatures, enzymes work ___*___ with increased temperature, but at higher temperatures they become inactivated or ___*___. This effect is ___*___.

17. Enzymes are also affected by changes in pH. Most enzymes have an ___*___ (most favourable) pH of just over 7 (the pH inside cells). Enzymes that work in the gut work outside cells and may have pH optima very different from 7.

18. The rate at which an enzyme molecule can work depends on the concentration of the ___*___, since it can only handle one ___*___ molecule at a time. At high ___*___ concentrations enzyme cannot work any faster, so is unaffected by ___*___ concentration.

19. The activity of many enzymes depends on the presence of another molecule or ion, called a ___*___-___*___.

20. Enzymes are also prevented from working by ___*___. For example cyanide inhibits a key enzyme in ___*___.

ISBN: 9780170214094

19 Cells and Energy, Respiration and Fermentation

Cells use energy for processes such as the following:

- *Movement*, for example:
  - daughter chromosomes are dragged apart in mitosis,
  - the beating of cilia and flagella,
  - the contraction of muscle.

- *Active transport* (Chapter 21).

- *Anabolism*, or the making large molecules from small ones, such as proteins from amino acids.

In addition, certain cells use energy for:

- *Bioluminescence*, or the production of light, for example the New Zealand 'glow worm'.

- *Homeothermy* (maintaining body temperature) in mammals and birds.

Sources of energy

Cells obtain energy by the breakdown of carbohydrate, fat or protein. The energy yields of different fuels are shown in Table 19.1.

Fat yields nearly two and a half times as much energy per gram as carbohydrate. This is linked to the fact that it contains little oxygen, and hence has to undergo more oxidation, with release of more energy. Storing energy as fat thus saves weight, which is particularly important in flying animals and in wind-dispersed seeds.

The complete breakdown of glucose can be summarised as follows:

Substrate	Energy value (kJ per gram)
Carbohydrate	17
Fat	39
Protein	17

Table 19.1 Energy values of different fuels

$$C_6H_{12}O_6 + 6O_2 \longrightarrow 6CO_2 + 6H_2O + \text{large amount of energy}$$

The trouble is that when a glucose molecule is broken down in this way, it releases far more energy than a cell can use efficiently. If we liken energy to money, a glucose molecule is the equivalent of a $100 bill. When a cell 'spends' energy, it doesn't get any 'change' — any that is not used is lost as heat. This would not only be wasteful, but would damage the cell.

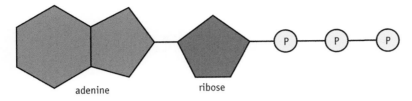

adenine   ribose

Fig. 19.1 Simplified structure of ATP

Cells get over this problem by converting the large 'packets' of energy in glucose and other fuel molecules into small packets of 'loose change', in the form of a substance called **adenosine triphosphate**, or **ATP**. This consists of adenine (one of the bases in DNA), ribose (a five-carbon sugar) and three phosphate groups (Fig. 19.1).

When ATP is broken down to adenosine diphosphate or **ADP**, it releases a small amount of energy. In the below equation, P represents inorganic phosphate:

$$ATP + H_2O \longrightarrow ADP + P + \text{small amount of energy}$$

If you think of ATP as a 'charged' battery, ADP is a 'flat' battery. To 'recharge' ADP back to ATP requires energy, which comes from the breaking down of glucose and other fuels.

In cells, the complete oxidation of a glucose molecule yields enough energy to make about 30 molecules of ATP:

$$C_6H_{12}O_6 + 6\,O_2 + 30\,ADP + 30\,P \longrightarrow 6\,CO_2 + 6\,H_2O + 30\,ATP + \text{waste heat}$$

In this way each large glucose energy 'packet' is split up into about 30 smaller ATP 'packets'. The yield of 30 ATP molecules represents an efficiency of about 45%, compared with about 20% for a car engine. The remaining energy is lost as heat. In mammals and birds this is not altogether wasted, as it helps keep the body warm.

ATP is continuously being produced as fast as it is used. An average human body contains about 30 grams of ATP, but uses about 40–70 kg every 24 hours. This is only possible because each ATP molecule is recycled very rapidly (Fig. 19.2).

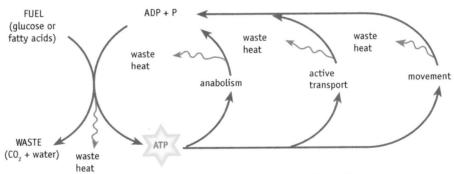

Fig. 19.2 ATP as the energy carrier in cells

The chemical processes in cells are collectively called **metabolism**. Some are involved in the building up of large molecules from smaller ones and are called **anabolism**. Those in which larger molecules are broken down into smaller ones are called **catabolism**.

Since ATP is continuously being 'spent' to drive cell processes, it must be continuously produced. In most eukaryotes (animals, plants and fungi) ATP is produced in two processes which occur in sequence (Fig. 19.3).

- **Glycolysis**, which occurs in the cytosol (Ch. 17), and does not use oxygen. In glycolysis each glucose molecule is converted to two **pyruvic acid** molecules and a little ATP is produced. In glycolysis there is no *net* oxidation. When glycolysis occurs in the absence of oxygen, it is followed by a process called **fermentation**, in which pyruvic acid is converted either to lactic acid or to ethanol and CO_2.

- **Respiration**, in which pyruvic acid is oxidised to CO_2 and water with the production of *many* molecules of ATP. Except in some bacteria, respiration is an *aerobic* process (requires oxygen), and occurs in the mitochondria. Because it produces much more ATP than glycolysis, respiration is much more efficient.

In yeast and in skeletal muscle, fermentation may occur instead of respiration (see below).

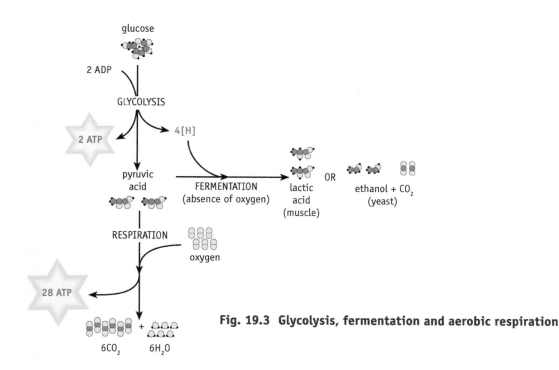

Fig. 19.3 Glycolysis, fermentation and aerobic respiration

Fermentation

Although fermentation is often (incorrectly) described as 'anaerobic respiration', the two processes are quite different (see Extension for explanation).

In fermentation the *oxidation* (removal of hydrogen) in glycolysis is followed by the *reduction* of pyruvic acid by addition of hydrogen. There is thus no *net* oxidation, as *is* the case in both aerobic and anaerobic respiration.

One of the best known kinds of fermentation occurs in the souring of milk and in active skeletal muscle, with the production of *lactic acid*:

$$\text{pyruvic acid} + 2[H] \longrightarrow \text{lactic acid}$$

In an alternative kind of fermentation, yeast converts pyruvic acid to *ethanol* and CO_2.

Since glycosis releases a small amount of energy it is inefficient, so the 'fuel' is used up much more quickly. To begin with this does not matter in, say, a bottle of milk since the 'fuel' is plentiful, but eventually the 'fuel' becomes scarce, slowing growth. Another disadvantage is that lactic acid and ethanol are slightly toxic.

Demonstrating respiration

The simplest way of showing that an organism is respiring is to detect the CO_2 produced. Fig. 19.4 shows a simple method.

Limewater turns milky in the presence of CO_2. If the liquid in tube A changes but the liquid in tube B (the control) does not, then the change *must be due to the animal*. With only small animals you will need to leave the apparatus for several hours to see any change. Alternatively, you could use an indicator, because CO_2 dissolves in water to form a weak acid. A suitable indicator is bromothymol blue, which is blue in neutral conditions and yellow in acidic solution.

To show that a green plant respires, you must keep it in darkness to prevent photosynthesis. Also, most plants respire more slowly than animals do, so it takes longer to detect.

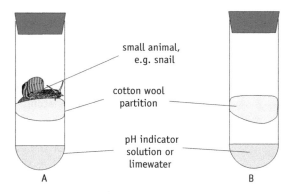

Fig. 19.4 Demonstrating CO_2 production by a small animal

Measuring the rate of respiration

The experiment described above is a *qualitative* one — it only tells us *whether* an organism is making CO_2. To make it into a *quantitative* investigation would mean measuring the *rate* of respiration.

The rate of respiration is most easily measured in terms of the amount of oxygen absorbed in a given time, using a **respirometer**. Fig. 19.5 shows a simple one. Although the animal cannot be prevented from producing CO_2, the soda lime absorbs it as fast as it is produced. As a *control*, a similar apparatus is set up, but without an animal. This records any changes that are not caused by the animal, such as expansion of the air as a result of warming. To minimise any warming by body heat, try to avoid handling the barrel of the syringe when you are setting the apparatus up.

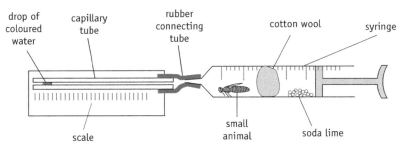

Fig. 19.5 A simple respirometer

After assembling the apparatus and the control, leave for about ten minutes to ensure that both are at the same temperature as the surroundings. Next, introduce a small drop of coloured water into the end of the capillary, and draw it up a centimetre or so by *gently* pulling out the plunger of the syringe. Any uptake of oxygen by the animal would result in the slow movement of the water drop toward the syringe, which is measured using the scale. When the water reaches the end of the scale you can return it to the beginning using the plunger (taking care not to touch the barrel of the syringe).

If the liquid in the control moves slightly, there has been a small change in either room temperature or atmospheric pressure, and you will have to make a correction. Suppose for example that over the period of measurement the drop moves 2 mm *towards* the syringe in the control and 22 mm towards the syringe in the apparatus with the animal. The change due to the animal would thus be 22 – 2 = 20 mm. If on the other hand the control drop had moved 3 mm *away* from the syringe, the change due to the animal would be 22 + 3 = 25 mm.

Of course, a big animal uses more oxygen in a given time than a small one. To compare the rate of oxygen uptake by a snail and a centipede, you must take account of the different body sizes. You would do this by expressing the rate of respiration as the volume of oxygen absorbed per minute *per gram of animal*. It is calculated as follows:

$$\frac{\text{volume of oxygen absorbed}}{\text{time in minutes} \quad \times \quad \text{mass of animal}}$$

For example, suppose a mouse with a mass of 40 grams uses 4 cm³ oxygen in two minutes. In one minute the animal uses 2 cm³, but each gram of mouse must be using 1/40 of 2 cm³, or 0.05 cm³ per minute. Thus the rate of respiration of mouse *tissue* is 0.05 $cm^3min^{-1}g^{-1}$.

Factors affecting the rate of respiration

Temperature

Enzymes are very sensitive to temperature and, consequently, so are the cell processes dependent upon it. Organisms except birds and mammals have body temperatures that depend at least to some extent on that of the environment. In warmer conditions metabolism speeds up, and fuel and oxygen are used more quickly. For every 10° C rise in temperature, metabolic rates roughly double. If the temperature is raised through 20° C metabolic rates increase four times. At higher temperatures enzymes become denatured and death results.

Oxygen

Oxygen is needed for the efficient use of fuel to provide energy. In its absence, organisms such as yeast and tissues such as skeletal muscle can switch to fermentation. Some organs (such as the brain) cannot survive without oxygen.

Poisons

Respiration is inhibited by cyanide. This prevents oxygen from combining with hydrogen at the end of the electron transport chain (see below), and brings ATP production to a halt.

Extension: Respiration in more depth

Oxidation and reduction

Some of the key reactions in glycolysis and respiration are *oxidation* and *reduction* reactions. Oxidation is the removal of an electron; reduction is the addition of an electron. If an electron is transferred from X to Y, X becomes oxidised and Y is reduced, so all oxidation reactions also involve a simultaneous reduction.

Sometimes an electron is removed accompanied by a proton, in the form of a hydrogen atom.

Oxidation can also occur when a substance combines with oxygen. The oxygen is so 'greedy' for electrons that it tends to take more than its 'fair' share. As a result the substance with which the oxygen has combined loses part of its share of electrons and is thus oxidised.

A substance that takes electrons is an *oxidising agent*, and a substance that gives them up is a *reducing agent*.

One of the key substances involved in the transfer of electrons is called **NAD** (**n**icotinamide **a**denine **d**inucleotide), which is produced from vitamin B$_3$. NAD is an example of a *coenzyme*. In its oxidised state (NAD$^+$) it is short of an electron. In several stages in the oxidation of the fuel, two hydrogen atoms are removed. NAD$^+$ takes the two electrons and one of the protons to form NADH, the other proton remaining in solution:

$$NAD^+ + 2[H] \longrightarrow NADH + H^+$$

Glycolysis

The overall result of this process is the conversion of glucose to **pyruvic acid** (a 3-carbon compound) and the net production of two molecules of ATP (Fig. 19.6). Though glycolysis does not require the presence of oxygen, one of the reactions of glycolysis is an oxidation, in which the fuel is oxidised by NAD$^+$, forming NADH + H$^+$.

Glycolysis can only continue if there is a supply of NAD$^{+=}$, which means that the NADH must be able to unload its electrons to re-form NAD$^+$. How this happens depends on whether oxygen is present or not.

In the presence of oxygen the NADH is oxidised to NAD$^+$ in the mitochondria in *respiration* as explained below. If oxygen is lacking, the electrons are got rid of in fermentation, dealt with later in this chapter.

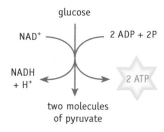

Fig. 19.6 Summary of glycolysis

Respiration

Respiration is the second stage of energy production from glucose. Unlike glycolysis it can only occur in the presence of oxygen. It can be divided into three stages:

- Formation of acetyl co-enzyme A.
- The citric acid cycle (also called the Krebs cycle after its discoverer).
- Oxidative phosphorylation.

The first two occur in the matrix of the mitochondria and the third occurs in the inner membrane of the mitochondria.

Prelude to the citric acid cycle: Formation of acetyl co-enzyme A

This involves a substance called *co-enzyme A* (coA), which is made from vitamin B$_5$. This reaction results in the removal of hydrogen and CO$_2$ from pyruvic acid to form a two-carbon compound called *acetyl coA*:

$$coA + pyruvic\ acid + NAD^+ \longrightarrow acetyl\ CoA + NADH + H^+ + CO_2$$

The CO$_2$ is a waste product, but the NADH is a fuel for the final stage of respiration. The two-carbon compound then enters the citric acid cycle.

The citric acid cycle (Fig. 19.7)

Its overall result is the conversion of the two-carbon fragment to CO_2 and NADH:

$$acetyl\ coA + 4NAD^+ \longrightarrow 2CO_2 + 4NADH + 4H^+ + coA$$

For each molecule of acetyl coA broken down, a molecule of ATP is also produced. Notice two important points:

- It is *strictly aerobic*, meaning that it cannot occur without oxygen. This might seem strange, because oxygen takes no direct part (see later for the explanation).

- It yields only one molecule of ATP for each pyruvic acid molecule produced (two for every glucose used).

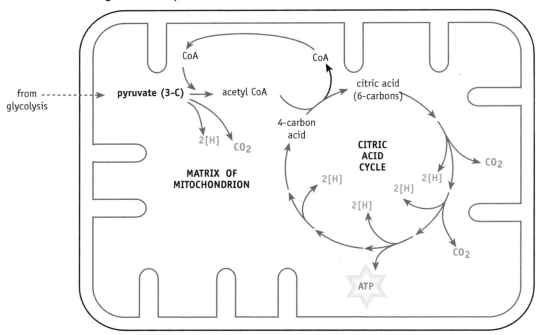

Fig. 19.7 Outline summary of the citric acid cycle

Oxidative phosphorylation

The citric acid cycle yields NADH and a *little* ATP. Much more ATP is produced when the NADH is oxidised in the inner membrane of the mitochondria in oxidative phosphorylation. The process can be divided into two stages:

- Electron transport and proton pumping.

- ATP formation.

It is easier to understand if we deal with them in reverse order.

ATP formation

ATP is produced in the inner mitochondrial membrane by a process that is essentially *active transport* (Chapter 21) *in reverse*. In active transport, ions or molecules are pumped across a membrane from a low concentration to a higher one, using the energy supplied by breakdown of ATP. In the inner membrane, the tendency for hydrogen ions (protons, H^+) to diffuse down a concentration gradient is used to generate ATP.

An analogy is the difference between an electric motor and a dynamo, in which each is the reverse of the other. In an electric motor electrical energy is converted to mechanical energy, and in a dynamo mechanical energy is converted to electrical energy. In ATP formation the energy of a concentration gradient is used to make ATP, whereas in active transport ATP energy is used to create a concentration gradient.

ISBN: 9780170214094

The concentration of protons is much lower in the matrix than in the intermembrane space, so protons diffuse in. This occurs through special protein 'gates' in the membrane. These channels pass through tiny mushroom-like structures that project into the matrix. Each contains a complex of proteins called **ATP synthase**, that catalyses the production of ATP.

Fig. 19.8 shows a mechanical model of ATP formation, and can be likened to waterwheel generating electricity.

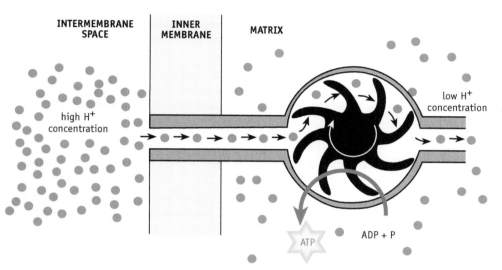

Fig. 19.8 Waterwheel analogy showing how diffusion of H⁺ ions powers the production of ATP

Proton pumping

Obviously, protons can only continue to diffuse into the matrix if the concentration gradient is maintained. This is achieved by pumping protons out of the matrix (Fig. 19.9). The energy for proton pumping comes from the oxidation of NADH in electron transport, to be described next.

Proton pumping and ATP synthesis both involve transport of protons across the inner membrane. The area of the membrane is greatly increased by folding into cristae (Fig. 19.10).

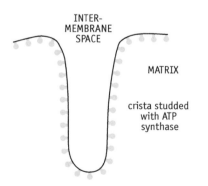

Fig. 19.10 A crista studded with ATP synthase

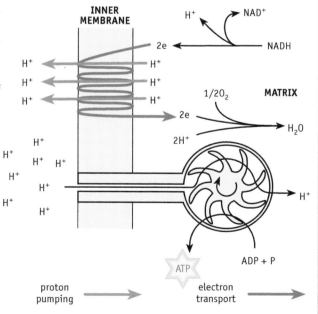

Fig. 19.9 Proton pumping and ATP synthesis

Electron transport

Electron transport is a complex process in which NADH loses two electrons to a series of carriers, collectively called the **respiratory chain**. Each carrier (A, B, C and D in Fig. 19.11) picks up an electron from the one higher up the chain and gives it to the one further down. Some of the electron transfers release enough energy to pump a proton out of the matrix against the concentration gradient. At the end of the chain is oxygen, which combines with the two electrons and two protons to form water.

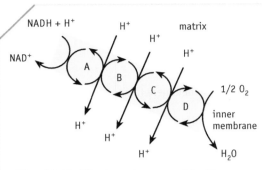

Fig. 19.11 Electron transport in the mitochondrion

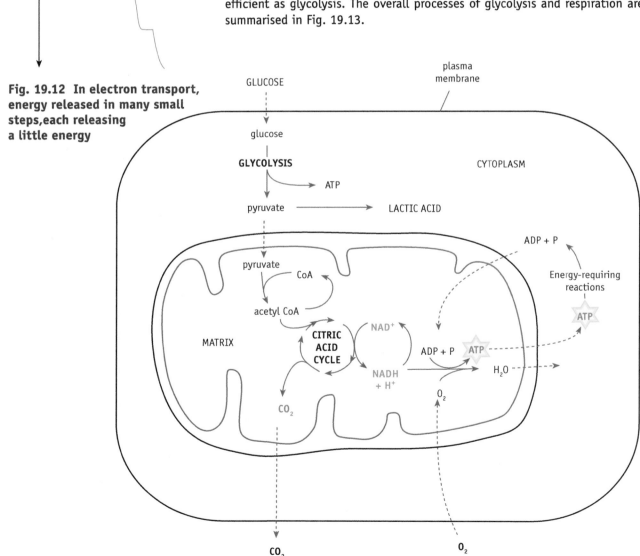

Fig. 19.12 In electron transport, energy released in many small steps, each releasing a little energy

The overall result can be summarised as follows:

$$NADH + H^+ + oxygen \longrightarrow NAD^+ + water + energy$$

Why so many carriers? By dividing up the energy release into many small steps, less is wasted (Fig. 19.12).

By giving up two electrons, NADH is converted back into NAD^+, enabling it to pick up more electrons in the citric acid cycle. Oxygen is thus necessary for the regeneration of NAD^+. Without oxygen there is no NAD^+ to remove hydrogen in the citric acid cycle reactions, so the citric acid cycle stops.

The last of the electron carriers is inactivated by cyanide, which is why it is such a deadly poison. Cyanide brings the whole electron transport chain to a halt, and thus stops most of the ATP production.

The ATP yield

For each original glucose molecule, a total of about 30 ATP molecules are produced (earlier estimates were 36–38). Of these, respiration yields about 28, glycolysis a net yield of two. Respiration is therefore about 14 times as efficient as glycolysis. The overall processes of glycolysis and respiration are summarised in Fig. 19.13.

Fig. 19.13 Summary of the main stages of energy metabolism

ISBN: 9780170214094

Obtaining energy without oxygen: anaerobic metabolism

Many animal tissues (especially skeletal muscle), plant cells, and fungi such as yeast can make ATP without using oxygen, in *anaerobic metabolism* (Fig. 19.14). The first stage of anaerobic metabolism is the formation of pyruvic acid in glycolysis, as in aerobic metabolism. This is followed by *fermentation*, the details of which depend on the organism. In plants and yeast, the pyruvic acid is converted to ethanol and CO_2, and in animals and the bacteria that cause milk to go sour, it is converted to lactic acid.

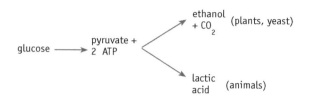

Fig. 19.14 Summary of anaerobic metabolism

Fermentation has two drawbacks:

- Lactic acid and ethanol are slightly poisonous.

- Only two ATPs are produced for each glucose used, compared with 28 in respiration.

Anaerobic metabolism is therefore 2/28 = 1/14 as efficient as respiration. To produce ATP at the same rate, the fuel must therefore be used up 14 times faster. This does not matter to a tapeworm surrounded by digested food, or to a yeast cell in rotting fruit.

In a 200-metre sprint, muscles are expending energy about 300 times as fast as when they are at rest, yet the blood supply can increase only about 20 times. The difference is accounted for by the fact that skeletal muscles store **glycogen** (a polysaccharide), which can rapidly be converted into glucose. Because the muscles have their own store of carbohydrate they can spring into violent action without having to 'wait' for the blood system to deliver extra glucose.

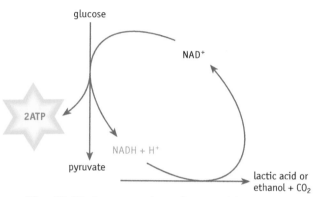

Fig. 19.15 Regeneration of NAD without the use of oxygen

In a sprint, the muscles rapidly use up glucose and accumulate lactic acid. During the recovery the lactic acid is reconverted back to pyruvic acid and then to glucose. This requires oxygen, which explains why you continue to breathe heavily *after* a sprint. What is happening is that, while your muscles are working anaerobically, they develop an *oxygen debt*, which is 'repaid' during the recovery.

The conversion of pyruvic acid to lactic acid or ethanol does not in itself produce ATP, so what is the 'point' in their production? The answer lies in the fact that glycolysis can only continue in the presence of NAD^+. This means that the NADH + H^+ produced must somehow be oxidised back to NAD^+. In aerobic conditions this happens in electron transport in the mitochondria. When oxygen is not sufficiently available, the NADH is oxidised by the pyruvic acid itself (Fig. 19.15).

Fermentation, aerobic and anaerobic respiration

In fermentation the fuel is oxidised by NAD^+ and then reduced by NADH, so there is no *net* oxidation. In respiration the fuel is oxidised by a substance obtained from the environment. In aerobic respiration this is oxygen. In anaerobic respiration the fuel is oxidised by some other substance, such as nitrate (as in denitrifying bacteria in the soil):

Aerobic respiration:

pyruvic acid + oxygen $\xrightarrow{\text{oxidation}}$ CO_2 + water

Anaerobic respiration by denitrifying bacteria:

pyruvic acid + nitrate $\xrightarrow{\text{oxidation}}$ CO_2 + water + nitrogen gas

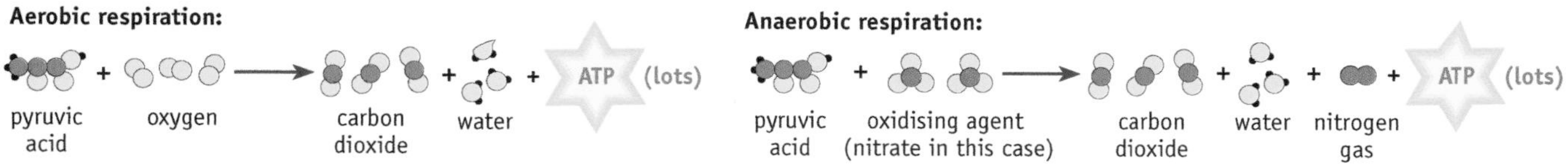

Aerobic respiration:

pyruvic acid + oxygen → carbon dioxide + water + ATP (lots)

Anaerobic respiration:

pyruvic acid + oxidising agent (nitrate in this case) → carbon dioxide + water + nitrogen gas + ATP (lots)

Fig. 19.16 Pictorial representation of aerobic and anaerobic respiration. The equations are more complex than shown and are not intended to balance.

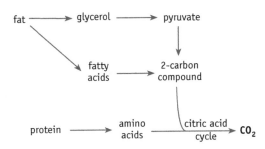

Fig. 19.17 How fat and protein are used in respiration

Respiration using fat or protein

Some of the fuel used in respiration in humans is fat and protein. Both are complex foods and are first broken down to simpler units (Fig. 19.17).

Fat is first broken down to fatty acids and glycerol. Glycerol can be converted into pyruvic acid, which then enters the citric acid cycle in the mitochondria. Fatty acids undergo a complex process by which two-carbon fragments are repeatedly 'pruned off' and enter the citric acid cycle. Since glycolysis is the only route in which fuel can be broken down anaerobically, fatty acids and amino acids can only be used in the presence of oxygen.

Proteins are broken down to amino acids, which are then relieved of their amino groups by a process called **deamination**. The resulting acids then enter the citric acid cycle.

Summary of key facts and ideas in this chapter

○ Glycolysis, which occurs in the cytosol and yields a small quantity of ATP.

- Respiration, which occurs in the mitochondria and yields a large quantity of ATP.

○ Test your basics

Copy and complete the following sentences. In some cases the first letter of a missing word is provided.

1. All cells use energy by the breakdown of organic molecules. The most energy-rich fuel is ___*___, which yields more than twice as much energy on oxidation as ___*___ or p___*___.

2. The form in which all cells use energy is a substance called ___*___. When this is converted to ___*___ and inorganic phosphate, a small amount of energy is released for use by the cell.

3. Cells have two ways of obtaining energy from organic molecules. In ___*___ the fuel is oxidised and converted to CO$_2$ and ___*___, yielding a large amount of energy in the form of ___*___. In ___*___ ___*___ there is no net oxidation and the fuel is converted from one kind of organic substance to another, such as e___*___ or ___*___ acid, and yielding a relatively small amount of energy.

4. The breakdown of glucose to CO$_2$ and water occurs in two stages. The first occurs in the ___*___ and is called ___*___. It results in the conversion of glucose to ___*___ acid. The second is called ___*___ and takes place in the ___*___. It results in the complete oxidation of ___*___ acid to ___*___ and w___*___.

5. In anaerobic metabolism in muscle during intense exercise, glucose is converted to ___*___ acid and a small amount of ___*___. Though it is inefficient, it can occur extremely rapidly, generating ATP at a high rate. Yeast can carry out a similar process, except that in this case ___*___ is converted to e___*___ and ___*___. A disadvantage of both these processes is that the waste products are slightly t___*___.

6. The rate of respiration is roughly ___*___ for a 10°C increase in temperature, until at higher temperatures (above about 40°C in humans), it begins to slow due to the thermal ___*___ of enzymes.

7. Respiration is ___*___ by cyanide, which prevents the final step in the process, the formation of ___*___.

20 Light to Chemical Energy: Photosynthesis

Green plants (and certain bacteria) are the only organisms that do not depend on other organisms for their energy supply. Instead of feeding on organic matter made by other living things, plants make their own by *photosynthesis*. They are therefore **autotrophic** ('self-feeding'). Animals, fungi and most bacteria cannot make their own organic substances from inorganic materials and are said to be **heterotrophic**.

In photosynthesis, CO_2 and water are converted to sugar and oxygen. The energy for the process is *light*, which is trapped by the green pigment **chlorophyll**. The entire process occurs in the **chloroplasts**. The following simple equation shows the raw materials and end products:

$$CO_2 + \text{water} + \text{light energy} \xrightarrow{\text{chloropyhll}} \text{sugar} + \text{oxygen}$$

What happens to the sugar made in photosynthesis?

Eventually, all the sugar made in photosynthesis is used in one of two ways (Fig. 20.1):

1. Some is used to make other organic molecules such as cellulose, lipids and amino acids.

2. Some is used in respiration to supply *energy*.

Not all the sugar made in photosynthesis is used immediately — some is usually stored temporarily as starch in the chloroplasts, but at night this is broken down into sugar again and transported to other parts of the plant. Many plants (such as potato and kumara) store starch in large underground tubers.

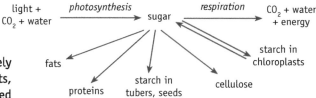

Fig. 20.1 What happens to the sugar made in photosynthesis

Chemistry of photosynthesis

Photosynthesis is really the conversion of light energy into chemical energy stored in carbohydrate and oxygen. Although it consists of many reactions, photosynthesis can be divided into two main stages, only one of which requires light:

- The light-dependent stage or *photo stage*, in which light energy is used to split water into hydrogen and oxygen. This is called **photolysis** and can be summarised thus:

$$\text{light} + 2H_2O \xrightarrow{\text{chloropyhll}} 4[H] + O_2$$

ISBN: 9780170214094

As explained below, the hydrogen is not produced as gas, but is combined with a hydrogen carrier. This is not shown, but is indicated by the square brackets [H].

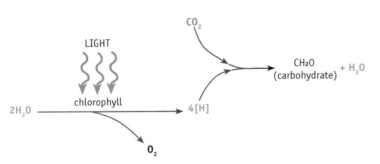

Fig. 20.2 Summary of photosynthesis

- The light-independent, or *synthetic stage*. This is sometimes called the 'dark stage', because under certain laboratory conditions it can occur in darkness. The term is misleading, however, because in nature it *does not* occur in the dark. In this stage the hydrogen made in the photo-stage is used to convert CO_2 to sugar [simplified to (CH_2O)]:

$$4H + CO_2 \longrightarrow (CH_2O) + H_2O$$

The overall result is summarised in Fig. 20.2.

The photo stage

The light-dependent stage occurs in the **thylakoids** of the chloroplasts. These are flattened, hollow sacs stacked in piles or **grana** (singular, *granum*). The walls of the thylakoids consist of lipid bilayers and contain the light-absorbing pigments.

Light energy absorbed by the chlorophyll is used in two ways:

- Photolysis, or the use of light energy to split water into hydrogen and oxygen. Instead of appearing as hydrogen gas, the hydrogen is picked up by a hydrogen carrier called **NADP$^+$**. This is very similar to the NAD$^+$ in respiration, except that it has an extra phosphate group.

$$\text{light} + 2H_2O + 2NADP^+ \longrightarrow 2NADPH + 2H^+ + O_2$$

- ATP production. Whereas in respiration ATP is made from energy released in chemical reactions, in photosynthesis it is produced directly from light trapped by chlorophyll:

$$\text{light} + ADP + P \longrightarrow ATP$$

The overall result of the photo-stage is therefore the conversion of light into chemical energy in the form of two energy-rich compounds, NADPH and ATP. These are then used to 'drive' the energy-requiring light-independent stage. Fig. 20.3 summarises the relationship between the two stages.

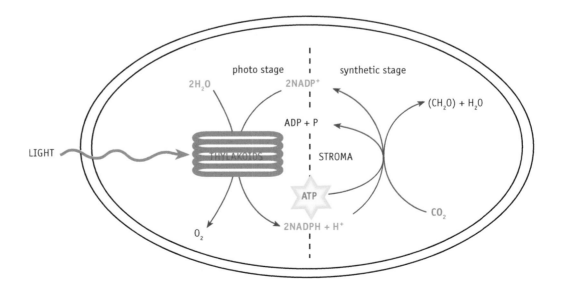

Fig. 20.3 Interrelationship between the two stages of photosynthesis

ISBN: 9780170214094

The synthetic stage

This occurs in the **stroma**, or liquid part of the chloroplasts. In this stage the energy-rich products of the photo stage are used to convert CO_2 to carbohydrate. It can be summarised as follows:

$$CO_2 + NADPH + H^+ + ATP \longrightarrow carbohydrate + NADP^+ + ADP + P$$

The process is much more complicated than this, and takes the form of the **Calvin cycle** (after one of its discoverers).

The absorption of light: Chloroplast pigments

Before light can be used, it must first be absorbed. This is the function of the four lipid-soluble pigments in the membranes of the grana. These belong to two chemical families:

- **Chlorophylls**. A chlorophyll molecule is tadpole-shaped with a 'head' containing magnesium, and a long hydrocarbon 'tail'. The tail is non-polar and lies embedded in the lipid bilayer of the thylakoid membranes. In the plant kingdom there are two kinds of chlorophyll — chlorophyll *a* (bottle green) and chlorophyll *b* (olive green).

- **Carotenoids**, which are yellow or orange, and do not contain a metal. In land plants there are two kinds of carotenoid — **carotene** (orange) and **xanthophyll** (yellow). They are normally masked by chlorophyll but in deciduous trees they are revealed as the chlorophyll breaks down in the autumn, giving the familiar autumn colours. Like the chlorophylls, carotenoids are embedded in the thylakoid membranes.

Separating chloroplast pigments

All the photosynthetic pigments are in the lipid bilayer of the chloroplast membranes. They can be removed by grinding up a leaf in ethanol or propanone (acetone). The resulting solution is a mixture of the four pigments, which can easily be separated by *paper chromatography* (Fig. 20.4).

First, make a pencil line about a centimetre from one end of a strip of chromatography paper. Next, use a cocktail stick to streak a tiny drop of the extract along the pencil line. After the solvent has evaporated, streak another drop, and so on until a medium-dark green streak has been produced. Finally, dip the tip of the strip into a suitable solvent (10 parts petrol ether : 12 parts propanone). The solvent runs up the paper, which acts like a wick.

As the solvent creeps up the paper the pigments slowly become separated. Those pigments that are least strongly attracted to the paper move fastest. By the time the solvent has reached the top of the paper the pigments have become separated into distinct spots.

While the chromatogram is 'running', it is best left in dim light or darkness because chlorophyll bleaches very quickly in sunlight. In the intact chloroplast this does not happen because it is in some way 'protected' by the carotenoids.

After separation by chromatography, the light-absorbing properties of each pigment can then be investigated. Light of different wavelengths is passed through a solution of the pigment. For each wavelength, the fraction of the light that is absorbed is measured. When the results are plotted on a graph, an *absorption spectrum* is obtained. Fig. 20.5 shows the absorption spectra for chlorophylls *a* and *b*.

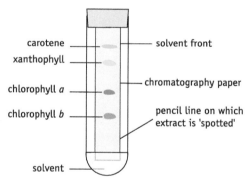

Fig. 20.4 **Chromatographic separation of chloroplast pigments**

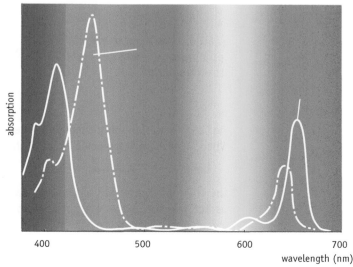

Fig. 20.5 **Absorption spectra of chlorophylls**

Notice that neither pigment absorbs much green light, most of which therefore passes straight through. The green colour we see is that of the light that is *least* absorbed. Since red and blue light are strongly absorbed, these colours are the most useful in photosynthesis.

Extension: More about the photo stage

The essential feature of the photo stage is the conversion of light energy into chemical energy. A major part of this is the splitting of water to release oxygen. Oxygen is a powerful *oxidising agent*, meaning that it has a very strong tendency to remove electrons from other substances. NADPH is a strong *reducing agent*, meaning that it 'wants' to give electrons to other substances.

In respiration the electrons move 'downhill' from NADH (which is a strong reducing agent) to oxygen. The result is the release of lots of energy and the formation of water.

In photosynthesis the reverse happens. Light energy is used to move electrons from the oxygen in water (which fiercely 'tries' to hold on to them) to NADP+ (which doesn't 'want' them). The result is the formation of oxygen and NADPH.

The energy needed to transfer an electron from water to NADP+ is more than the energy content of a single photon of light. What happens is that two photons are used, in two different reactions. Photosystem I (PS I) reduces NADP+ to NADPH, while Photosystem II (PS II) oxidises water by removing electrons from it.

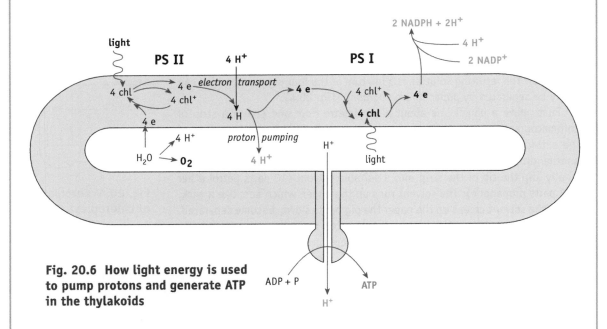

Fig. 20.6 How light energy is used to pump protons and generate ATP in the thylakoids

Fig. 20.6 shows a highly simplified scheme of what happens.

When a chlorophyll molecule in PS II absorbs a photon, it ejects an electron:

$$light + chl \longrightarrow chl^+ + e^-$$

The chlorophyll now has a positive charge and thus has a strong tendency to attract electrons. It has actually become an even stronger oxidising agent than oxygen, and is consequently able to remove electrons from water:

$$4\ chl^+ + 2H_2O \longrightarrow 4\ chl + O_2 + 4H^+$$

ISBN: 9780170214094

The fate of the electron ejected from the chlorophyll in PS II will be dealt with later.

PS I, meanwhile, also absorbs photons and emits electrons. These have enough energy to reduce $NADP^+$ to NADPH:

$$\text{light} + \text{chl} \longrightarrow \text{chl}^+ + e^-$$

$$4 \, e^- + 2 \, NADP^+ + 4 \, H^+ \longrightarrow 2 \, NADPH + 2H^+$$

The chlorophyll now has a 'hole' or positive charge, and thus has a strong tendency to gain electrons. These electrons come from the electrons ejected from PS II, and whose fate we have conveniently ignored up till now. Instead of travelling directly from PS II to PS I, however, the electrons pass along a series of carriers, using the released energy to pump protons into the thylakoids. The tendency for protons to diffuse back out of the thylakoids is then used to generate ATP.

Thus in both mitochondria and chloroplasts, ATP is generated by the tendency of protons to diffuse down a gradient established by pumping them across a membrane. In both, the energy for proton pumping is released as electrons pass along a chain of carriers. In both organelles the area available for pumping in greatly increased — in mitochondria by cristae, and in chloroplasts by the thylakoids. They differ in the direction of pumping: in mitochondria protons are pumped *out* of the matrix into the intermembrane space (which becomes alkaline), whereas in chloroplasts they are pumped *into* the thylakoids (which become acidic).

Measuring the rate of photosynthesis

The rate of photosynthesis is most easily expressed as the amount of:

- oxygen given off per hour, or

- CO_2 taken in per hour.

If different plants are being compared, then the rate must be expressed as the amount of photosynthesis per hour *by each cm² leaf area* (or by each gram of plant).

True and apparent photosynthesis

When the rate of photosynthesis is measured, it is not the true rate. This is because of *respiration*, which occurs all the time. This idea is explained more fully in Chapter 10.

Interdependence between the two stages

Though the light-dependent and light-independent stages have been explained separately, *each depends on the other*. As Fig. 20.3 shows, the light-independent stage can only occur in the presence of ATP and NADPH, which are only made in the light. It is therefore misleading to call the light-independent stage the 'dark stage', because it normally only occurs in the light. For similar reasons, the light stage can only occur if CO_2 is present. If CO_2 is in short supply, ATP and NADPH are only slowly used up, so ADP and $NADP^+$ are only slowly re-formed. Without adequate ADP and $NADP^+$ much of the light is therefore wasted.

Environmental factors affecting the rate of photosynthesis

Photosynthesis is affected by a variety of external factors:

- Light intensity
- Light wavelength
- CO_2 concentration
- Temperature
- Water availability
- Mineral supply.

The effect of these factors on photosynthesis is complicated because they do not act separately — the effect of changing the supply of one factor may depend on the level of another.

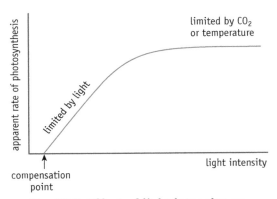

Fig. 20.7 Effect of light intensity on photosynthetic rate

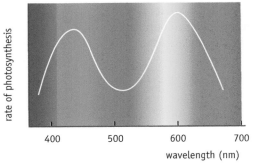

Fig. 20.8 Effect of light wavelength on photosynthesis

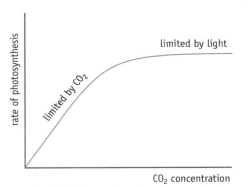

Fig. 20.9 Effect of CO_2 concentration on photosynthesis

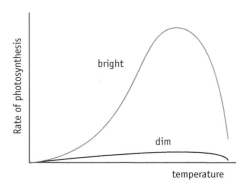

Fig. 20.10 Effect of temperature on photosynthesis

Light intensity

This is shown in Fig 20.7.

- In dim light the graph is a straight line, so the rate of photosynthesis is *directly proportional* to light intensity — doubling the light intensity doubles the rate of photosynthesis. At low intensities, light is in shorter supply than any other factor. We say that the rate of photosynthesis is *limited* by light. In dim light, therefore, increasing the temperature or CO_2 concentration makes little difference to photosynthetic rate.

- Above a certain light intensity, making the light brighter makes no difference to photosynthetic rate. This is because some other factor is in short supply, such as CO_2. Since more of the light is wasted in very bright light, photosynthesis is actually less *efficient* under these conditions than in dim light.

- At a certain low light intensity, called the **compensation point**, the plant appears not to be photosynthesising at all, because photosynthesis is exactly balanced by respiration. At the compensation point the plant is neither gaining nor losing organic matter.

Light wavelength

Since chlorophyll absorbs little green light (Fig. 20.5), this is much less useful in photosynthesis than blue or red light (Fig. 20.8). Plants living on the forest floor therefore have a double problem. Not only is the light dimmer, but the light that does reach them contains a high proportion of the less useful green wavelengths.

Actually, green light is not completely wasted. Some is absorbed by the orange carotenoid pigments and the energy passed on to chlorophyll *a*.

CO_2 concentration

The effect is similar to that of light — above a certain level, CO_2 concentration has little effect because light or temperature is limiting the overall rate (Fig. 20.9).

Temperature

Since the reactions of the Calvin cycle are catalysed by enzymes, photosynthesis would be expected to be very sensitive to temperature. Fig. 20.10 shows what happens.

In bright light and with good supplies of CO_2, photosynthesis is very sensitive to temperature. At higher temperatures the rate slows because enzymes are being denatured. In dim light, temperature has little effect because the Calvin cycle is held up by a shortage of ATP and NADPH.

Water supply

Since water is a raw material in photosynthesis you might think that its supply would affect photosynthesis. So it does, but not *as a raw material*. This is because even when a plant is on the point of death from dehydration the cells still contain abundant water. The explanation is that when a plant is short of water the stomata close to reduce water loss, which reduces the uptake of CO_2.

ISBN: 9780170214094

Mineral salts

Besides non-metals such as nitrate, phosphate and sulphate, plants need certain metals such as potassium, calcium, magnesium and iron. One of these, magnesium, is needed because it forms part of the chlorophyll molecule.

Plants growing in soils that are deficient in magnesium tend to have a pale yellow-green colour, a condition that is called *chlorosis*. Iron is also needed for chlorophyll synthesis, even though it is not a constituent of chlorophyll.

Summary of key facts and ideas in this chapter

○ Plants make their own organic compounds from CO_2 and water in the process of *photosynthesis*, summarised thus:

   CO_2 + water + light energy $\xrightarrow{\text{chloropyhll}}$ sugar + oxygen

○ The entire process occurs in organelles called *chloroplasts*.

○ Actual photosynthetic rate is always greater than the net photosynthetic rate by the rate of respiration:

   net photosynthesis = true (gross) photosynthesis — respiration

○ Photosynthesis can be divided into two stages:
   - A light-dependent *photo stage*, which occurs in the *thylakoids* of the chloroplasts where the chlorophyll is located. In this stage light energy is used to split water, yielding oxygen, and hydrogen; ATP is also produced.
   - A light-independent *synthetic stage*, which occurs in the *stroma* of the chloroplasts. In this stage CO_2 is reduced to carbohydrate using hydrogen and ATP produced in the photo stage. The synthetic stage is catalysed by enzymes and is thus very sensitive to temperature.

○ The two stages are mutually dependent, since each uses the products of the other. For example, the photo stage uses ADP produced in the synthetic stage, and the synthetic stage uses ATP produced in the photo stage.

○ In the thylakoid membranes are four pigments: *chlorophyll* a (bottle-green), *chlorophyll* b (lime-green), -*carotene* (orange) and *xanthophyll* (yellow).

○ The rate of photosynthesis can be expressed in terms of the amount of CO_2 absorbed per hour, the amount of oxygen produced per hour, or the amount of organic matter produced per hour.

○ The rate of photosynthesis is affected by the light intensity, CO_2 supply, and temperature. Water supply affects the rate indirectly, as a plant that is short of water closes its stomata, reducing the supply of CO_2.

○ The rate of photosynthesis is also affected by the wavelength ('colour') of light. Chlorophyll does not absorb green light to any significant extent, so green light is much less useful than red or blue.

○ The environmental factor that limits the rate of photosynthesis is the one that is in shortest supply relative to the others. On a sunny winter's day, temperature is likely to be the limiting factor. On a warm summer's day, CO_2 is likely to be rate-limiting. Near dawn and dusk, light limits the rate.

Test your basics

Copy and complete the following sentences. In some cases the first letter of a missing word is provided.

1. In photosynthesis, ___*___ from the air and ___*___ from the soil are converted to c___*___ and ___*___, using ___*___ energy from the ___*___. The ___*___ is trapped by the green pigment ___*___, and the entire process takes place in organelles called ___*___.

2. Photosynthesis occurs in two main stages, which occur in different parts of the ___*___. The photo stage takes place in the t___*___ of the chloroplast, which are piled into stacks called ___*___. In this stage light energy is used to split water, releasing ___*___ gas. The hydrogen is not released as gas but is picked up by a hydrogen carrier called ___*___. The other useful product of the photo stage is the energy carrier ___*___.

3. The synthetic stage does not require light but uses the NADPH and ___*___ produced in the photo stage. This stage occurs in the ___*___ of the chloroplast and results in the reduction of ___*___ to ___*___.

4. The thylakoid membranes contain chlorophylls ___*___ and ___*___, and also ___*___ (orange) and ___*___ (yellow). These can be separated using the technique of ___*___.

5. The colour of chlorophyll is because ___*___ light is ___*___ (passes straight through), whereas b___*___ and ___*___ are ___*___.

6. The rate of photosynthesis of a given plant under different conditions can be compared by measuring the amount of ___*___ given off per hour or the amount of ___*___ taken in per hour. To compare photosynthetic rates of different plants, we need to take into account the total leaf ___*___ or alternatively, the mass of plant material.

7. On a warm day in bright sunlight, the rate of photosynthesis is most likely to be limited by ___*___ supply. On a cold, sunny day it is most likely to be limited by ___*___. On the forest floor in summer it is likely to be limited by ___*___.

8. When a plant is short of water, photosynthesis is likely to be limited by the closure of the ___*___, thus restricting the uptake of ___*___.

9. Chlorophyll contains ___*___, and shortage of this metal ion in the soil leads to ___*___, in which the leaves are pale ___*___.

21 How Cells Absorb and get rid of Materials

Cells are constantly importing raw materials and exporting waste products. An animal cell for example, takes in oxygen and gets rid of CO_2, while a photosynthesising cell does the reverse. Fig. 21.1 shows some examples of these chemical exchanges.

An important function of the plasma membrane is to regulate this two-way traffic. It does this in two quite different ways:

- *Passively*, by acting as a molecular 'sieve', enabling small particles to pass through while restricting the movement of larger ones.

- *Actively*, by expending energy to transfer molecules and ions that cannot pass through passively.

Substances enter and leave cells by several mechanisms:

- Diffusion
- Osmosis
- Active transport
- Facilitated diffusion
- Endocytosis and exocytosis

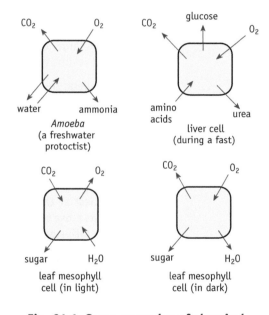

Fig. 21.1 Some examples of chemical exchanges in cells

Diffusion

Diffusion is the movement of a substance from where it is more concentrated to where it is less concentrated. It depends on the fact that in a liquid or gas, molecules and ions are in a constant, random movement (Fig. 21.2).

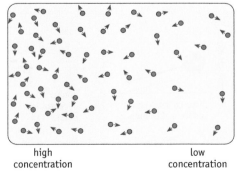

high concentration　　　low concentration

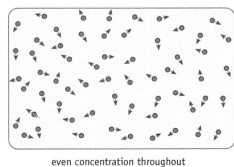

even concentration throughout

Fig. 21.2 Diffusion of randomly moving particles

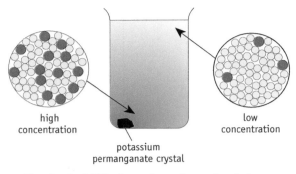

concentration

low
concentration

potassium
permanganate crystal

Fig. 21.3 Diffusion of a coloured substance in water

Fig. 21.3 shows what happens when a crystal of potassium permanganate is dropped into a beaker of still water. The deep purple slowly colour spreads outwards until it is equally concentrated throughout the solution (diffusion means *spreading*).

The situation in living cells is different from that in Fig. 21.3. An animal cell, for example, is continuously using oxygen and producing CO_2, so neither becomes evenly concentrated.

The energy for diffusion is not supplied by the organism but comes from the movement (kinetic energy) of the molecules or ions involved. Also, diffusion is a *net* movement; when oxygen diffuses into a cell, some oxygen molecules are moving out of the cell, but more are moving inwards.

The speed of diffusion depends on four things (the first two being by far the most important):

- The kind of medium through which it occurs. Diffusion is about 10 000 times faster in a gas than in solution. This is because molecules in a liquid change direction much more frequently in a liquid since they are much closer together.

- The steepness of the concentration gradient (see below).

- The size of the particles. At a given temperature, small molecules move faster (on average) than larger ones. Molecules of CO_2 are about 1.4 times bigger than molecules of oxygen (O_2), and it diffuses about 1.4 times more slowly.

- Temperature. Molecules move faster at higher temperatures, so diffusion is correspondingly faster. The effect is actually small, since it depends on the *absolute* temperature (a rise from 10° C to 20° C is actually an increase from 283 K to 293 K).

Concentration gradients

Imagine CO_2 diffusing into two plant cells, one with a vacuole and one without. The concentration of CO_2 is highest at the cell surface and lowest in the chloroplasts. Because the vacuole displaces the chloroplasts close to the cell wall, the diffusion path is much shorter. The change in CO_2 concentration along the diffusion pathway can be thought of as a 'gradient', which can be shown as a graph (Fig. 21.4). The gradient is steeper in the cell with the vacuole (concentration changes faster with distance from the cell surface), so diffusion is faster.

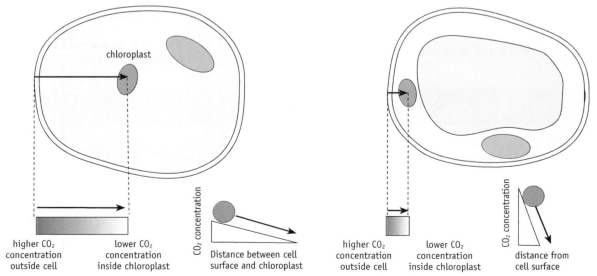

chloroplast

| higher CO_2 concentration outside cell | lower CO_2 concentration inside chloroplast |

CO_2 concentration

Distance between cell surface and chloroplast

Cell without vacuole, shallow gradient, slow diffusion

| higher CO_2 concentration outside cell | lower CO_2 concentration inside chloroplast |

CO_2 concentration

distance from cell surface

Cell with vacuole, steep gradient, rapid diffusion

Fig. 21.4 Model showing effect of concentration gradient on diffusion rate of CO_2 in two imaginary plant cells – one with a vacuole and one without (only one chloroplast shown)

ISBN: 9780170214094

Osmosis

Osmosis is a process of fundamental importance in living organisms. Here are some examples of the effects of osmosis:

- Most marine fish die if transferred to freshwater.
- When a drop of blood is mixed with distilled water, the blood cells burst.
- Living plant tissues that have lost water become firm when supplied with water.

When applied to animal cells, osmosis is relatively straightforward. To make sense of osmosis in plant cells, we have to introduce the idea of **water potential**.

First, some definitions:

- A *partially permeable membrane* is a kind of molecular sieve. It allows small molecules (such as water) to pass through but not larger ones (such as sugar).
- The *water potential* of a solution is the tendency of water to *leave* that solution. Water *always* tends to move from a higher to a lower water potential.
- *Osmosis* is the movement of water through a partially permeable membrane.

Fig. 21.5 illustrates the idea of water potential. First, we need to appreciate that when sugar is added to water, the sugar reduces the concentration of the water. A dilute solution of sugar thus has a higher *water concentration* than a concentrated sugar solution.

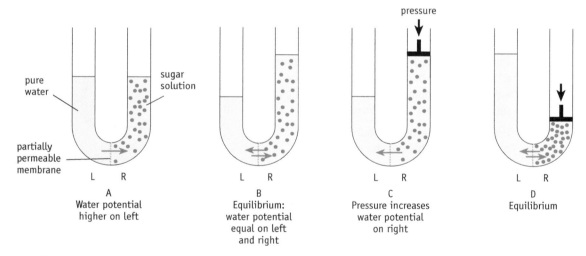

Fig. 21.5 Model showing osmosis and the idea of water potential

In A, pure water is separated from sugar solution by a partially permeable membrane. Water thus diffuses across the membrane from the higher to the lower water concentration. The level of the sugar solution rises, and so therefore does its *pressure*. Eventually the level in the right hand limb (R) stops rising, as in B, *despite the fact that the water is more concentrated in the left hand limb* (L). The additional pressure in R is thus enough to oppose the effect of the concentration difference. When there is no further change in levels, the water potential in L and R are equal.

The water potential of a solution therefore depends on two things *added together*:

- its water concentration
- its pressure.

Now suppose downward pressure is exerted by a piston, as in C. This raises the water potential in R, so water now moves into L. In doing so, *it is moving from a lower to a higher water concentration*. If the piston is pushed hard enough, the level in L may rise above that in R, as in D. Eventually, a new

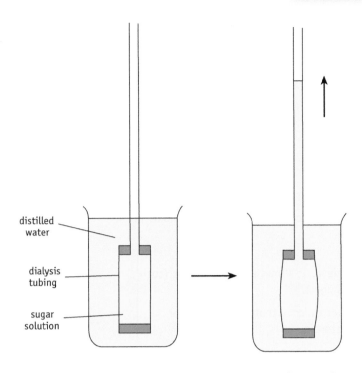

distilled water

dialysis tubing

sugar solution

Fig. 21.6 Simple demonstration of osmosis

equilibrium is established and the water potentials in L and R are equal again.

Fig. 21.6 shows a simple laboratory demonstration of osmosis. The dialysis tubing is partially permeable. To begin with, the levels of the sugar solution and water are equal, so there is no pressure difference. The water is more concentrated in the beaker, so the water potential in the beaker is higher than in the tube. Water therefore enters the dialysis tubing and the level begins to rise, raising its pressure. If the tube were high enough and the dialysis tubing and seal were strong enough, the sugar solution would rise many metres. When it eventually stopped rising, the water potential in the beaker and tube would be equal, though the water concentration would still be higher in the beaker.

It is important to realise that as the level in the tube rises, water molecules are actually moving across the membrane in both directions simultaneously. It is just that at a given moment more are moving in one direction than the other. Osmosis is thus a *net* movement of water molecules.

Osmosis in plant cells

So far as osmosis is concerned, a plant cell has three important parts (Fig. 21.7):

- The *cell wall,* which is fully permeable, allowing all molecules to pass through. Because the wall is strong enough to resist tension, a plant cell can develop a high internal pressure.

- The *cytoplasm,* which is bounded by the partially permeable *plasma membrane*.

- The *vacuole,* which contains a watery *cell sap,* a solution of salts and other dissolved substances. The vacuole is separated from the cytoplasm by another partially permeable membrane, the *tonoplast*.

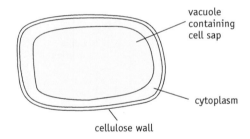

vacuole containing cell sap

cytoplasm

cellulose wall

Fig. 21.7 The osmotically important parts of a plant cell

You can see osmosis in plant cells by mounting a plant tissue such as a strip of rhubarb epidermis ('skin') in a concentrated solution such as molar sucrose. After a few minutes the vacuoles lose water by osmosis, causing the cytoplasm to pull away from the cell wall. This process is called **plasmolysis** (Fig. 21.8A). In a plasmolysed cell the wall is limp so there is no internal pressure, and the cell sap has the same water concentration as the external solution.

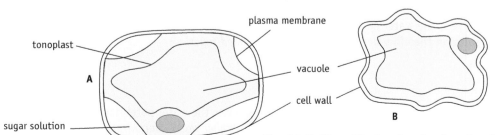

tonoplast

plasma membrane

A

vacuole

cell wall

sugar solution

B

Fig. 21.8 The effects of water loss in plant cells by osmosis (A) and by evaporation (B)

Plasmolysis only occurs when cells are surrounded by strong *solutions* and rarely, if ever, occurs under natural conditions. When a plant cell plant loses water by *evaporation* the result is quite different. Instead of the cytoplasm contracting away from the wall, the cytoplasm pulls the wall in with it so the whole cell caves in (Fig. 21.8B).

Where osmosis is concerned, the important thing is the number of solute particles per dm³ ('litre') of solution. Although 1 dm³ of 10% sucrose solution contains the same *mass* of sugar as a 10% solution of glucose, it contains just over half the *number* of sugar molecules, since sucrose molecules are almost twice as big as glucose molecules.

Because of this, solute concentrations are best expressed in terms of *molarity* rather than percentages. A molar solution is one in which 1 dm³ of solution contains the molecular mass of the substance in grams. Thus 1 dm³ of molar glucose solution contains 180 grams of glucose and 1 dm³ of molar sucrose contains 342 grams of sucrose. For any substances whose particles do not dissociate into ions, *solutions of equal molarity have the same number of solute particles in a given volume of solution*. The higher the molarity of a solution, the lower its water concentration.

Because it is strong enough to resist stretching, the cellulose wall enables plant cells to develop high internal pressures. When water enters a plant cell, the pressure inside it rises and eventually becomes sufficient to prevent any more water entering. The cell is like a tightly inflated football and is said to be **turgid**, and its internal pressure is called *turgor pressure*. A cell that has low internal pressure is said to be **flaccid** or limp. In young, soft parts of plants, turgor helps provide support, as shown when a leaf wilts as the cells lose water.

Fig. 21.9 shows what happens when a flaccid plant cell is placed in distilled water. Water enters the cell and its pressure rises, like blowing up a football. Eventually the pressure is sufficiently high to prevent any more water entering.

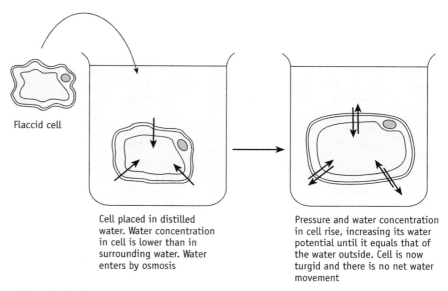

Flaccid cell

Cell placed in distilled water. Water concentration in cell is lower than in surrounding water. Water enters by osmosis

Pressure and water concentration in cell rise, increasing its water potential until it equals that of the water outside. Cell is now turgid and there is no net water movement

Fig. 21.9 What happens when a plant cell is placed in distilled water

In this turgid state the water potential in the cell equals that of the surrounding water, and there is no net water movement. *This is despite the fact that water is less concentrated inside than outside.*

Osmosis in animal cells

Osmosis in animal cells is simpler than in plants because there is no cell wall, so there is never any internal pressure. The effect of osmosis can be seen in animal cells by diluting blood with excess distilled water (Fig. 21.10). Normally, blood cells are surrounded by plasma that has the same water concentration as the cytoplasm. When the plasma is diluted, water enters the blood cells by osmosis, causing them to swell and burst, leaving the empty cell membranes. If concentrated salt solution is added to blood, water leaves the cells by osmosis, causing them to shrink.

Fig. 21.10 Effect of different solutions on red blood cells

concentrated / salt solution — cells shrivel — normal blood cells — dilution with water — cells swell — further dilution — cells burst

ISBN: 9780170214094

Osmosis in single-celled organisms

Paramecium is a single-celled organism that lives in ponds and ditches (Fig. 21.11). It is large for a single-celled organism and is just visible to the naked eye. It swims using thousands of cilia, which it also uses to feed by filtering bacteria and other single-celled organisms from the water.

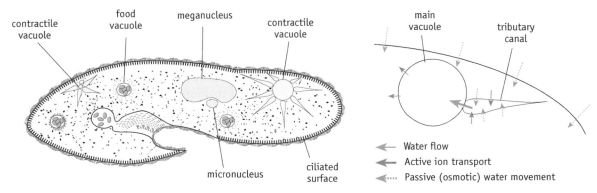

Fig. 21.11 Left: *Paramecium*, a single-celled freshwater organism Right: Small part of the body showing one tributary canal of contractile vacuole

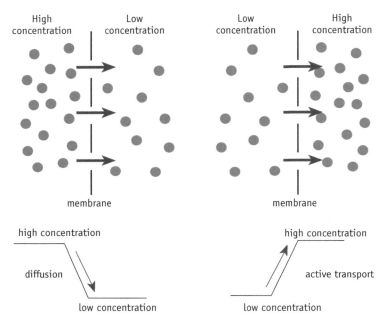

Fig. 21.12 Passive and active transport as 'downhill' and 'uphill' processes

Fresh water is a very dilute environment, and as a result water is constantly entering the body by osmosis. At the same time, water is actively pumped out from the body by two **contractile vacuoles**. Each consists of several 'feeder' vacuoles which drain into a larger central vacuole. The central vacuoles slowly fill and then empty at a pore on the surface. The two vacuoles fill and empty alternately, so when one is nearly full the other has only just started filling.

The liquid that is expelled is very dilute, with the result that water is expelled from the body at the same rate as it enters. In this way the water concentration in the body is regulated at a constant level. This process is called **osmoregulation**. It is an example of **homeostasis** — the maintenance of near-constant conditions inside the body.

The water concentration in the cytoplasm is higher than in the vacuole, so water must be moving in the opposite direction to that of osmosis. The organism must therefore be actively 'pumping' water into the vacuoles from the cytoplasm.

That the process requires energy is shown by the fact that if a trace of cyanide is added to the water, the organism swells and bursts. Cyanide inhibits respiration and therefore prevents the active transport that pumps water out.

The mechanism of water pumping depends on *active transport* (see below) of ions into the feeder canals. Water follows by osmosis, and subsequently the ions are pumped out of the central vacuole, leaving the water.

Active transport

Even if the lipid bilayer were permeable to glucose, amino acids and ions, these could only diffuse across the plasma membrane if the concentration gradient were the 'right way around'. However,

ISBN: 9780170214094

movement of these substances is often 'uphill', or against a concentration gradient (Fig. 21.12). Since this kind of movement across a membrane requires energy it is called *active transport*.

Since it occurs in *all* living cells, active transport is one of the fundamental characteristics of living organisms. Examples are the absorption of glucose by the small intestine and kidney tubules, and salt uptake by the roots of plants, and amino acid uptake by fungi.

At rest, most animal cells use about a third of their resting energy 'budget' in active transport — in nerve cells it may be two thirds. Without active transport the inside of a cell would slowly become like its surroundings. For example, in all animals the concentration of potassium ions inside the cells is much higher than outside. Very slowly, potassium ions leak out by diffusion, but this is counteracted by active uptake. This ability to maintain near-constant internal conditions is another example of *homeostasis*.

In some animal tissues specialised for active transport, the area of the plasma membrane is greatly increased by thousands of **microvilli** (Fig. 21.13). Cells involved in active transport also have abundant mitochondria, which generate the necessary ATP.

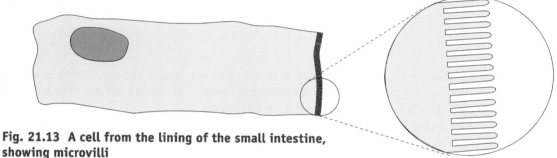

Fig. 21.13 A cell from the lining of the small intestine, showing microvilli

Evidence for active transport

How can we tell if transport is active or passive? It is active if:

- the material is being moved from a lower to a higher concentration.

- it is inhibited by cyanide (which stops respiration) or by lack of oxygen (Fig. 21.14).

- it is highly sensitive to temperature (diffusion is only slightly temperature-sensitive).

How does active transport work?

Since active transport occurs across a cell membrane, the 'pumps' must be located within the membrane itself, so they cannot be much bigger than large molecules. In two respects the mechanism resembles that of enzyme action:

- Active transport mechanisms are *specific*. For example, certain substances inhibit the transport of one kind of molecule without interfering with the transport of other substances. This suggests that different 'pumps' transport different kinds of molecule.

- The mechanism can become 'saturated'. Fig. 21.15 shows the effect of increasing the concentration of potassium ions on its rate of uptake by carrot tissue. As the external potassium concentration is increased, so does its rate of uptake, *but only up to a point*. Beyond this, further increases in concentration have no effect. This is readily explained by supposing that each carrier molecule briefly combines with a potassium ion, just as an enzyme combines with its substrate.

Both of these effects parallel the action of enzymes, and are strong evidence that the 'pumps' are globular proteins in the membrane. Fig. 21.16 shows the kind of mechanism probably involved.

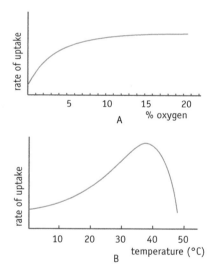

Fig. 21.14 Effect of oxygen supply on rate of uptake of bromide ions by barley roots (A) and effect of temperature on rate of uptake of potassium ions by carrot discs (B)

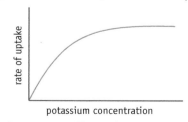

Fig. 21.15 Effect of external potassium concentration on its rate of uptake by carrot tissue

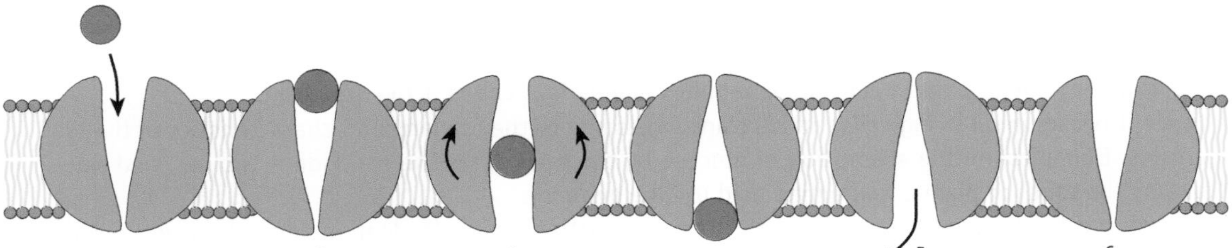

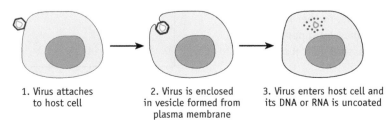

1	2	3	4	5	6
	molecule combines with membrane protein	membrane protein changes shape, discharging molecule to inner side of membrane		molecule released on other side	Using ATP energy, protein regains former shape

Fig. 21.16 Possible mechanism of active transport

The ion or molecule to be transported first combines with the carrier protein. The protein then changes its shape, discharging the molecule or ion to the other side of the membrane. To transport another molecule the protein must first regain its original shape, which requires energy in the form of ATP.

Facilitated diffusion

Like 'ordinary' diffusion, this process is passive in that it does not involve the expenditure of energy and can only occur from a higher to a lower concentration. However, facilitated diffusion can only take place through special protein channels in the plasma membrane.

Like the carrier proteins involved in active transport, channel proteins are specific for particular molecules or ions. The process is passive because substances can only move through channel proteins from a higher to a lower concentration. In mammals, most cells absorb glucose by facilitated diffusion. One of the effects of insulin is to activate glucose transporter channels, enabling the cells to absorb glucose. If insulin is deficient, most of the cells cannot absorb glucose and have to use fat for respiration.

Cytosis

In addition to individual molecules, many cells can transport much larger packages across their membranes by *cytosis*. This involves the folding and fusion of membranes, and since this is a form of movement, it must involve energy expenditure. There are two main kinds — *endocytosis* and *exocytosis*.

Endocytosis

Endocytosis is the taking in of bulk material from the environment by an intucking of the plasma membrane around it, forming small 'bags' or vesicles. One kind of endocytosis is **phagocytosis** ('cell eating') when, for example, a white blood cell engulfs a bacterium. The particle is enclosed in a food vacuole (or phagosome), which then fuses with a lysosome containing digestive enzymes (Chapter 17). The food is then digested and the products absorbed through the membrane of the vacuole.

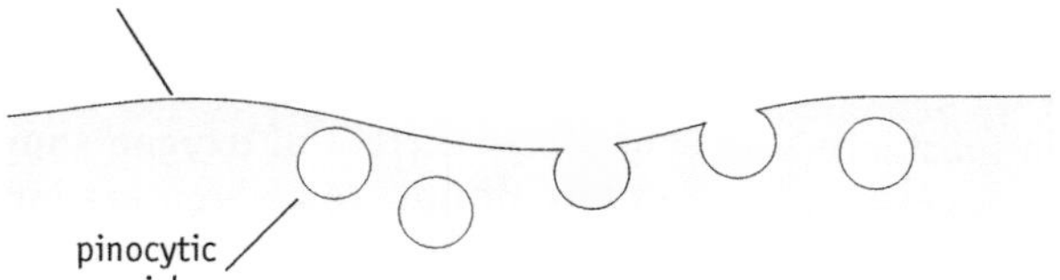

Phagocytosis is triggered by the presence of the food particle. For example, white cells engulf bacteria much more readily if the bacteria are coated with antibodies (defence proteins made by the body in response to the presence of foreign substances). This suggests that membranes can 'recognise' specific kinds of chemical.

Whereas only certain cells show phagocytosis, virtually all eukaryotic cells exhibit **pinocytosis** ('cell drinking'). In this process the plasma membrane continually forms small intuckings, enclosing vesicles of the surrounding fluid (Fig. 21.17).

plasma membrane

pinocytic vesicle

Fig. 21.17 Pinocytic vesicles forming

1. Virus attaches to host cell

2. Virus is enclosed in vesicle formed from plasma membrane

3. Virus enters host cell and its DNA or RNA is uncoated

Fig. 21.18 Pinocytic vesicles forming

An interesting case of endocytosis is the entry of virus particles into animal cells. Each kind of animal virus can only attack certain types of cell in certain species of animal. For example the polio virus attacks certain nerve cells in monkeys, apes and humans. For a virus to enter an animal cell, the proteins in the outer coat of the virus must join with certain 'receptor' proteins in the plasma membrane. The plasma membrane then tucks in and encloses the virus in a vacuole (Fig. 21.18). The protein coat is then digested away to release the nucleic acid.

Exocytosis

The reverse of endocytosis is exocytosis, and it occurs when a cell secretes a substance for use elsewhere (Chapter 17).

Exocytosis adds membrane to the cell surface and endocytosis removes it. Since the area of plasma membrane remains more or less constant, the rates of these two processes must be in balance.

Summary of key facts and ideas in this chapter

○ Test your basics

Copy and complete the following sentences. In some cases the first or last letters of a missing word are provided.

1. The energy for diffusion comes from the ___*___ energy of the particles.

2. At a given temperature, substances with smaller particles diffuse ___*___ than substances with larger particles.

3. Diffusion is much slower in ___*___ than in a ___*___ because the particles ___*___ and change ___*___ much more frequently.

4. The rate of diffusion between two points depends on the steepness of the ___*___ ___*___. This in turn depends on the ___*___ between the two points and the ___*___ in concentration between them.

5. Osmosis is the movement of ___*___ through a ___*___ ___*___ membrane .

6. The water potential of a solution depends on both its water c___*___ and its ___*___.

7. In plant cells the cell wall is strong enough to resist t___*___ so the cell can withstand internal p___*___. The cell wall is also fully ___*___ to dissolved substances.

8. The cytoplasm of a plant cell is surrounded by the ___*___ membrane, which is ___*___ ___*___.

9. Most of the volume of a plant cell is occupied by a large sap-filled space, the ___*___, containing a high concentration of s___*___es.

10. When a plant cell is placed in distilled water, it absorbs water until the internal ___*___ rises enough to prevent any more water entering. A cell in this state is said to be ___*___, and in this state the ___*___ ___*___ inside and outside the cell are ___*___.

11. When a plant cell is immersed in a solution more concentrated than its own cell sap, water leaves the cell until the plasma membrane pulls away from the ___*___ ___*___. The cell is said to be ___*___ed.

12. Movement of a substance across a membrane against a concentration gradient is called ___*___ ___*___. Unlike diffusion it is very sensitive to ___*___.

13. Facilitated diffusion resembles active transport in depending on protein ___*___ in cell ___*___. It differs in that it can only occur ___*___ a concentration gradient.

14. Another mechanism of transport across membranes is ___*___, in which a membrane folds and becomes pinched off, enclosing a portion of material from the other side of the membrane. In ___*___, substances are transported into the cell, for example when a white blood cell engulfs a bacterium, and when a virus enters an animal cell. In ___*___, substances are exported out of the cell, for example the secretion of digestive enzymes by cells lining the gut.

Excellence in Biology Level 2

ISBN: 9780170214094

22 Chromosomes and Cell Division

All sexually reproducing eukaryotes begin life as a fertilised egg or **zygote**. This is formed by the joining together of two cells called **gametes** — a large female gamete or **egg**, and a much smaller male gamete or **sperm** (Fig. 22.1). The zygote then divides into two cells, and then each of these cells divides, and so on, eventually producing a body consisting of millions of cells.

130 µm

50 µm

Fig. 22.1 A human egg (left) and sperm (right) compared

Chromosomes

To develop into a many-celled organism, a fertilised egg must have *information*. When the zygote divides, the two daughter cells each receive a complete set of this information. This is carried in the form of a chemical called **d**eoxyribo**n**ucleic **a**cid, or **DNA** as it is usually known. It is present in thread-like bodies called **chromosomes**, which are in the nucleus. A zygote contains two sets of chromosomes, one set from the egg and one set from the sperm (Fig. 22.2).

Chromosomes only appear as separate structures during cell division, and even then they must first be stained if ordinary microscopes are used (*chromo* means 'colour', and *soma* means 'body'). In between divisions

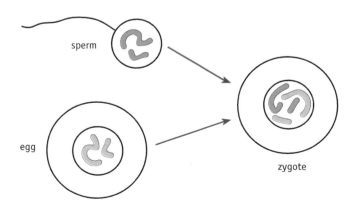

sperm

egg

zygote

Fig. 22.2 Chromosomes differ in their size and structure, and occur in pairs. Gametes have one of each pair; the zygotes and body cells derived from it have both members of each pair.

ISBN: 9780170214094

the chromosomes are too long, thin and tangled to be seen. In the years after their discovery in the late nineteenth century, biologists discovered a number of interesting things about them:

- With very few exceptions, all the body cells of any given species have the same number of chromosomes. For example, human cells have 46, kiwi have 80, dogs have 78, mice have 40, kowhai have 18 and rimu have 20. *Chromosome number is therefore a characteristic of a species.*

- In a given species the chromosomes differ in length and in certain other characteristics, such as the position of the *centromere* (see below). Each chromosome also has a characteristic pattern of banding after treatment with special stains. Individual chromosomes can thus be distinguished, and each given a number.

- In any gamete there is only *one* of each kind of chromosome. Gametes are said to have the **haploid** number of chromosomes (*n*). In humans, *n* = 23.

- In all the body cells of an animal or a plant, each chromosome has a 'partner' or **homologue** that looks just like it, with the same length, banding pattern and centromere position. There are thus *two* chromosomes of each kind in the nucleus of a body cell, one member of each pair being inherited from each parent. The two together are called a **homologous pair**. Chromosomes inherited from the male parent are called *paternal* chromosomes, those inherited from the female parent are called *maternal* chromosomes. Cells with two of each kind of chromosome have the **diploid** number (2*n*). In humans, 23 of the chromosomes are of maternal origin and 23 are of paternal origin.

Using special staining techniques, each chromosome has been shown to have a specific banding pattern, rather like a supermarket bar code. Each chromosome has a banding pattern identical with its homologue.

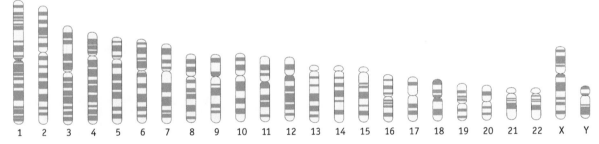

1 2 3 4 5 6 7 8 9 10 11 12 13 14 15 16 17 18 19 20 21 22 X Y

Fig. 22.3 The human karytoype. Each chromosome is shown single-stranded, but in cells used for these preparations they have replicated and are double-stranded.

Each species can thus be distinguished from every other species by its chromosomes. All these chromosomal characteristics together form an organism's **karyotype**. When photographs of the individual chromosomes are cut out and arranged in pairs and in order of size, the result is a **karyogram** (Fig. 22.3). Chromosomes are numbered, starting from the largest.

Genes and chromosomes

The complete set of genes in an organism is its **genome**. The human genome has about 23 000 genes. These are arranged on 23 pairs of chromosomes, so each chromosome has an average of about 1000 genes. Each gene occupies a particular position or **locus** on a particular chromosome. For example, the gene for the ABO blood groups is near one end of chromosome number 9.

Excellence in Biology Level 2

ISBN: 9780170214094

Kinds of division

The nucleus of a eukaryotic cell (see Chapter 17) can divide in two different ways (Fig. 22.4):

- By **mitosis**. In this kind of division the daughter nuclei are genetically identical to the parent nucleus. Mitosis is concerned solely with increasing the *number* of cells, and in humans is part of growth ('repair' involves localised growth).

- By **meiosis**. Meiosis is an essential part of sexual reproduction, and is concerned with producing genetic *change*. While mitosis occurs throughout life, in most multicellular organisms meiosis occurs only in the adult. Meiosis is concerned with producing genetic variation and is dealt with in Chapter 14.

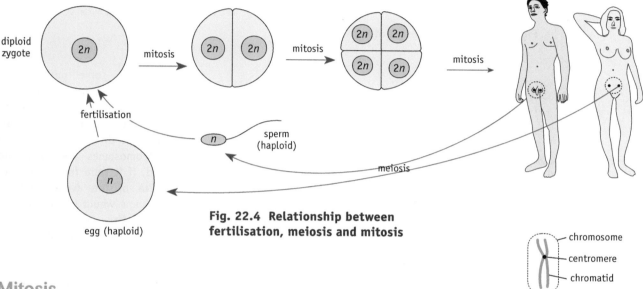

Fig. 22.4 Relationship between fertilisation, meiosis and mitosis

Fig. 22.5 Structure of a chromosome in the early stages of mitosis

Mitosis

In humans and other multicellular organisms, mitosis is an important part of *growth*. In single-celled organisms it is the basis of asexual reproduction. It can occur in diploid or in haploid cells — for example mosses are haploid, and as they grow they increase the number of cells by mitotic division.

Early in the development of multicellular organisms, cell division occurs throughout the entire body. In adults it occurs in certain parts only, such as the skin and bone marrow of mammals and the root tips and shoot tips of flowering plants.

Before a cell can divide mitotically the DNA of each chromosome replicates. The result is two identical threads or **chromatids**, attached at a point called the **centromere** (Fig. 22.5).

As a result of mitosis each of the two daughter nuclei receives one of the two chromatids of each chromosome. If the cell is going to divide again, each chromosome replicates again. During successive cell divisions, each chromosome goes through a cycle of replication, separation of chromatids, replication, and so on (Fig. 22.6).

The essential results of mitosis are illustrated in Fig. 22.7.

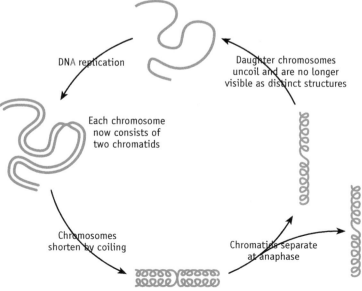

Fig. 22.6 Chromosome replication and mitosis

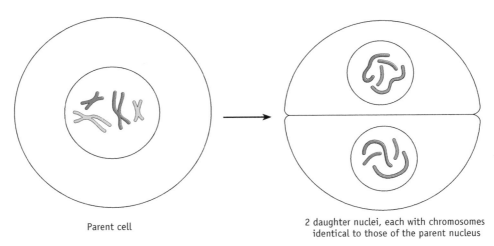

Parent cell

2 daughter nuclei, each with chromosomes identical to those of the parent nucleus

Fig. 22.7 Essential result of mitosis

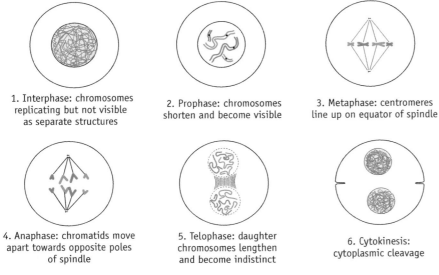

1. Interphase: chromosomes replicating but not visible as separate structures

2. Prophase: chromosomes shorten and become visible

3. Metaphase: centromeres line up on equator of spindle

4. Anaphase: chromatids move apart towards opposite poles of spindle

5. Telophase: daughter chromosomes lengthen and become indistinct

6. Cytokinesis: cytoplasmic cleavage

Fig. 22.8 The main events of mitosis in an animal cell, shown in more detail. Though not part of mitosis, interphase and cytoplasmic cleavage are shown for convenience.

Now for some more detail (Fig. 22.8). Though mitosis is a continuous process, for convenience it is divided into different stages. Between cell divisions (interphase) the chromosomes are long and tangled. If it were fully extended, the DNA of the average human chromosome would be about 4 cm long. In this tangled state it would be mechanically impossible for the chromatids to move apart to form two daughter groups. The early stage (prophase) of mitosis is concerned with the coiling up of the chromosomes. This makes them short and compact enough for the chromatids to separate. In this state they are visible under the microscope when stained.

- **Prophase** The chromosomes shorten and thicken by coiling up, and become visible when stained. The end of prophase is marked by the formation of the **spindle**, a barrel-shaped system of thousands of protein fibres. At the same time the nuclear envelope breaks down.

- **Metaphase** Each chromosome becomes attached to the spindle fibres at its centromere. The chromosomes become arranged on the equator of the spindle (the imaginary plane between the two ends or poles).

- **Anaphase** The chromatids of each chromosome move apart to each pole of the spindle.

- **Telophase** This is the opposite of prophase — the spindle disappears and a nuclear envelope develops round each group of daughter chromosomes, forming two nuclei. The chromosomes elongate and become indistinct again.

Cytoplasmic division

Towards the end of mitosis the cytoplasm begins to split into two. In animal cells a furrow develops between the two nuclei, pinching the cytoplasm into two. In plant cells a new cell wall develops between the two daughter nuclei (Fig. 22.9). At first, the two daughter cells are both enclosed by the wall of the old cell. However this is soon broken down at the 'join', separating the daughter cells from each other.

The cell cycle

In cells that are repeatedly dividing, mitosis is just the more easily observed and dramatic period in the process of cell growth and division. Between divisions, when not much seems to be happening, the cell is a hive of activity in which mitochondria are dividing, chromosomes are replicating and proteins are being synthesised. The entire sequence of events is called the *cell cycle* (Fig. 22.10).

Growth in plant cells

Because a plant cell wall is non-living, it cannot grow actively but has to be stretched by pressure from within. A plant cell enlarges by the osmotic uptake of water. Though this is a passive process, it can only be maintained if the difference in water concentration is maintained by active uptake of salts, which *is* an active process (Fig. 22.11).

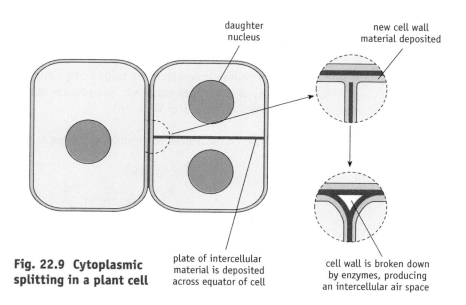

Fig. 22.9 Cytoplasmic splitting in a plant cell

daughter nucleus

new cell wall material deposited

plate of intercellular material is deposited across equator of cell

cell wall is broken down by enzymes, producing an intercellular air space

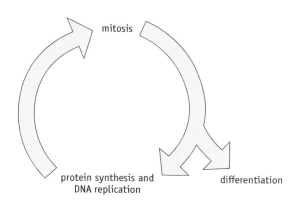

mitosis

protein synthesis and DNA replication

differentiation

Fig. 22.10 Key events in the cell cycle

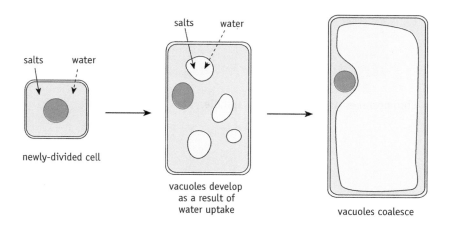

salts water

newly-divided cell

salts water

vacuoles develop as a result of water uptake

vacuoles coalesce

Fig. 22.11 How a plant cell enlarges. Solid arrows indicate active transport; dotted arrows indicate osmotic uptake.

Extension: Stages of mitosis in an onion root tip

The following photomicrographs were taken using a microscope with a magnification of approximately 1000x. Names of the stages are given, but you are not expected to learn them for NCEA. (Courtesy Dr Barry O'Brien, University of Waikato.)

Interphase: Chromosomes not visible; the large dark blob is a nucleolus.

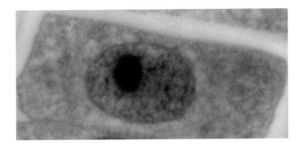

Early prophase: Chromosomes have become distinct, nucleolus still visible.

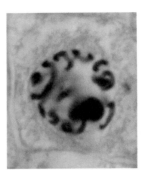

Late prophase: Nucleolus no longer visible.

Metaphase: Chromosomes on equator of the spindle.

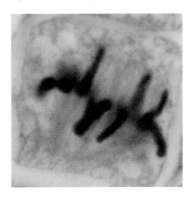

ISBN: 9780170214094

Anaphase: Chromatids dragged apart to opposite poles of the spindle.

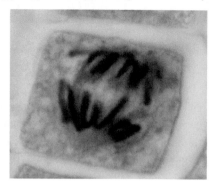

Late anaphase: Daughter chromosomes collect at poles of spindle.

Early telophase: Each set of daughter chromosomes forms a compact cluster.

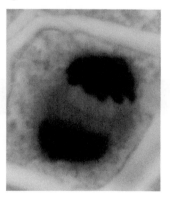

Late telophase: Chromosomes begin to become less distinct, and nucleolus re-forms. Though not part of mitosis, the new cell wall begins to form between the daughter nuclei.

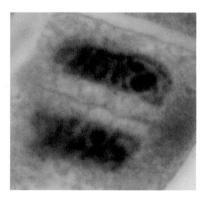

ISBN: 9780170214094

Summary of key facts and ideas in this chapter

○ Test your basics

Copy and complete the following sentences.

1. In eukaryotes, the genetic information is carried in the ___*___, on threads called ___*___.

2. Every chromosome has a region called the ___*___, which is the part that attaches to the ___*___ in cell division. Chromosomes also differ in length and in the pattern of ___*___ when stained.

3. Body cells of animals and flowering plants have ___*___ of each kind of chromosome, so they have the ___*___ number. Chromosomes thus exist in ___*___ pairs. In a gamete there is only ___*___ of each pair, so gametes have the ___*___ number of ___*___. Each species can be distinguished by the characteristics of its chromosome set, called the ___*___.

4. In ___*___ the daughter nuclei are genetically identical to each other and their chromosome number is the ___*___ as that of the parent nucleus. In ___*___ the daughter nuclei are genetically different from each other and have ___*___ as many chromosomes as the parent nucleus.

5. Between divisions of the nucleus, the ___*___ is replicated, mitochondria ___*___, and proteins and other cell constituents are ___*___. In plants, cell growth involves the active uptake of ___*___, accompanied by the osmotic uptake of ___*___. This stretches the ___*___ ___*___, but its thickness is maintained by deposition of new ___*___.

23 DNA: Its Structure and Replication

'It has not escaped our notice that the specific pairing we have postulated immediately suggests a possible copying mechanism for the genetic material.' (James Watson and Francis Crick, in a letter to *Nature*, 25 April 1953)

Deoxyribonucleic acid or DNA

Though DNA was first extracted (from pus cells) as early as 1868, it remained a grey, fibrous material on chemists' shelves until the mid-twentieth century, when it was finally proved to be the stuff of which genes are made.

This realisation started a race to work out its structure and how it worked. The actual double-helical structure was worked out by Watson and Crick in 1953, using X-ray data obtained by Rosalind Franklin. As a result, Watson and Crick received the Nobel Prize for Physiology or Medicine. Franklin had died in 1958, so prize was shared with Maurice Wilkins, director of the laboratory in which Franklin had worked.

The information in DNA serves two purposes:

- It enables cells to join amino acids into the correct sequences to make particular proteins.

- It enables the DNA to be *replicated*, or copied. This happens every time a cell divides mitotically; each daughter nucleus receives an identical copy of the DNA in the parent nucleus.

DNA is one of two classes of compound called **nucleic acids**. The other is **RNA** (ribonucleic acid), which plays a central role in protein synthesis (Chapter 24).

DNA structure

A DNA molecule consists of two **polynucleotides** wrapped around each other in a spiral or *helix*. A polynucleotide is a polymer consisting of a chain of sub-units called **nucleotides**. Each individual nucleotide consists of three smaller parts (Fig. 23.1):

- a *phosphate* group;
- a pentose (5-carbon) *sugar*, **deoxyribose**;
- a nitrogenous (nitrogen-containing) *base*.

ISBN: 9780170214094

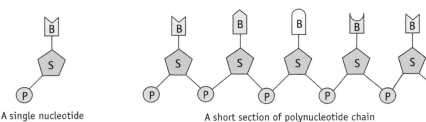

A single nucleotide

A short section of polynucleotide chain

Fig. 23.1 A single nucleotide and a short section of polynucleotide chain. B = base, S = sugar, P = phosphate. The four kinds of base are represented by different shapes and colours.

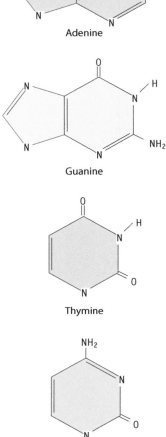

Adenine

Guanine

Thymine

Cytosine

Fig. 23.3 The four DNA bases

Whereas the sugar and the phosphate are the same in all of the nucleotides in DNA, the base can be one of four kinds — **adenine (A)**, **guanine (G)**, **thymine (T)** or **cytosine (C)**.

Four kinds of base means there are *four kinds of nucleotide*. It is the *order* of the four kinds of nucleotide in the polynucleotide chain (or rather, the order of their bases since it is the bases that are different) that represents the information for making a protein. The base sequence in DNA is thus a kind of message in code, rather like the order of dots and dashes in Morse code.

A DNA molecule actually consists of *two* chains of nucleotides, arranged in a double helix, rather like a twisted ladder (Fig. 23.2).

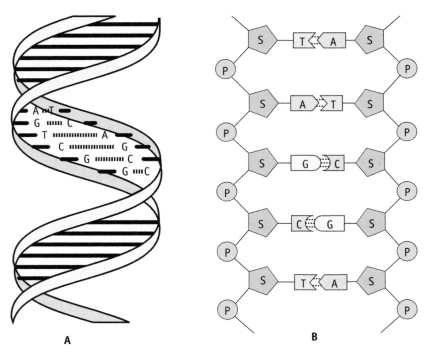

A

B

Fig. 23.2 (A) A simple representation of the DNA double helix. (B) The helix straightened out to show base pairing between the two chains.

The DNA double helix, as it became known, has the following important features:

- The two nucleotide chains are cross-linked by hydrogen bonds between the bases. The bases are thus in *pairs,* each pair forming a 'rung' of the ladder. Since hydrogen bonds are weak, the two strands can be easily separated.

- Adenine and guanine are two-ringed **purines**, whereas cytosine and thymine are single-ringed **pyrimidines** (Fig. 23.3). Since a purine always pairs with a pyrimidine, a large base always pairs with a small one. This means that the 'rungs' of the ladder are all the same width, and the two sides are parallel.

ISBN: 9780170214094

How does DNA carry information?

A protein consists of one, or sometimes several, polypeptide chains. A polypeptide is a chain of amino acids, linked together in a specific order. To make a particular polypeptide, a cell must have the information to join the amino acids in the correct sequence. This information is contained in a sequence of DNA bases called a **gene**.

DNA replication

In between cell divisions, when the chromosomes are too long and thin to be seen, the DNA molecule in each chromosome makes another copy of itself.

First, the two strands separate locally to expose the bases, rather like the opening of a zip fastener. Under the influence of the enzyme **DNA polymerase**, new nucleotides pair up with the exposed bases of the existing half-strands. The nucleotides are then linked together lengthwise. Because adenine pairs with thymine and guanine pairs with cytosine, each new half-strand is identical to one of the original half-strands. The result is two identical DNA molecules, each consisting of an 'old' and a 'new' half. Because each replication involves keeping one of the existing strands, DNA replication is said to be **semiconservative** (*semi* = 'partial', *conservative* = 'keeping'), shown in Fig. 23.5.

To replicate its DNA, a cell needs four things:

- *Raw materials* in the form of new nucleotides.

- *Energy* supplied by respiration.

- *Information* in the sequence of bases in each of the two strands.

- An *enzyme*, DNA polymerase.

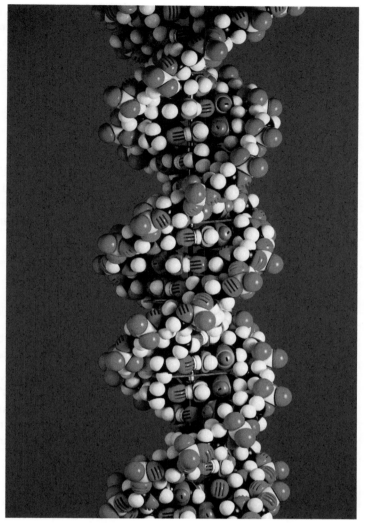

Fig. 23.4 A model of the DNA double helix

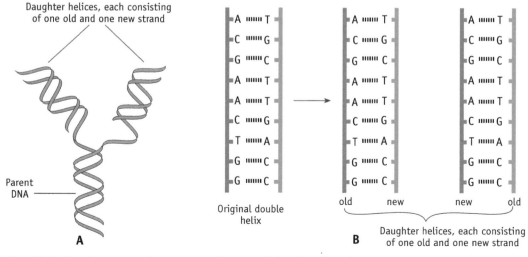

Fig. 23.5 Semiconservative DNA replication (A) Pictorial view, no details of base pairing (B) Shown with helix straightened out

Very rarely, DNA does not replicate exactly, resulting in a change in the genetic information called a **mutation**. This can change the sequence of amino acids in a protein encoded in a gene, but the code itself remains unchanged.

Extension: The two ends of a polynucleotide chain are different

An important property of polynucleotide chains is that the two ends are different, and are called the 3' (spoken as three-prime) and 5' ends. If we liken a polynucleotide chain to a line of elephants linked trunk-to-tail, no matter how many nucleotides ('elephants') are at one end, the 'head' of the line always ends in a trunk (3' end), and the other in a tail (5' end), as shown in Fig. 23.6. In the double helix, the two chains are *antiparallel*; the 5' end of one chain being opposite the 3' end of the other.

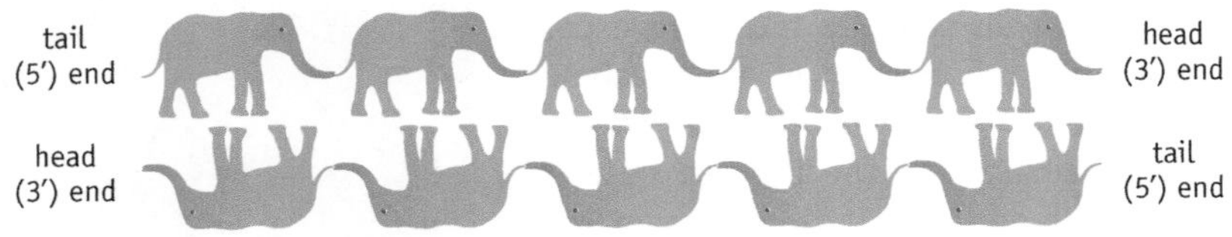

tail (5') end   head (3') end

head (3') end   tail (5') end

Fig. 23.6 Two pentanucleotide (five nucleotides) chains represented by elephants linked trunk to tail

The basis for the 5' and 3' naming of the two ends depends on the numbering of the carbon atoms in the sugar (Fig. 23.7).

When two nucleotides join together, the phosphate of one nucleotide joins with the OH group on carbon number 3 of the other nucleotide. No matter how many nucleotides in the chain, one end (the 3' end) has a free OH group and the other (the 5' end) has a free phosphate group (Fig. 23.8).

Fig. 23.9 shows the antiparallel arrangement of two DNA strands.

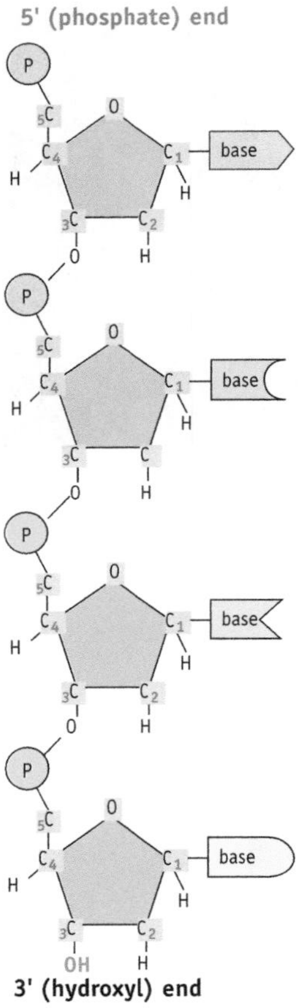

5' (phosphate) end

3' (hydroxyl) end

Fig. 23.8 A tetranucleotide, showing how the 5' and 3' ends can be distinguished. New nucleotides can only be added to the 3' end.

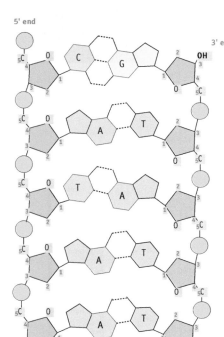

5' end

3' end

3' end

5' end

Fig. 23.9 How the two DNA strands are antiparallel

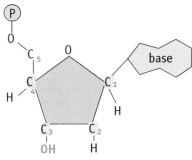

Fig. 23.7 Numbering of carbon atoms in deoxyribose sugar

Experimental evidence for the double helix

A key prediction of Watson and Crick's model was that DNA replication is *semiconservative*, half the old strand being conserved in each of the daughter DNA molecules, and half being new. In 1958, five years after Watson and Crick announced their double helix model, supporting evidence came in the form of an experiment by Matthew Meselsohn and Franklin Stahl using a heavy isotope of nitrogen, ^{15}N.

Compounds containing ^{15}N are slightly heavier than those containing the normal isotope, ^{14}N, and can be separated using a technique called *density-gradient centrifugation*. The extracted DNA is placed in a tube containing a solution whose density increases with depth. After spinning at very high speed, the DNA settles out at the level corresponding to its density, 'heavy' DNA settling out lower down than 'light' DNA.

E. coli bacteria were grown in a medium in which the nitrogen was ^{15}N. After many generations all the nitrogen-containing compounds made by the bacteria, including DNA, were 'heavy'. The ^{15}N nutrient was then replaced by ^{14}N. At intervals corresponding to the time for one generation, bacteria were removed and the DNA extracted and its density determined in the centrifuge. The results, together with the explanation, are shown in Fig. 23.10.

Before the change from ^{15}N to ^{14}N, all the DNA was 'heavy' and occupied a single band containing only ^{15}N. As predicted by the Watson-Crick model, after one generation the DNA occupied a single band that was slightly 'lighter' than the $^{15}N^{15}N$ DNA and presumably consisted of 'hybrid' ($^{15}N^{14}N$) DNA. After two generations the DNA occupied two bands, a hybrid band with the same density as in the previous generation, and a lighter ($^{14}N^{14}N$) band.

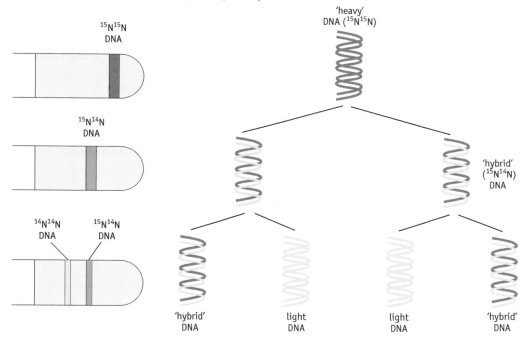

Fig. 23.10 Experimental proof that DNA replication is semiconservative

Extension: DNA replication is actually very complicated

DNA replication is more complicated than above, in several ways:

- Fig. 23.5 might suggest that DNA replication begins at the end of a chromosome and works its way along to the other end. In reality it occurs at many points simultaneously (Fig. 23.11). The shortest human chromosome contains about 50 million base pairs, and in human cells DNA replication occurs at a rate of about 50 per second. If replication were to start at one end and work its way along the whole length, it would take about 11 days to complete.

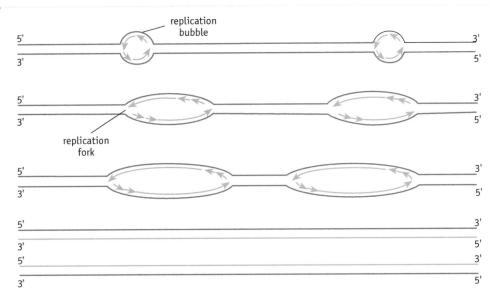

Fig. 23.11 In eukaryotes DNA replication begins simultaneously at many points on each chromosome (only two shown here), the replicated regions eventually joining up

- It is *discontinuous* at the 5' end. Polynucleotides can only increase in length at the 3' end. Since the two DNA chains are antiparallel, one end of each new strand is produced in sections. These are called *Okazaki fragments* after the scientist who first postulated their existence. Each fragment grows at the 3' end before being joined by an enzyme (*DNA ligase*) to the previously-formed fragment (Fig. 21.11 and Fig. 21.12). Okazaki fragments are much longer than shown in Fig. 23.12. The continuously growing end of a growing polynucleotide chain is called the *leading* strand and the discontinuously growing end is called the *lagging* strand.

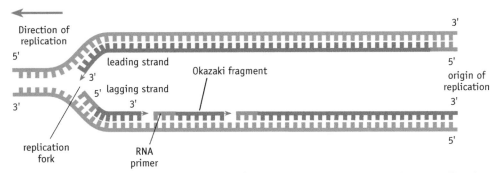

Fig. 23.12 Discontinuous replication of DNA. (The other side of the now long replication bubble is not shown, but lies to the right.)

- An RNA *primer* is needed. New DNA nucleotides can only be added to the 3' end of an *existing* polynucleotide strand. To get the process started, a strand of RNA is laid down as a primer, and then DNA is added to this (Fig. 23.12).

- Energy and raw materials are supplied as a combined package. The raw materials for DNA synthesis are actually *nucleoside triphosphates*. These are very similar to ATP except that the base can be any of four kinds and the sugar is deoxyribose instead of ribose. As it joins to the existing chain, each nucleoside triphosphate has its two terminal phosphate groups split off, releasing enough energy to 'drive' the process. The nucleoside triphosphates thus act as both raw materials and a source of energy.

- DNA polymerase acts as a 'proofreader'. Though the frequency with which the 'wrong' bases are incorporated into new DNA is very low, it is still much higher than the spontaneous mutation rate. The reason for the discrepancy is that besides linking nucleotides together, DNA polymerase checks for mismatched bases, and removes any incorrect ones.

The cell cycle

Cells that are constantly dividing go through a cycle of events, of which mitosis is but the most dramatic. After mitosis, the cell regains its original size by synthesis of new proteins, lipids and other cell constituents. Then, still during interphase, it makes a 'decision' as to whether to divide again or not. If is destined to divide again, the cell enters a period of DNA replication, after which there is a further period of synthesis of proteins and other materials (Fig. 23.13).

The whole sequence of events, including mitosis, is called the **cell cycle**, of which DNA synthesis occupies about a third. The duration of the cycle depends on temperature and on the organism and, in multicellular organisms, the kind of tissue.

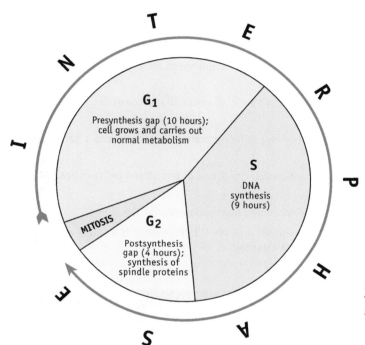

Fig. 23.13 The cell cycle. Times are for human cells in tissue culture.

The DNA content of a nucleus can vary four-fold

While cells can be diploid or haploid, their DNA content can vary *four-fold*. This is because a chromosome can be single-stranded or double-stranded, depending on whether the cell is destined to divide again. Fig. 23.14 shows how the DNA content of cells changes during the life of an animal.

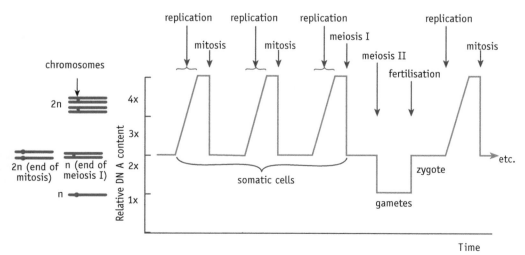

Fig. 23.14 DNA content of cells at different stages of an animal life cycle

Summary of key facts and ideas in this chapter

○ DNA is one of two classes of *nucleic acids*, the other being RNA.

○ DNA consists of two strands wrapped around each other in a *double helix* (spiral).

○ Each strand is a *polynucleotide*, consisting of many nucleotides joined together in a chain.

○ Each nucleotide consists of a five-carbon sugar (deoxyribose), a phosphate, and a base.

○ Each base can be one of four kinds: *adenine*, *guanine*, *cytosine* or *thymine*.

○ Adenine and guanine are two-ringed *purines*; cytosine and thymine are single-ringed *pyrimidines*.

○ The two polynucleotide strands are held together by *hydrogen bonds* between the bases.

○ Bases are paired according to a specific rule: adenine pairs with thymine and guanine pairs with cytosine. This is called *complementary base pairing*.

○ As a result of complementary pairing, the sequence of bases in each strand contains the information for building the other strand.

○ DNA replication is *semiconservative*, each daughter strand consisting of one original ('old') and one 'new' strand.

○ Every polynucleotide chain has two different ends (like a 'head' and a 'tail'), called the 3' ('head') and the 5' ('tail') end.

○ The two DNA strands are *antiparallel*, the 5' end of one strand being opposite the 3' end of the other strand.

○ Polynucleotide chains 'grow' by adding new nucleotides to the 3' end.

○ DNA replication occurs between cell divisions ('interphase'). If a cell is not going to divide again, the chromosomes remain single-stranded.

Test your basics

Copy and complete the following sentences. In some cases the first or last letters of a missing word are provided.

1. A nucleotide consists of a 5-carbon ____*____, a nitrogenous ____*____ and a ____*____ group.

2. A DNA molecule consists of two poly___*___ chains wrapped around each other in a double ___*___, like a twisted ladder. The sides of the 'ladder' are alternating p___*___ and ___*___, and the 'rungs' consist of pairs of bases linked together by ___*___ bonds.

3. The bases in DNA are of ___*___ kinds. ___*___ and g___*___ are double-ringed ___*___, and t___*___ and ___*___ are ___*___. In the two strands, a___*___is paired with ___*___ and g___*___ is paired with ___*___.

4. When DNA is replicated, the two strands ___*___, exposing the bases. Under the influence of the enzyme ___*___ ___*___, new nucleotides are added by pairing with the exposed bases. Each of the 'daughter' DNA molecules consists of one old and one new strand; this is called ___*___ replication.

5. Rarely, DNA replication is not exact, in which case a 'mistake' or mutation occurs.

24 What Do Genes Do?
Gene Expression

Early in the twentieth century, genes were very mysterious. No one knew what a gene was made of, what it did or how it did it. Even in those early days, however, a few far-sighted biologists suspected that a gene controls the production of an enzyme.

One of these was Sir Archibald Garrod, a physician who took an interest in certain hereditary diseases. Among these was **alkaptonuria** ('black urine disease'). In this condition the urine contains *homogentisic acid*, which turns black on exposure to air. Garrod noted that out of 11 cases, three were the result of inbreeding (matings between relatives). After consulting a geneticist, he concluded that the condition was due to a recessive allele.

Garrod also took an interest in other inherited conditions, such as **albinism** (inability to make the pigment melanin). He suggested that all these 'inborn errors of metabolism', as he called them, were caused by a failure of body chemistry due to the inability to make a certain enzyme (Fig. 24.1).

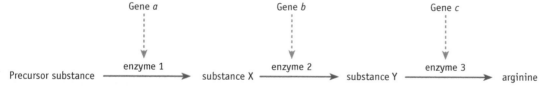

Fig. 24.1 A generalised metabolic pathway and genes controlling it. If any one of the enzymes is defective, the entire pathway will be non-functional.

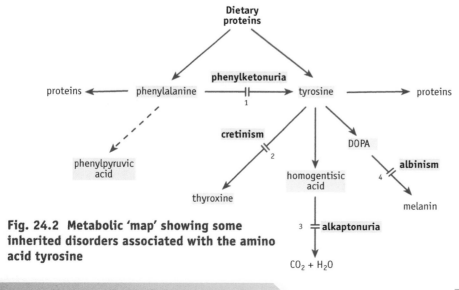

Fig. 24.2 Metabolic 'map' showing some inherited disorders associated with the amino acid tyrosine

ISBN: 9780170214094

It only needs one enzyme on the pathway to be defective for the whole pathway to be blocked. If the end product is essential for life, then deficiency of the enzyme would be lethal.

Decades later, in the 1940s, scientists showed that enzymes are indeed under the control of genes. Fig. 24.2 shows the location of these blocks on the 'map' of enzymes involved in the metabolism of the amino acid *tyrosine*.

A case study: Phenylketonuria

Since Garrod's work, many other examples of the effect of defective genes on metabolism have been found. Among them is **phenylketonuria (PKU)**, a condition resulting from deficiency of the enzyme that converts the amino acid *phenylalanine* into another amino acid, *tyrosine*. As with any other metabolic block, it has two distinct effects:

- Accumulation of phenylalanine. This is converted into phenylpyruvic acid, which is toxic and severely affects brain development. Before birth, phenylalanine does not accumulate because it passes into the mother's blood. After birth, an elevated blood phenylalanine can be easily detected, and testing of newborn babies is standard practice in New Zealand. If PKU is diagnosed, the child must be fed on a diet low in phenylalanine for normal intellectual development.

- Since tyrosine cannot be produced from phenylalanine, it must be obtained from the diet and may be slightly deficient. A result of this is insufficient production of melanin, the pigment of skin and hair. People with PKU thus tend to be very fair-haired.

How does a gene control the production of an enzyme?

Before they could even begin to answer this question, scientists had to know something about the structure of genes and the structure of enzymes.

In the early twentieth century enzymes were as mysterious as genes, until it was shown (in 1926) that enzymes are *proteins*. Proteins were known to consist of polypeptides, or chains of *amino acids*. It was also known that proteins differed with regard to the proportions of their amino acid constituents.

Then, in 1951, Frederick Sanger worked out the structure of the hormone *insulin*. He showed that it contains 51 amino acids, linked *in a particular order*.

Thus the problem became: How does a gene contain information for joining amino acids in a specific sequence?

As explained in the previous chapter, the structure of DNA was worked out in 1953 by James Watson and Francis Crick. In some way the sequence of amino acids must be represented by the sequence of bases in DNA.

Information – the role of DNA and RNA

Since there are 20 kinds of amino acid in proteins but only four kinds of base, each amino acid cannot be specified by just one base. Nor could a sequence of two bases, since this would only give 4 x 4 = 16 possible sequences:

AA	AT	AC	AG
TA	TT	TC	TG
CA	CT	CC	CG
GA	GT	GC	GG

If we use a third base, then for each of these 16 possibilities there would be four more possibilities, giving 4 x 4 x 4 = 64 possibilities:

AAA	AAT	AAC	AAG
ATA	ATT	ATC	ATG
ACA	ACT	ACC	ACG
AGA	AGT	AGC	AGG
TAA	TAT	TAC	TAG
TTA	TTT	TTC	TTG
TCA	TCT	TCC	TCG
TGA	TGT	TGC	TGG
CAA	CAT	CAC	CAG
CTA	CTT	CTC	CTG
CCA	CCT	CCC	CCG
CGA	CGT	CGC	CGG
GAA	GAT	GAC	GAG
GTA	GTT	GTC	GTG
GCA	GCT	GCC	GCG
GGA	GGT	GGC	GGG

So the minimum requirement is for a triplet (three 'letter') code, but of course this does not mean that it *is* triplet code. In the 1960s it was shown that it is indeed a triplet code.

Thus a sequence of three bases represents an amino acid, just as in the old Morse code, a sequence of three dots specified the letter 'S' and three dashes specified the letter 'O'. Thus to represent a protein consisting of 100 amino acids we would expect a sequence of 100 x 3 = 300 bases.

Directly or indirectly, making proteins involves many parts of the cell

Making proteins is a complex process and involves several structures cooperating together. Before getting into detail, it's worth revising the parts concerned, dealt with in Chapter 17. Fig. 24.3 shows the various structures involved.

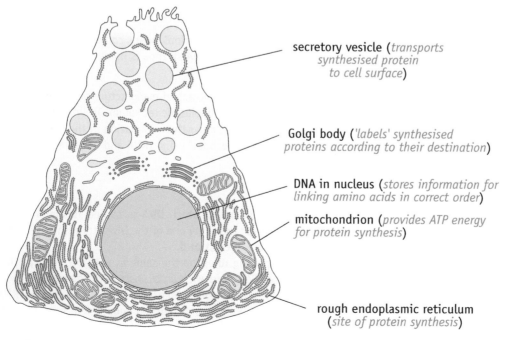

secretory vesicle (*transports synthesised protein to cell surface*)

Golgi body (*'labels' synthesised proteins according to their destination*)

DNA in nucleus (*stores information for linking amino acids in correct order*)

mitochondrion (*provides ATP energy for protein synthesis*)

rough endoplasmic reticulum (*site of protein synthesis*)

Fig. 24.3 A pancreatic cell, showing organelles involved in the production and secretion of digestive enzymes

ISBN: 9780170214094

Proteins are made in the cytoplasm

Experiments using radioactively labelled amino acids have shown that proteins are produced in the cytoplasm. Since most of the DNA is the nucleus, information must somehow be transferred from the nucleus to the cytoplasm.

The transfer of the coded information and its subsequent decoding involve another kind of nucleic acid, called **RNA (ribonucleic acid)**. This differs from DNA in three key ways:

- The sugar is **ribose** rather than deoxyribose.
- One of the pyrimidine bases is **uracil** instead of thymine.
- It is *single-stranded* (though it may be bent back on itself).

There are three kinds of RNA:

- **Messenger RNA**. Like DNA, this does not 'do' anything, but simply carries stored information. This form of RNA is called **messenger RNA (mRNA)** because it transfers information from the DNA in the nucleus to the 'factory floor' in the cytoplasm. In doing so it acts as a temporary, working copy of a gene.
- **Transfer RNA (tRNA)**. This brings each amino acid into association with the mRNA triplet specifying it. Hence there must be at least 20 kinds of tRNA, one for each amino acid.
- **Ribosomal RNA (rRNA)**. Together with proteins, rRNA is a major constituent of the **ribosomes**. These are the tiny granules that actually 'read' the message in mRNA and use it to assemble a polypeptide. Each ribosome consists of two parts, which only come together when 'reading' mRNA.

Protein synthesis consists of two main stages

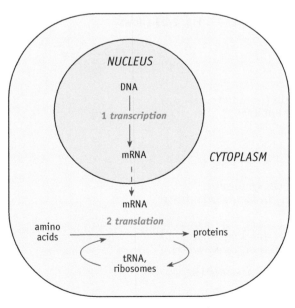

Fig. 24.4 Outline of the two main stages of protein synthesis

Protein synthesis is a complex process, so before delving into the details, here's a brief overview (Fig. 24.4).

There are two main stages (in eukaryotes there is a third stage, which is beyond the scope of this book).

1. **Transcription**. 'Transcription' means *copying*. In this stage a copy of one of the DNA strands is made in the form of a strand of messenger RNA.

2. **Translation**. This is the actual manufacture of the protein and occurs in the endoplasmic reticulum. To link the amino acids in the correct order the information in the mRNA has to be 'read' and 'decoded' using tRNA and ribosomes.

Transcription

The basic mechanism by which DNA is copied is similar to its replication, except that only one of the DNA strands is used; this is called the **template strand**.

The base pairing rules are the same as with DNA except that the *adenine of DNA pairs with uracil instead of thymine*.

First, the two DNA strands separate to form a 'transcription bubble' (Fig. 24.5). The enzyme **RNA polymerase** then moves along the template strand, linking ribonucleotides together to form RNA. As with all polynucleotides, the RNA grows by adding nucleotides to the 3' end, so the RNA chain grows in the 5' ⟶ 3' direction. Since the two DNA and mRNA strands are *antiparallel*, the RNA polymerase moves along the template strand in the 3' ⟶ 5' direction.

ISBN: 9780170214094

Although RNA polymerase moves along the template strand, it actually produces a copy of the *non-template* strand (except that U replaces T). For this reason the non-template strand is referred to as the **coding strand** or **sense strand**, and its 5′ and 3′ ends point the same way as those of the RNA. Since the RNA grows in the 5′ ⟶ 3′ direction, the 5′ end is said to be the 'upstream' end. This is the same polarity as in the coding strand of the DNA, so the 5′ end of the gene is considered to be the 5′ end of the coding strand.

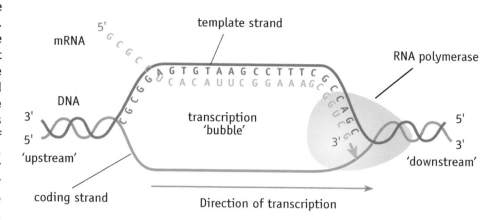

Fig. 24.5 How DNA is transcribed into RNA

As RNA polymerase moves along a gene, it unwinds the DNA ahead of it and rewinds it behind. Thus a 'transcription bubble' moves along the DNA.

How does the RNA polymerase 'know' which of the two DNA strands to use as a template? This depends on a specific nucleotide sequence called the **promoter**, to which the RNA polymerase binds.

Having 'recognised' and bound to the promoter, the RNA polymerase moves along the template strand in the 3′ ⟶ 5′ direction, making RNA (in the 5′ ⟶ 3′ direction) as it does so. The promoter sequence is not copied.

In addition to a supply of energy, transcription thus requires three things:

- *raw materials*, four kinds of ribonucleotide;

- an *enzyme*, RNA polymerase;

- *information*, in the form of one of the two DNA strands.

Translation

Before the mRNA can be used to make a protein, it has to leave the nucleus via the pores in the nuclear envelope (these are not simple holes, but are very complex structures).

As explained above, each amino acid in the protein is represented by a triplet of three bases in the mRNA. Hence the mRNA coding for an octapeptide (eight amino acids) would consist of 8 x 3 = 24 nucleotides. Each triplet of bases in mRNA is called a **codon**.

The code is usually represented using mRNA triplets (Fig. 24.6). For example, using the standard symbols G, A, U and C, the triplet GAG represents or 'stands for' the amino acid glutamic acid and GUG represents valine.

		Second base of codon				
		U	C	A	G	
First base of codon (5′ end)	U	phe	ser	tyr	cys	U
		phe	ser	tyr	cys	C
		leu	ser	**STOP**	**STOP**	A
		leu	ser	**STOP**	trp	G
	C	leu	pro	his	arg	U
		leu	pro	his	arg	C
		leu	pro	gluN	arg	A
		leu	pro	gluN	arg	G
	A	ile	thr	aspN	ser	U
		ile	thr	aspN	ser	C
		ile	thr	lys	arg	A
		met + START	thr	lys	arg	G
	G	val	ala	asp	gly	U
		val	ala	asp	gly	C
		val	ala	glu	gly	A
		val	ala	glu	gly	G

(right column header) Third base of codon (3′ end)

Fig. 24.6 The genetic code 'dictionary'

General features of the genetic code

- It is a *triplet* code, meaning that three bases specify or represent each amino acid.

- It is *degenerate*, meaning that several amino acids are specified by more than one codon. In fact only two amino acids are specified by only one codon.

- Where several codons represent the same amino acid, the alternative codons usually differ with respect to the last letter.

- Because of the polarity of polynucleotide chains, opposite sequences are read differently. For example, CUG is read differently from GUC.

- In almost all cases it is *non-overlapping*, each block of three bases is read separately, rather than the last base of one codon acting as the first base of the next.

- Three codons act as 'stop' signals, meaning 'end of message'.

- One codon acts both as 'start' and also codes for the amino acid methionine.

- With minor and rare exceptions the code is *universal*, each amino acid being specified by the same codons in bacteria, fungi, plants, animals. An important consequence of this is that genes can be transferred between widely diverse species and still function. For example, bacteria can be used to make human proteins.

N.B. The term 'code' is often misused to mean 'message' or 'information'. They are quite different, since one code can be used to send an almost infinite number of messages. It is quite incorrect to describe mRNA codons as 'codes', or to say that mRNA 'carries the code to the cytoplasm'. mRNA carries *information* to the cytoplasm. The genetic code is not a sequence of bases but a *relationship* between base sequence and amino acid sequence.

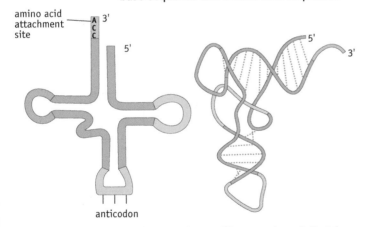

Fig. 24.7 A tRNA molecule shown 'flattened out' (left), and more accurately on the (right)

Transfer RNA (tRNA)

To link amino acids together in the order specified by the sequence of mRNA codons, each amino acid must be brought into association with the codon specifying it. This is the function of tRNA, which thus acts as a kind of 'adaptor'.

Since there is a different kind of tRNA for each amino acid, there must be at least 20 kinds of tRNA. tRNA is a relatively small molecule, consisting of 70–90 nucleotides. Like other RNAs it is single-stranded, but is folded back on itself in various places by complementary base pairing to form a complex three dimensional shape. When flattened out, it resembles a cloverleaf (Fig. 24.7).

Each tRNA molecule combines with 'its' amino acid to form a tRNA-amino acid complex. The process requires an enzyme and ATP:

amino acid + tRNA + ATP $\longrightarrow$ tRNA-amino acid complex + AMP + 2 phosphates

To 'recharge' AMP to ATP requires the energy equivalent of two ATPs. The 3' end of the tRNA is the end that joins with the amino acid has the same base sequence (CCA) in all tRNAs. How, then, does each tRNA recognise 'its' amino acid? The answer lies in the way the tRNA is folded, which is slightly different for each tRNA. The folding pattern depends on the pairing between the bases along different regions of the chain, and hence on the base sequence. Each tRNA is 'recognised' by the enzyme that catalyses its combination with the amino acid.

At one point on the tRNA there are three unpaired bases called an **anticodon**. They are complementary to a codon on the mRNA so can combine with it by complementary base pairing. In this way each amino acid can be linked up with the mRNA codon specifying it.

Like mRNA, tRNA is made on a DNA template. In eukaryotes there are many copies of each tRNA gene.

The role of the ribosomes

Unlike the words on a page, which are separated by spaces, there are no spaces separating adjacent codons in mRNA. Consider the following message:

THECATATETHEBIGFATRAT

To make sense, the words need to be read in threes, but there are three possible *reading frames*, depending on which letter we begin with:

If we begin with the first letter we have: THE CAT ATE THE BIG FAT RAT
Beginning with the second letter: HEC ATA TET HEB IGF ATR AT
Beginning with the third letter: ECA TAT ETH EBI GFA TRA T

Only one of these makes any sense. It is the job of ribosomes to ensure that codon-anticodon pairing occurs using the correct reading frame. Its ability to do this depends on two facts:

* Codon-anticodon pairing can *only* occur in the presence of a ribosome.

* A ribosome can recognise the beginning of the message (the 5' end of a mRNA molecule) and can move along it *one triplet at a time*. Thus the site where codon-anticodon pairing can occur moves along one triplet at a time, starting at the beginning.

Each ribosome consists of two subunits, one larger than the other. When not making a protein they are separate from each other, but they come together during protein synthesis.

Ribosomal RNA is produced on a DNA template. The regions of the chromosomes involved are clustered together to form darkly-staining **nucleoli** (singular, *nucleolus*). The ribosomal proteins are made in the cytoplasm. These then enter the nucleus where the ribosomes are assembled.

The process begins when the two subunits of a ribosome come together on the 5' end of the mRNA (Fig. 24.8). Near the 5' end is the codon AUG, which means 'begin here'. This also codes for the amino acid methionine, which is removed after the polypeptide is completed. The ribosome moves along the mRNA one codon at a time. Each time it does so, an amino acid is added to the growing polypeptide chain. When the ribosome reaches a 'stop' codon, it breaks free and can begin 'reading' another mRNA molecule.

Normally, several to many ribosomes move along at any one time (Fig. 24.9).

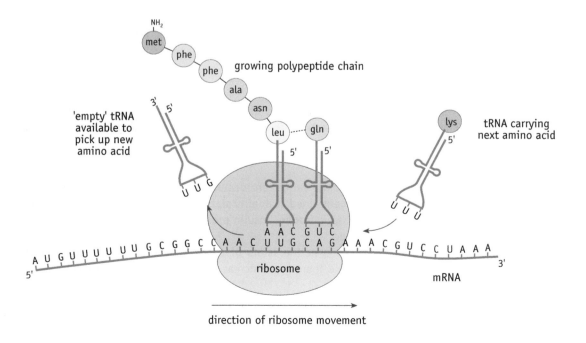

Fig. 24.8 A ribosome 'reading' mRNA. Each amino acid is shown by a different colour.

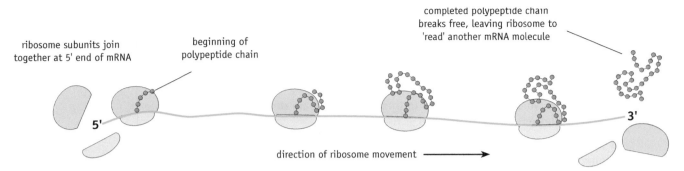

ribosome subunits join together at 5' end of mRNA

beginning of polypeptide chain

completed polypeptide chain breaks free, leaving ribosome to 'read' another mRNA molecule

5'

3'

direction of ribosome movement ⟶

ribosome breaks up into subunits

Fig. 24.9 Several ribosomes 'read' an mRNA molecule simultaneously

Genes can be switched on or off

Lactose intolerance

Milk contains the disaccharide sugar *lactose*. All newly-born mammals produce the enzyme *lactase*, which digests lactose to glucose and galactose. In many human populations the lactase gene is switched off after weaning, thus saving energy by not making an enzyme that is not needed. However, in populations that have traditionally used animals as a source of milk, the gene remains active. People whose lactase gene has been inactivated cannot digest lactose and suffer intestinal symptoms after drinking it. This is because the bacteria in their large intestine feed on the lactose, producing large amounts of gas.

Among Australian aborigines, 85% are lactose-intolerant, and in certain Asian societies 95% may be intolerant. In Northern European societies the figure is usually less than 5%. Since lactose intolerance is genetic, there must have been a mutation in a regulator gene controlling lactase production. Milk has only been a normal constituent of the human diet for less than 10 000 years, so the genetic change must have occurred fairly recently.

Foetal and adult haemoglobin

A molecule of adult haemoglobin is a *tetramer*, consisting of four polypeptide chains, two   (alpha) chains and two   (beta) chains. In the foetus the haemoglobin is slightly different; instead of the two -chains there are two   (gamma) chains. The significance of this is that foetal haemoglobin is more 'greedy' for oxygen, thus favouring the transfer of oxygen from maternal blood to foetal blood in the placenta. At birth, the gene for -globin is switched off and the gene for -globin is switched on.

Hormones and genes

In the examples above, genes are permanently switched off or on. There are many cases, however, in which a gene is not simply on or off, but its activity is constantly varying according to ever-changing circumstances. The production of the protein hormone insulin is an example.

Insulin and blood glucose

Despite the fact that the intake of carbohydrate varies considerably throughout the day and night, the concentration of glucose in the blood is maintained at a fairly stable level. One of the chief hormones involved in the regulation of blood glucose is *insulin*. Insulin promotes the utilisation of glucose by cells, and hence its removal from the blood.

After a meal rich in carbohydrate, cells in the Islets of Langerhans in the pancreas step up their secretion of insulin. At the same time there is an increase in the activity of the insulin gene, increasing insulin production.

Pleiotropy

At the molecular level, the effect of a gene is the production of a polypeptide. Yet at the level of the whole organism, many genes have multiple, or **pleiotropic**, effects. An example is the 'frizzle'

ISBN: 9780170214094

condition in fowls, mentioned in Chapter 14. Frizzle fowls have trouble maintaining their body temperature. The heart becomes enlarged, as do the crop, gizzard, pancreas, adrenal glands and kidneys — presumably as a response to the need for increased supply of nutrients and oxygen to the cells, to maintain a higher respiratory rate.

Pleiotropy is far from a special case; many genes have secondary effects that seem quite unrelated to their presumed primary one. For instance, people with blood group O are more likely to develop gastric ulcer, and group A people are more susceptible to stomach cancer.

Pleiotropy may have played a part in evolution. Natural selection can only act on phenotypic characters than are, at least to some extent, already in existence. It may act to increase the size or shape of a structure, or the speed of a process, but this still leaves the problem of new characteristics arise in the first place.

The answer may be that structures arise as a result of pleiotropic effects. For example in fish, a tiny fold of skin may have arisen as a secondary effect of genes that were primarily concerned with other functions. If such a rudimentary skin fold had a slightly beneficial effect on stability, the genes responsible would be selected for. Similarly, protobirds may have developed tiny skin outgrowths as pleiotropic effects. If these rudimentary feathers helped retain heat, the genes responsible would be selected for and the feathers would increase in size, permitting the evolution of high body temperature and later, flight.

Extension: Evolution from the perspective of molecular biology – The Central Dogma

The modern, neo-Darwinian view of how evolution has occurred is that changes in the base sequence of DNA are random and that if the result is a better phenotype, it is simply good luck (Fig. 24.10).

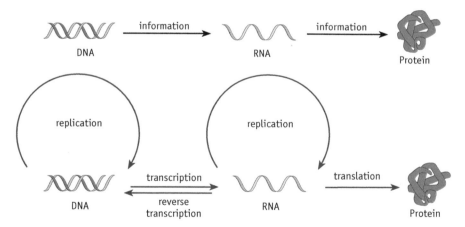

**Fig. 24.10 Above: the 'Central Dogma' in its original form.
Below: revised form of the 'Central Dogma'.**

This can be stated another way:

- Information can flow from DNA to RNA and from RNA to protein, but not in the reverse direction.

This view of how information flows was first articulated by Francis Crick, who called it **the Central Dogma** of molecular biology. What Crick was saying was that genes provide the information for building bodies, but that the body cannot directly influence the plan that helped to build it.

This, in molecular terms, is the essence of the neo-Darwinian view of evolution. A follower of Lamarck, on the other hand, would maintain that the body responds adaptively to changes in the environment, and that these changes in the body are inherited because they result in changes in the DNA. At the molecular level, this would mean that changes in the amino acid sequence of proteins must somehow result in changes in the base sequence of DNA. Not only is there no evidence for this idea, it is very difficult to see how it could happen.

Since Crick stated the Central Dogma, it has had to be modified as a result of two discoveries:

- In some viruses the genetic material is RNA rather than DNA.

- In retroviruses (e.g. HIV, the virus that causes AIDS), the RNA of the virus is used to make DNA, which is then used to make more viral RNA.

The revised view of the Central Dogma can thus be restated:

- Information can flow from nucleic acid to protein, or from one kind of nucleic acid to another, but not from protein to nucleic acid or from protein to protein.

Genotype, phenotype and environment

The information for building an organism — the *genotype* is stored in coded form in DNA. During growth and development, this information is used to build a complex body.

At the molecular level, the phenotype is expressed as proteins. But the actions of individual proteins are but the first links in a chain of effects that culminate in the development of a complex body. However, even if we knew the structure and actions of every protein produced by an organism, we would still not be able to predict all the details of the phenotype.

One reason for this is that the environment can have major effects on the development of the phenotype. These effects may be non-adaptive or adaptive. The difference can be illustrated by the effects of shortage of magnesium and light on plant growth.

When a plant is short of magnesium, the leaves become pale due to lack of chlorophyll, a condition called *chlorosis*. This is a *direct* effect of nutrient shortage, and is also non-adaptive in that it does not help the plant deal with the shortage.

On the other hand, when seedlings are grown in darkness, the stems are pale and elongated and the leaves remain small (Fig. 24.11C). In this case the effect of lack of light is *indirect*. It is also adaptive because by growing longer, it increases the chances of reaching the light. Stem elongation is further helped by producing smaller leaves, diverting more resources to stem extension.

Other examples of adaptive environmetal effects

- In trees and shrubs, wood produced in the spring usually has wider water-conducting vessels than autumn wood. This enables the xylem to deliver more water to the leaves when conditions favour rapid transpiration.

- In forest trees, leaves that develop in bright light usually have a higher density of stomata and more strongly developed palisade tissue than leaves developing in the shade (Fig. 24.11A).

- In response to decreasing day length, grizzly bears lay down large quantities of fat, in anticipation of the demands of hibernation. Changing photoperiod also stimulates many birds to lay down fat reserves prior to migration.

- The sex of many reptiles depends on the temperature at which the eggs are incubated. In tuatara, lower temperatures favour the development of females and higher temperatures favour the production of males. In Nile crocodiles, temperatures above 34.5 °C or below 31.7 °C favour the development of females, whereas intermediate temperatures tend to produce males.

- Female bee larvae fed on 'royal jelly' throughout their development grow into queens, which are fertile. Grubs that are fed on royal jelly for only the first part of development grow into workers, which are sterile.

ISBN: 9780170214094

- Termites are social insects. In a termite colony there is division of labour into several *castes*: queen, a king, workers and soldiers (Fig. 24.11B). Workers and soldiers are all offspring of the queen and king and can be of either sex. Every egg laid by the queen can differentiate into worker or soldier. What, then, determines which caste a young termite develops into? It depends on the balance between various *pheromones* (chemicals used to communicate between members of the same species). In some way, then, these pheromones must trigger the expression of different genes.

- In the reef-dwelling clownfish there is only one female in each social group. On death of this female, the most dominant male changes into a female (Fig. 24.11D).

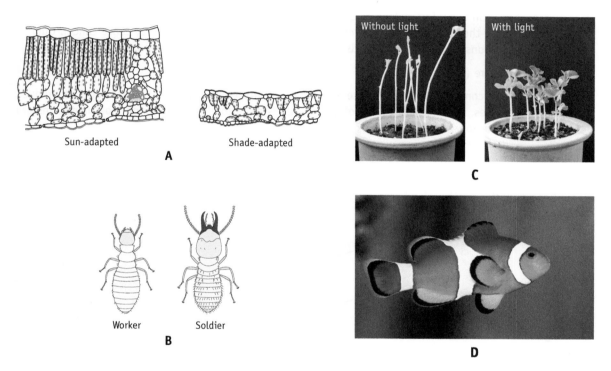

Fig. 24.11 Examples of the effect of environment on gene expression: (A) Sections through sun-adapted and shade-adapted leaves of maple. (B) Worker and soldier caste of Prorhinotermes simplex, a termite. (C) Effect of light on pea seedlings. (D) Clownfish, which begin as males but may become females when the 'resident' female dies.

Gene mutation

A gene mutation is a change in the base sequence of a gene, resulting in a new allele. Mutations are also relatively stable and potentially long-lasting. *Recombination* on the other hand, simply reshuffles existing alleles into new but relatively short-lasting combinations.

N.B. A mutation is not a change in the genetic code. With minor exceptions the code is universal and does not change. Mutation results in a change in the 'message' — not the code.

Characteristics of gene mutations

Gene mutations have the following characteristics:

- They are *random*, meaning that they do not result from *need*. Any advantage is purely due to chance. Although the *rate* of mutation may be influenced by environmental change (e.g. it rises with temperature), the *kind* of mutation bears no relation to the kind of environmental change.

- Most are *harmful* under existing environmental conditions. Most alleles exist because they have passed the test of natural selection, so any random change is more likely to be harmful than beneficial (a random change to a watch mechanism is unlikely to improve its timekeeping). Under new environmental circumstances, a minority of gene mutations *may* be beneficial.

- They are *rare*. In fruit flies for instance, a new mutation is typically present in one in 30 000–50 000 gametes.

- In diploid organisms most are *recessive*, and so are only expressed in homozygotes. Most are present in heterozygotes and are thus protected from natural selection by dominant alleles. All diploid organisms are said to carry a *genetic load* of harmful, recessive alleles.

- They can occur in any kind of cell. Mutations in body cells are called *somatic mutations* and, if they occur in genes involved in the control of cell division, may cause cancer. Only mutations in gametes or in cells ancestral to gametes (the germ line) can be inherited.

- They are (rarely) reversible.

Though recombination can produce a virtually infinite number of genotypes, it doesn't produce any new alleles. The ultimate source of new alleles is mutation. Thus, although most mutations are harmful, mutation is essential for evolution to occur.

There are fundamentally two types of gene mutation

Gene mutations, otherwise known as *point mutations*, can be of two fundamentally different kinds: base *substitutions* and *frameshifts*.

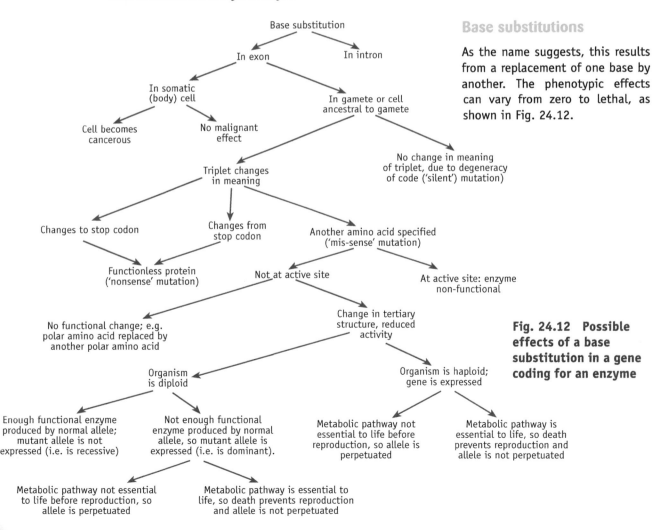

Base substitutions

As the name suggests, this results from a replacement of one base by another. The phenotypic effects can vary from zero to lethal, as shown in Fig. 24.12.

Fig. 24.12 Possible effects of a base substitution in a gene coding for an enzyme

Frameshifts

Frameshifts occur when a base is inserted or deleted, resulting in a change in the 'reading frame'. Consider the following sentence, with spaces between the words to make it easier to read:

> THE FAT CAT ATE THE RAT

If we insert a letter after 'CAT' to make it 'CATS', the triplets after the insertion are completely changed:

> THE FAT CAT SAT ETH ERA T

Deleting a letter ('F' in this case) would result in an equally meaningless message:

> THE ATC ATA TET HER AT

Unless a frameshift mutation occurs near the end of a gene, it is likely to result in a functionless protein, for two reasons:

- In the resulting protein, most or all of the amino acids 'downstream' of the change will be changed.

- Since the triplets 'downstream' of the insertion or deletion will now be changed randomly, there will be a chance that any given triplet will be converted into a 'stop' triplet. Since there are three 'stop' codons, the probability that any given triplet will be converted to a stop codon is 3/64.

If two frameshifts of opposite kind occur, the second will restore the reading frame:

> THE ATC ATS ATE THE RAT

So, if an insertion and a deletion are sufficiently close together, the affected region of the polypeptide chain may be short and the protein could still be functional. Most cases of reversed mutation are probably of this kind, though an exact reversal of the original mutation would be very unlikely, as would reversal of a base substitution.

A gene can mutate to produce many different mutant alleles

When we say that a gene exists as two alleles, *A* and *a*, this is often not strictly true. Just as there are many different reasons why a car will not start, there are many ways a gene can mutate to produce a non-functional product. For example, over 1500 different mutant forms of the gene for cystic fibrosis are known, some of which are less serious than others.

Similarly, the mutant gene causing the disease phenylketonuria (PKU) has over 400 known variants. Depending on the kind of mutation, the activity of the enzyme involved can vary between complete inactivity to 50% of normal.

What causes mutations?

Mutations may arise for two reasons:

- Some are *spontaneous*, meaning that they arise without any apparent external cause. The rate of spontaneous mutation does, however, significantly increase with *temperature*.

- Most are *induced* by external agents called **mutagens**. Since cancer is a result of somatic mutation, mutagens are also carcinogens (cancer-producing agents).

There are three classes of mutagen: radiation, chemicals and viruses.

Radiation

Certain forms of radiation have enough energy to break chemical bonds, producing ions and other highly reactive particles. Examples of such *ionising* radiation are *gamma ( )-rays, X-rays, alpha ( )-rays, beta ( )-rays* and *neutrons*. -rays and X-rays are the most energetic forms of *electromagnetic radiation*, which also includes ultraviolet, visible light, and radio waves (Fig. 24.13).

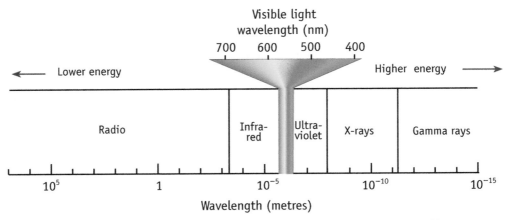

Fig. 24.13 The elecromagnetic spectrum, of which light is but a very small part

Particles produced when ionising radiation smashes molecules are so reactive that they combine with the first organic molecule they collide with. If this happens to be DNA, then one base may be changed into another. The action of ionising radiation on DNA is therefore *indirect*.

Ultraviolet (UV) radiation, on the other hand, is strongly absorbed by DNA and specifically damages it. The poor penetrating power of UV gives no protection to microorganisms, for which UV is a powerful mutagen and is sometimes used to sterilise surfaces.

Chemicals

Thousands of chemicals can act as mutagens and therefore as carcinogens. For example in parts of Asia there is a high incidence of liver cancer caused by *aflatoxin*, a mutagen produced by certain moulds growing on nuts and grain stored in damp conditions. Benzopyrene in cigarette smoke is another mutagen.

Viruses

Some viruses have genetic material that includes *oncogenes*, or cancer-causing genes. The virus causing hepatitis B may eventually lead to liver cancer, and human papilloma virus may lead to cervical cancer.

What is a safe dose?

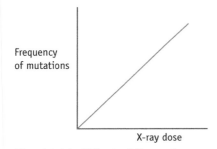

Fig. 24.14 Effect of X-ray dose on frequency of induced mutations in *Drosophila*

One of the first people to show that X-rays cause mutations was Hermann Muller. In the 1920s he showed that in *Drosophila* the frequency of gene mutations was directly proportional to the dose (Fig. 24.14). Even in very low 'background' radiation, some mutations occur. A medical X-ray is said to be 'safe' only because the risks are very small compared with the chance of leaving a possibly serious condition undiagnosed.

While it is often possible to say that in a population a certain *proportion* of observed mutations is due to a mutagen, we can never be certain of the cause in any individual case. This is because a certain low proportion are spontaneous (occur without any external cause).

Mutation and cancer

Cancer is a group of diseases in which cells divide uncontrollably and spread around the body. Cell division and the ability of cells to remain in their proper location are both under genetic control. When these controls break down, cancer may be the result.

A key event in the formation of a cancer cell is the mutation of a *proto-oncogene* to an *oncogene*. Proto-oncogenes are normal genes that regulate cell division. An oncogene is therefore a mutant product of a normal gene. Over 20 oncogenes have been discovered.

ISBN: 9780170214094

It is believed that two or more oncogenes must be present for cancer to develop. Cancer thus involves a series of mutation events. Some people inherit an oncogene, and for these people only one more mutation need occur. This explains why certain cancers, such as breast cancer and colon cancer tend to 'run in families'. In such cases it is even more important to avoid lifestyle risk factors.

| Radiation | Chemicals | Viruses |

Fig. 24.15 The major classes of mutagen

Summary of key facts and ideas in this chapter

○ All the organic substances of the body are either proteins or are made by enzymes, which are proteins.

○ A protein consists of one or more chains of amino acids called *polypeptides*.

○ The information for making a polypeptide is stored in a length of DNA called a *gene*.

○ Some proteins consist of more than one polypeptide, and so are coded for by more than one gene.

○ The amino acid sequence in a polypeptide is specified by the base sequence in DNA.

○ The genetic code is a *triplet code*, meaning that each amino acid is specified by a sequence of three nucleotide bases.

○ Proteins are produced on particles called *ribosomes*, which occur in large numbers in the *rough endoplasmic reticulum*.

○ After synthesis, many proteins are transported through the ER to the *Golgi complex*, where they are further modified or chemically 'labelled' according to their destination.

○ Besides DNA, protein synthesis also involves *ribonucleic acid* (RNA).

○ RNA is similar to DNA except that
  - it is single-stranded,
  - the sugar is ribose rather than deoxyribose, and
  - the base uracil replaces thymine.

○ In eukaryotes, protein synthesis consists of two stages, the first occurring in the nucleus:
  - *Transcription*, which produces a mRNA copy of one of the DNA strands.
  - *Translation*, in which the information in the mRNA is 'decoded' to produce a polypeptide.
    · Translation involves two other forms of RNA called tRNA and rRNA.
    · Each amino acid in mRNA is represented by a sequence of three bases in the mRNA called a codon.
    · There are 20 kinds of tRNA molecule, each with three unpaired bases called an *anticodon*. This is complementary to a codon in mRNA.
    · Ribosomes move along the mRNA in the 5' ⟶ 3' direction, one triplet at a time, adding a new amino acid each time.

○ Though the amino acid sequence of a polypeptide is under exclusively genetic control, environmental factors can have major effects on the level of activity of genes.

○ Though the primary product of a gene is a polypeptide, a non-functional or partially functional gene may have many secondary or tertiary phenotypic effects, called *pleiotropism*.

○ *Mutations* are the ultimate source of new alleles; recombination merely reshuffles existing genes.

○ All genetic variation — whether by mutation or recombination — is random, and is simply good luck if it is advantageous.

○ Mutations may be *spontaneous*, or *induced* by environmental factors.

○ So far as mutagens are concerned, there is no safe dose.

Test your basics

Copy and complete the following sentences. In some cases the first letter of a missing word is provided.

1. A gene carries the information for producing a ___*___. Some proteins consist of more than one kind of ___*___, so these proteins require more than one gene to specify them.

2. Each ___*___ ___*___ in a polypeptide is encoded by a sequence of ___*___ bases in DNA called a ___*___. There are ___*___ kinds of ___*___ ___*___ in proteins, but there are ___*___ possible sequences of ___*___ DNA bases. Hence some ___*___ may be specified by more than one DNA ___*___.

3. The first stage of protein synthesis occurs in the ___*___ and involves the production of a m___*___ copy of a gene. This is similar to ___*___ except that it is ___*___-stranded, the base ___*___ replaces ___*___, and the sugar is ___*___ instead of ___*___.

4. In the production of mRNA, the two ___*___ strands separate, exposing the bases. The bases of one of the two strands, the ___*___ strand, are used to pair up with ribonucleotides to form m ___*___. The complementary base pairing rules are the same as in DNA except that t___*___ in DNA pairs with ___*___ in RNA. Each RNA triplet is called a ___*___.

5. The enzyme responsible for RNA synthesis is ___*___ ___*___. It 'recognises' the DNA strands to use by a specific base sequence called a ___*___, to which it binds.

6. The second stage of protein synthesis occurs in the ___*___ and is called ___*___. In this stage, ___*___ ___*___ are joined together in the sequence encoded in the m___*___. This stage is carried out by the ___*___, using t___*___ to bring the amino acids to them. Though ___*___-stranded, t___*___ is folded back on itself into a 3-dimensional shape like a cloverleaf.

7. The m___*___ is ___*___ to the template strand and is thus actually a *copy* of the DNA strand that is not used. This is called the c___*___ or ___*___ strand.

8. If a triplet in the coding strand is TAG, the corresponding RNA codon is ___*___, and the DNA triplet in the template strand is ___*___.

9. Although the RNA bases are 'read' in threes, they are not arranged in threes — there are no gaps between codons. There are thus three possible 'reading ___*___' in the mRNA, depending on which base is taken as the starting point. The function of the ribosomes is to ensure that the mRNA is 'read' in the correct reading ___*___.

Glossary

Note: Unless otherwise stated, references to animal structures apply to humans or mammals in general.

A

Abscisic acid Plant growth substance that plays an important role in the regulation of stomatal aperture.

Absorption spectrum Graph showing how the absorption of light by a substance varies with wavelength.

Acclimatisation A purely phenotypic change in an individual organism in a way that increases its chances of survival.

Active site The part of an enzyme molecule that takes part in the reaction.

Active transport Movement of molecules or ions across a membrane against a concentration gradient.

Adaptation A feature of an organism (structure, physiology, behaviour or biochemistry) that increases its chances of survival and reproduction.

Aestivation A state in which an animal survives a dry period by entering a state of dormancy.

Age structure Relative proportions of each age group in a population.

Air capillaries Microscopic air-filled tubes in bird lungs; the site of gas exchange.

Alimentary canal Tube where food is digested by an animal.

Allele One of two or more alternative forms of a gene at a particular locus.

Allele frequency The proportion of a particular allele in a gene pool.

Allelopathy Production by a flowering plant of a substance that inhibits growth of potential competitors.

Amino acids The building units of proteins.

Amylopectin One of two constituents of starch, consisting of branched chains of glucose

Amyloplast A starch-storing organelle.

Amylose One of two constituents of starch, consisting of unbranched chains of glucose.

Anabolism The building of larger molecules from simpler ones in metabolism.

Anaphase Stage of nuclear division when the centromeres of the chromosomes are dragged towards the poles of the spindle.

Androecium Collective name for the stamens of a flower.

Angiosperm Member of the group consisting of flowering plants.

Annual Plant with a life cycle lasting one year.

Anther Structure containing pollen sacs, within which pollen grains are produced.

Antibiosis Production of chemicals by an organism such as a fungus, which inhibits the growth of potential competitors.

Anticodon Sequence of three bases (triplet) in tRNA that pairs with a codon in mRNA.

Aorta Large artery taking blood out of the left side of the heart.

Apoplast Water pathway through plant consisting of spaces in cell walls together with the vessels and tracheids of the xylem.

Archegonium Female gamete-producing organ of non-seed plants.

Arteriole Microscopic artery.

Artery Vessel that takes blood away from the heart.

Atheridium Male gamete-producing organ of non-seed plants.

ATP (adenosine triphosphate) The immediate source of energy for cells.

Autotrophic Able to produce organic compounds from CO_2 and water.

B

Biennial Plant with a life cycle lasting two years.

Bilayer The basic framework of cell membranes, consisting of two back-to-back layers of phospholipid.

Bile Greenish yellow alkaline fluid secreted by the liver; emulsifies fats.

Binomial system System of naming organisms designed by Linnaeus, each organism having a two-part name.

Biogeochemical cycle A nutrient cycle involving geological as well as biological processes.

Biotic factor Environmental factor that is due to other organisms.

Blood The liquid in which substances are transported around the body

Boundary layer Layer of still air next to the epidermis of a plant.

Breathing centre Brain centre that controls rate and depth of breathing.

Bronchial tree Branching system of air-filled tubes in the lungs of a mammal.

Bronchiole Microscopic air-filled tube in the lungs, not supported by cartilage.

Bronchus An air-filled tube in the lungs, supported by cartilage.

C

Caecum Blind-ending outgrowth of the gut of animals such as many insects and mammals.

Calvin cycle The reactions of photosynthesis that do not require light and occur in the stroma of the chloroplasts.

Calyx Collective name for the sepals of a flower.

Canine Tooth behind incisors in mammals, typically long and pointed in carnivores.

Carbohydrate Compound containing carbon, hydrogen and oxygen, in which the hydrogen and oxygen are in the same ratio as in water.

Carnassials Last upper premolar and first lower molar teeth in many carnivores, which act like cutting blades.

Carnivore An animal that feeds on other animals. A primary carnivore feeds on herbivores; a secondary carnivore feeds on primary carnivores.

Carotenoids Group of orange-yellow pigments in chloroplasts and chromoplasts.

Carpel The female equivalent of a stamen.

Carrying capacity The maximum population density that an environment can sustainably support.

Catabolism The breaking down of large molecules to smaller ones in metabolism.

Cell cycle Cycle of events in a cell during and between divisions.

Cell sap Solution in the vacuole of a plant cell.

Cell wall Outer 'casing' outside plasma membranes of plants and fungi.

Cellulose Chief constituent of plant cell walls; consisting of unbranched chains of glucose units.

Cement Outer layer of the root of a tooth.

Centriole Minute structure present in cells of animals and some plants, and which organise the spindle in cell division.

Centromere The part of a chromosome that attaches to the spindle in cell division, and which holds chromatids together after chromosome duplication.

Chitin The chief constituent of the cuticle of arthropods and the walls of fungal hyphae.

Chlorophyll Green pigment that traps light in photosynthesis.

Chloroplast Organelle in which photosynthesis occurs in eukaryotic cells.

Cholesterol Steroid constituent of cell membranes.

Chromatid One of two identical chromosome strands, produced by replication, held together by a centromere.

Chromatography Technique for separating substances based on their different solubilities.

Chromoplast Plastid that gives colour to many flowers and fruits.

Chromosome Thread consisting of DNA and protein, present (in eukaryotes) in the nucleus.

Chyme Creamy white, semi-digested food produced by action of the stomach of humans and other mammals.

Cilia Hair-like extensions of plasma membrane of eukaryotic cells, which beat rhythmically, creating movement.

Cisterna Membrane-bound cavity of the endoplasmic reticulum.

Citric acid cycle (Krebs cycle) Process in the matrix of mitochondria in which pyruvic acid is completely oxidised to CO_2.

Climax community The final stage in succession, when there is no further change in a community, except that which is due to changes in climate.

Closed blood system System in which the blood is confined entirely within blood vessels, so it does not come into contact with the tissues it serves.

Coding strand DNA strand that is copied to make mRNA.

Codon Sequence of three bases in mRNA, specifying an amino acid.

Cohort A group of individuals starting life approximately the same time (e.g. in the same year).

Collagen Fibrous protein that is the chief constituent of tendon and ligament.

Commensalism Relationship in which one organism obtains benefit from another; neither benefits nor is harmed.

Community All the organisms that live in a defined area.

Companion cell Cell that transfers organic substances in and out of sieve tubes in the phloem.

Compensation point Light intensity at which photosynthesis equals respiration.

Competition State of affairs in which demand for a resource exceeds supply, so at least some organisms obtain less than they need.

Competitive Exclusion Principle Idea that no two species can indefinitely occupy the same niche in the same habitat.

Concentration gradient Rate at which the concentration of a substance changes with distance between two points.

Consumer Animal that feeds on other organisms, or parts of them, killing tissue in the process.

Contractile vacuole Osmoregulatory organelle common in many single-celled freshwater organisms.

Corolla Collective name for the petals of a flower.

Cortex Layer beneath epidermis in plant roots and stems.

Cotyledon Leaf of a plant embryo.

Crista Infolding of inner membrane of a mitochondrion.

Crop Gut region where food is temporarily stored prior to digestion in earthworm, weta and many other animals.

Crossing over Process in meiosis when chromatids of homologous chromosomes exchange sections.

Cuticle (animals) Material secreted by the epidermis of an arthropod, consisting of chitin and protein.

Cuticle (plants) Waxy layer secreted by the epidermis of a plant.

Cytoplasm The part of a eukaryotic cell outside the nucleus.

Cytosis Active process by which substances are transported in bulk across a plasma membrane.

Cytosol Part of the cytoplasm where organelles are suspended.

D

Deamination The removal of amino groups from amino acids with the production of ammonia.

Decomposer An organism that feeds on dead matter, breaking it down to inorganic materials in the process (also known as a saprobe or saprotroph).

Denaturation Loss of three-dimensional shape of a protein, resulting in its inactivation.

Denitrification Conversion of nitrate to nitrogen gas by anaerobic bacteria.

Dentine Material forming the bulk of a tooth.

Deoxyribose Five-carbon sugar in DNA.

Detritivore An animal that feeds on dead matter.

Diastema Large gap between front teeth and premolars in many herbivores.

Dichotomous key Used for rapid identification of an organism.

Diffusion The movement of atoms, molecules or ions from a region of higher concentration to a region of lower concentration.

Digestion The breaking down of complex food substances by enzymes into simpler substances.

Dihybrid cross A cross between true breeding parents differing with respect to two pairs of contrasting traits.

Diploid Having two sets of chromosomes.

Disaccharide Complex sugar consisting of two simple sugars linked together.

Discontinuous variation Variation in which individuals can be placed in distinct groups.

DNA The genetic material of all organisms (some viruses have RNA).

DNA polymerase Enzyme that catalyses the synthesis of DNA.

Dominant trait A trait that is expressed in heterozygotes and homozygotes; only one allele need be present for it to be expressed.

Dormancy A state of greatly reduced activity that enables organisms to survive adverse conditions.

Double circulation System in which blood travels twice through the heart in any complete circuit.

Dry mass The mass of all material minus water.

Duodenum The first part of the small intestine.

E

Ecdysis Process in which an arthropod sheds its cuticle during growth.

Ectoderm Outer layer of cells in a cnidarian such as a sea anemone or jellyfish.

Egestion The getting rid of faeces via the anus.

Egg A female gamete.

Electron transport Process in which electrons are transported along a series of electron carriers in the inner mitochondrial membrane, releasing energy that is used to make ATP.

Embryo (plants) The very young plant within a seed or developing seed.

Embryo sac The greatly reduced female gametophyte of flowering plants.

Emigration Organisms leaving an area.

Enamel The part of a mammalian tooth that contacts the food; the hardest material in a tooth.

Endocytosis Process by which substances are taken into a cell by folding of the plasma membrane around the material.

Endoderm Inner layer of cells in a cnidarian such as a sea anemone or jellyfish.

Endodermis Innermost layer of the cortex of a root.

Endoplasmic reticulum Complex system of membrane-bound spaces in the cytoplasm of eukaryotes.

Endosperm An energy-storage tissue in the seeds of many flowering plants.

Endothermic Able to keep body temperature above that of the surroundings by the production of metabolic heat.

Enteron The digestive cavity of a cnidarian, such as a sea anemone or jellyfish.

Environment All the factors in the surroundings of an organism that affect it.

Environmental resistance The cumulative effect of environmental factors tending to reduce population growth.

Enzyme co-factor Non-protein molecule or ion needed for the activity of an enzyme.

Enzyme-substrate complex Compound that is briefly formed during an enzyme-catalysed reaction.

Enzymes Organic catalysts; nearly all are proteins.

Ephemeral A plant that can complete several life cycles in a year.

Epigeal germination Germination in which the cotyledons are brought above ground.

Epiglottis Small flap-like structure at base of rear of tongue in mammals, which prevents food entering larynx when swallowing.

Eukaryote Organism with cells that have a distinct nucleus.

Eutrophic State in which a lake has become enriched by mineral nutrients, thus promoting growth of algae.

Evolution A directional genetic change in a population over successive generations.

Exine Outer layer of a wall of pollen grain.

Exocytosis Mechanism by which cells secrete large molecules in bulk.

Exoskeleton External skeleton, characteristic of arthropods.

Exploitation Relationship where one species obtains benefit from another, which is harmed.

Exponential growth Growth by a constant percentage each time unit (also called logarithmic growth).

F

F$_1$ generation The offspring of a cross between two organisms, true-breeding for different characters.

F$_2$ generation The offspring of inbreeding an F1 generation.

Facilitated diffusion Diffusion of a substance across a cell membrane by the action of specific protein carriers.

Faeces Undigested food expelled from the anus.

Fats Organic compounds consisting of glycerol combined with three fatty acids.

Fermentation Anaerobic process by which a cell's pyruvic acid is reduced to substances such as lactic acid or ethanol.

Flaccid Limp, lacking in internal pressure.

Flagella (eukaryotes) Hair-like extensions of plasma membrane, similar to cilia but longer and in smaller numbers per cell.

Flower Shoot highly modified for sexual reproduction in flowering plants.

Food chain A representation of feeding relationships where transfer of food is represented by arrows.

Food pyramid A representation of relative numbers, biomass or productivity of the organisms occupying the different trophic levels in a community.

Food web A representation of feeding relationships showing all the food chains in a community and how they are connected.

Founder effect Loss of genetic variation that occurs when a small group (in effect, a random sample) leaves a larger population.

Fruit A ripened ovary, containing seeds.

G

Gallbladder Stores bile.

Gamete A haploid cell that can only develop further if it fuses with another such cell.

Gametophyte Haploid, gamete-producing phase of plant life cycle.

Gas exchange The diffusion of oxygen and CO_2 in opposite directions across a surface.

Gene A length of DNA that codes for a polypeptide.

Genetic bottleneck A marked reduction of genetic variability resulting from a severe decrease of numbers in a population and the resulting inbreeding.

Genetic drift Random changes in allele frequency due to the effects of chance.

Genotype The genetic makeup of an organism.

Gills Gas exchange organs in fish and many other aquatic animals.

Gizzard Muscular part of foregut where food is ground up, in earthworm, weta and many other animals.

Globular protein Ball-shaped protein molecule.

Glycogen Storage carbohydrate in animals, consisting of branched chains of glucose.

Glycolysis Process in which cells convert glucose to pyruvic acid with the production of a small amount of ATP.

Golgi body Organelle in which newly-synthesised proteins are 'sorted' according to their destinations.

Gravitropism Growth response of a plant organ to gravity.

Grazing Relationship where an organism obtains food by eating part of many organisms, which are not killed.

Guard cells Cells bordering a stoma that by changing their shape, can open or close the stoma.

Gynoecium The female part of a flower.

H

Habitat The place where an organism lives.

Haemocoel Large body cavity containing blood in arthropods and most molluscs.

Haemocyanin Oxygen-carrying blood pigment in many crustaceans and molluscs.

Haemoglobin Red, oxygen-carrying protein in the blood.

Haploid Having one set of chromosomes.

Hepatic portal vein Vessel taking blood from small intestine to the liver.

Heterospory Production of spores of two kinds, giving rise to male and female gametophytes.

Heterotroph An organism that obtains its organic matter from other organisms.

Heterozygous Having two different alleles of a particular gene.

Hexose Sugar containing six carbon atoms per molecule.

Hibernation A state in which a mammal survives winter by entering a state of dormancy.

Homeothermic Able to regulate body temperature.

Homologous pair Two chromosomes, both of which occur in diploid cells but only one of which can normally exist in a haploid cell.

Homospory Production of spores of only one kind, as in most ferns.

Homozygous Having two identical alleles of a particular gene.

Hydrogen bond Weak force between a hydrogen atom and an electron-attracting atom such as oxygen.

Hydrophyte Plant adapted to conditions of excess water.

Hypha One of many threadlike structures that form the body of a fungus.

Hypocotyl Region of a plant embryo immediately between the cotyledons and radicle.

Hypogeal germination Germination in which the cotyledons remain below ground.

I

Immigration Organisms coming into an area.

Inbreeding Mating between close relatives or, in its extreme form, self-fertilisation.

Incisor Chisel-like tooth nearest the front of the mouth in a mammal.

Independent assortment Process in which different chromosome pairs segregate independently during meiosis I.

Indusium Membrane that protects a cluster of sporangia in ferns.

Inflorescence A cluster of flowers.

Instar A stage between successive ecdyses in an insect.

Integument Protective layer around an ovule.

Intine Inner layer of wall of pollen grain.

J

J curve Exponential increase in population followed by a crash in numbers.

K

Karyotype The chromosomal characteristics of an organism.

Keratin Fibrous protein that is the main constituent of hair, nails and feathers.

Kingdom The second largest taxonomic group used in classification, consisting of a number of related phyla.

L

Labium 'Lower lip' of insects, consisting of the fused third pair of mouthparts.

Lacteal Lymph capillary in a villus.

Lag phase Period after microorganisms are introduced to a new environment, during which the organism adapts its metabolism to a newly available kind of nutrient.

Larva Young stage of an insect that is structurally very different from the adult.

Lignin Complex material that is deposited in walls of many plant cells, making them rigid.

Lipase Digestive enzyme that breaks down fats to fatty acids and glycerol.

Lipids A group of water-insoluble, alcohol-soluble compounds containing fats and steroids.

Locus The position of a particular gene on a chromosome.

Lungs The gas exchange organs of mammals and other land vertebrates.

Lysosome Microscopic vesicle in cytotplasm of eukaryotic cells, containing digestive enzymes.

M

Mandible First of three pairs of mouthparts in an insect.

Maxilla Second of three pairs of mouthparts in an insect.

Megaspore Product of meiosis in seed-bearing and certain other plants, which gives rise to the female gametophyte.

Meiosis Process involving two successive nuclear divisions in which the chromosome number is reduced from diploid to haploid, the products differing genetically from one another; the four genetically different daughter nuclei have half the number of chromosomes as the parent nucleus.

Meniscus Curved water surface that develops in narrow, partially water-filled spaces, due to surface tension.

Mesophyll The photosynthetic tissue in a leaf.

Mesophyte Plant adapted to moderate supplies of water.

Metabolism The chemical reactions in cells.

Metaphase Stage of nuclear division when the centromeres of the chromosomes become arranged on the equator of the spindle.

Microfibril Bundle of cellulose molecules in plant cell walls.

Micrometre A millionth of a metre.

Micropyle Minute hole in the integuments protecting an ovule.

Microvilli Finger-like outgrowths of plasma membrane around cells lining the small intestine and certain other tubes in animals.

Mitochondria Organelles that carry out respiration in eukaryotic cells.

Mitosis Nuclear division resulting in production of genetically identical daughter nuclei.

Molars The only first teeth in the adult mammalian set; near the back of the jaw.

Monohybrid cross Cross between two organisms differing with respect to a single pair of contrasting charactersistics.

Mortality Death rate, expressed as the number of deaths per thousand per year.

Mutagen An agent causing mutations.

Mutation A change in the genetic material that persists unchanged over successive generations.

Mutualism Relationship between two organisms that both benefit from.

Mycelium Collective name for the thread-like hyphae making up the body of a fungus.

Mycorrhiza Relationship between fungus and roots of a plant, in which the fungus obtains carbohydrate and the plant obtains mineral salts.

N

NAD Electron carrier used in the oxidation of organic compounds in respiration.

NADP Electron carrier used in the reduction of CO_2 to carbohydrate in photosynthesis.

Nanometre A billionth of a metre.

Natality Number of births per thousand per year.

Natural selection Process by which some genotypes consistently make a greater contribution to the gene pool than others.

Niche The way of life of an organism; its 'profession'.

Nitrification Formation (in two stages) of nitrate from ammonia by bacteria.

Nitrogen fixation Conversion of the element nitrogen to a combined form such as ammonia (by bacteria) or nitrate (by lightning).

Nucellus Part of an ovule within which the female gametophyte and female gamete are produced.

Nuclear envelope Double layered membrane surrounding the nucleus.

Nucleic acids Polymers of nucleotides, e.g. DNA and RNA.

Nucleolus Darkly-staining region of the nucleus containing genes for making ribosomal RNA.

Nucleus Organelle in which most of a eukaryotic cell's DNA is contained.

O

Omnivore An animal that feeds on both animal and plant material.

Open blood system Blood system in which the blood bathes the tissues directly.

Operculum Gill cover in bony fish.

Organelle A membrane-bound structure within a cell, specialised for a particular function.

Osmosis The movement of water through a partially permeable membrane.

Ovary (plants) Structure that contains the ovules and develops into a fruit.

Ovule Structure that develops into a seed.

Oxidative phosphorylation Process in the inner membrane of a mitochondrion, in which most of the ATP in respiration is produced.

P

Pancreas Gland that secretes digestive juice into the duodenum.

Parabronchi Air-filled tubes in the lungs of bird, through which air flows unidirectionally.

Parasite Organism feeding off another organism without killing it.

Parasitoid Insect that feeds inside another over a prolonged period, eventually killing it.

Partially permeable membrane A membrane that allows passage of small molecules such as water but not larger molecules; a kind of molecular sieve. Also known as a semipermeable membrane.

Pentose Sugar containing five carbon atoms per molecule.

Pepsin Digestive enzyme secreted by the stomach; breaks down proteins to polypeptides.

Peptide bond The bond joining amino acids in proteins.

Perennial Plant that lives for many years.

Pericarp Wall of a fruit; a ripened ovary wall.

Peristalsis Waves of muscular contraction that propel liquid along the alimentary canal.

Petals Modified leaves that make the flower conspicuous to animal pollinators.

pH Measure of the acidity or alkalinity of a solution.

Pharynx The first part of the alimentary canal of many animals.

Phenotype The physical characteristics of an organism.

Phospholipid The major constituent of cell membranes.

Photolysis The splitting of water using light energy.

Photorespiration Process in many plants, in which organic matter is converted to CO_2 in hot, dry conditions.

ISBN: 9780170214094

Photosynthesis The use of light energy to produce carbohydrate from CO_2 and water.

Physical factor A non-living (or abiotic) environmental factor.

Phytoplankton Microscopic photosynthetic organisms that drift in the upper layers of the ocean or lakes.

Pioneer species The first plants to colonise an inhospitable habitat such as bare rock or sand dunes.

Pits Thin areas in plant cell walls, facilitating movement of substances between them.

Placenta (plants) The part of the ovary to which an ovule is attached.

Plasma The liquid part of the blood.

Plasma membrane Boundary membrane around the cytoplasm of a cell.

Plasmodesmata Thin strands of cytoplasm that run through plant cell walls, connecting adjacent cells.

Plasmolysis Process in which the cytoplasm of a plant cell pulls away from the cell wall as a result of loss of water, when the cell is placed in a solution with a higher solute concentration (i.e. lower water concentration) than the vacuole.

Plastid The general name for chloroplasts, amyloplasts and chromoplasts.

Pleiotropy Secondary effects of a gene on the phenotype.

Plumule The shoot of a plant embryo.

Pollen Single-celled structure (microspore) within which male gametes are produced in seed plants.

Pollen sac Structure where pollen grains are produced.

Pollen tube Structure produced by pollen grain, which carries male gametes to the female gamete in seed plants.

Pollination Transfer of pollen from an anther to a stigma in a flower of the same species.

Polynucleotide Chain of nucleotides.

Polypeptide Chain of amino acids.

Polysaccharide Complex carbohydrate consisting of many simple sugar units linked into a chain.

Predation Relationship where an organism obtains food from another, which is killed in the process.

Premolar Mammalian tooth behind the canine that, in humans, is used for crushing food.

Pressure flow hypothesis The most widely accepted mechanism for phloem transport in plants.

Primary consumer An animal that feeds off plants; a herbivore.

Prokaryote Unicellular organism in which the DNA is not enclosed in a nuclear envelope, so there is no distinct nucleus.

Promoter Base sequence in a DNA strand that indicates it is the strand to be used in transcription.

Prophase The first stage of nuclear division.

Protandrous Male parts of the flower ripen before the female parts.

Proteins Polymers of amino acids.

Protogynous Female parts of the flower ripen before the male parts.

Proton pumping Process in the inner mitochondrial membrane that generates the proton gradient used to produce ATP.

Pulmonary circuit System of vessels that carries blood to and from the lungs.

Pulp cavity The central part of a tooth, containing blood vessels and nerves.

Pupa A stage between larva and adult of an insect.

Purine Two-ringed nitrogenous base in DNA and RNA.

Pyrimidine Single-ringed nitrogenous base in DNA and RNA.

Pyruvic acid The end product of glycolysis and the raw material for respiration.

Q

Quadrat Small part of a habitat of known dimensions, used for counting a sample of organisms.

R

Radicle The root of a plant embryo.

Receptacle The tip of a flower stalk, to which flower parts are attached.

Recessive trait A trait that is expressed only in homozygotes; two alleles must be present to be expressed.

Reciprocal crosses A cross that is done both possible ways, e.g. male *AA* x female *aa*, and male *aa* x female *AA*.

Red cells (erythrocytes) Blood cells that carry oxygen.

Resolving power The ability to distinguish detail.

Respiration Process in mitochondria in which organic compounds are oxidised to CO_2 and water.

Respirometer Apparatus for measuring the rate of respiration.

Rhizoid Thread-like, anchoring structure produced by mosses and liverworts.

Ribosome Minute particle on which proteins are synthesised.

RNA polymerase Enzyme that joins ribonucleotides to make RNA.

Root hairs Thread-like extensions of epidermal cells of a root.

Root nodule Swelling on root of leguminous plant (and certain other kinds), containing nitrogen-fixing bacteria.

Root pressure Force generated in some plants that pushes water upwards.

Rumen First of four compartments in the foregut of sheep, cattle and deer, where cellulose is digested.

S

S-A node The heart's 'pacemaker'.

Saliva Digestive juice secreted into the mouth by salivary glands.

Saprobe Organism feeding on dead matter, such as many fungi and bacteria. The term 'saprophyte' is obsolete (*phyte* means 'plant', and fungi and bacteria are not plants).

Secondary consumer An animal that feeds off primary consumers; also known as a primary carnivore.

Segregation The separation of alleles or of homologous chromosomes during meiosis.

Semiconservative replication Replication (of DNA) in which each product molecule contains one original and one new strand.

Sense strand DNA strand that is copied to make mRNA.

Sepals Modified leaves that protect the flower in bud.

Sieve tube Conducting tube in the phloem.

Sigmoid growth 'S'-shaped graph of population growth.

Single circulation System in which blood travels once through the heart in any complete circuit.

Species The smallest taxonomic group used in classification; all members of a species are sufficiently alike to interbreed and produce fertile offspring.

Sperm A male gamete.

Sphincter Ring of muscle around a tube, preventing flow of liquid through it.

Spindle Barrel-shaped system of protein fibres that moves the chromosomes during cell division.

Spiracle Opening of a trachea at the body surface of an insect or myriapod.

Sporangium (plants) Structure in which spores are produced by meiosis.

Spore (plants) Haploid cell with the potential to develop into a haploid (gametophyte) plant.

Sporocyte Diploid cell that gives rise to spores by meiosis in plants.

Sporophyll Leaf specialised for production of spores rather than for photosynthesis.

Sporophyte Diploid, spore-producing phase of plant life cycle.

Stamen Structure that produces pollen in a flowering plant.

Starch Energy storage carbohydrate consisting of many glucose units.

Steroids Group of fat-soluble organic compounds consisting of four rings.

Stoma Microscopic pore in epidermis of a plant, by which rate of water loss can be regulated.

Stratification The layering of vegetation in a forest, each layer adapted at photosynthesis for a particular prevailing light intensity.

Stroma The colourless, liquid part of a chloroplast.

Substrate The substance an enzyme acts on.

Succession Process of slow change in the composition of a community, starting from originally bare habitat.

Survivorship curve Graph of number of survivors from a group born at the same time, plotted against time.

Symplast Water pathway through plant consisting of cytoplasm and plasmodesmata.

Systemic circuit System of vessels that carry blood to and from all parts of the body except the lungs.

T

Taxonomy The scientific study of how organisms are classified.

Telophase Last stage of nuclear division.

Template strand DNA strand that is used by complementary base pairing to make mRNA.

Testa Seed coat.

Testcross (backcross) A cross between an organism of unknown genotype with one that is homozygous recessive, in order to determine the former's genotype.

Thermocline Boundary between upper, warmer water and colder, deeper water in a lake or the ocean.

Thylakoids Flattened sacs in chloroplasts; the site of the light-dependent reactions of photosynthesis.

Tissue A group of usually similar cells specialised for a particular function.

Tolerance The range of conditions in which an organism can survive and reproduce.

Tonoplast The membrane around the vacuole of a plant cell.

Trachea Air-filled, branching tree-like tube in an insect of myriapod, along which oxygen and CO_2 diffuse. Also the main airway in mammals and birds leading to the lungs.

Tracheid Primitive, water-transporting cell of the xylem.

Tracheole One of the finest branches of the tracheal system of an insect; the site of gas exchange.

Transcription Making an RNA copy of a gene.

Transect The line in a habitat along which samples are taken.

Transfer RNA RNA that brings each amino acid to the codon specifying it.

Translation Joining amino acids in the sequence specified by the sequence of codons in mRNA.

Translocation Movement of substances in the phloem.

Transpiration The evaporation of water from the leaves and stems of a plant.

Transpiration pull The main force responsible for upward movement of water in vascular plants.

Trophic level A major feeding category in a community, e.g. herbivore, primary carnivore.

Turgor State in which a plant cell is tightly inflated with pressure

U

Uric acid Excretory product of insects, terrestrial reptiles and land snails.

V

Vacuole Large fluid-filled space in plant cells.

Vascular plants Plants with xylem and phloem; ferns, conifers and flowering plants.

Vein Vessel that takes blood toward the heart.

Venule Microscopic vein.

Vessel member The most specialised water-transporting cell of the xylem.

Villi Finger-like outgrowths of lining of small intestine.

W

Water potential The water potential of a solution is the tendency for water to move out of it; water always moves from a higher to a lower water potential.

X

Xerophyte Plant adapted to grow in very dry conditions.

Xylem The water-transporting tissue of vascular plants.

Z

Zonation Change in the composition of a community along a gradient of physical factors.

Zooplankton Tiny animals that drift in the upper layers of the ocean or lakes.

Zygomorphic (of flowers) Bilaterally symmetrical.

Zygote A fertilised egg.

Index

ISBN: 9780170214094

ISBN: 9780170214094

ISBN: 9780170214094

ISBN: 9780170214094

Excellence in Biology Level 2

ISBN: 9780170214094